Structural, Magnetic, Dielectric, Electrical, Optical and Thermal Properties of Nanocrystalline Materials: Synthesis, Characterization and Application

Structural, Magnetic, Dielectric, Electrical, Optical and Thermal Properties of Nanocrystalline Materials: Synthesis, Characterization and Application

Editors

Raghvendra Singh Yadav
Anju
Kottakkaran Sooppy Nisar

MDPI • Basel • Beijing • Wuhan • Barcelona • Belgrade • Manchester • Tokyo • Cluj • Tianjin

Editors

Raghvendra Singh Yadav
Centre of Polymer Systems
Tomas Bata University in Zlin
Zlín
Czech Republic

Anju
Centre of Polymer Systems
Tomas Bata University in Zlin
Zlin
Czech Republic

Kottakkaran Sooppy Nisar
College of Arts and Sciences
Prince Sattam bin Abdulaziz
Unversity
Al-Kharj
Saudi Arabia

Editorial Office
MDPI
St. Alban-Anlage 66
4052 Basel, Switzerland

This is a reprint of articles from the Special Issue published online in the open access journal *Crystals* (ISSN 2073-4352) (available at: www.mdpi.com/journal/crystals/special_issues/nanocrystalline_materials).

For citation purposes, cite each article independently as indicated on the article page online and as indicated below:

LastName, A.A.; LastName, B.B.; LastName, C.C. Article Title. *Journal Name* **Year**, *Volume Number*, Page Range.

ISBN 978-3-0365-3155-7 (Hbk)
ISBN 978-3-0365-3154-0 (PDF)

Contents

About the Editors

Raghvendra Singh Yadav

Dr. Raghvendra Singh Yadav is working as a "Senior Scientist/Researcher" at the Centre of Polymer Systems, Tomas Bata University in Zlin, Czech Republic.

He obtained his Ph.D. degree on the topic "Synthesis and Characterization of Nanophosphors (luminescent/optical nanoparticles)" from University of Allahabad, India. Dr. Yadav's current scientific activities are focused on "Lightweight, Flexible, Low-dimensional Electromagnetic Functional Nanocomposite Materials (MXene, MBene, Graphene, magnetic nanoparticles as nanofillers in a polymer matrix) and its Applications".

Anju

Anju is pursuing her Ph.D. degree at the Centre of Polymer Systems, Tomas Bata University in Zlin, Czech Republic. She received her Master Degree in Applied Physics from Amity University, India. She is working on the research topic "Preparation and Characterization of Advanced Spinel Ferrite Nanocomposites for Electromagnetic applications".

Kottakkaran Sooppy Nisar

Dr. Kottakkaran Sooppy Nisar is working as a Full Professor in Prince Sattam bin Abdulaziz University, Saudi Arabia. His current research interests are fluid dynamics, mathematical modeling, engineering problems, and numerical solutions for PDEs and ODEs.

Article

Zn Doped α-Fe₂O₃: An Efficient Material for UV Driven Photocatalysis and Electrical Conductivity

Suman [1], Surjeet Chahal [1], Ashok Kumar [1] and Parmod Kumar [2,*]

[1] Department of Physics, Deenbandhu Chhotu Ram University of Science and Technology, Murthal 131039, Haryana, India; sumanjangra1594@gmail.com (S.); Surjeetchahal.schphy@dcrustm.org (S.C.); ashokkumar.phy@dcrustm.org (A.K.)

[2] Department of Physics, J.C Bose, University of Science and Technology, YMCA, Faridabad 121006, Haryana, India

* Correspondence: kumarparmod@jcboseust.ac.in

Received: 18 March 2020; Accepted: 1 April 2020; Published: 4 April 2020

Abstract: Zinc (Zn) doped hematite (α-Fe₂O₃) nanoparticles with varying concentrations (pure, 2%, 4% and 6%) were synthesized via sol-gel method. The influence of divalent Zn ions on structural, optical and dielectric behavior of hematite were studied. X-ray diffraction (XRD) pattern of synthesized samples were indexed to rhombohedral *R3c* space group of hematite with 14–21 nm crystallite size. The lattice parameter (a and c) values increase upto Zn 4% and decrease afterwards. The surface morphology of prepared nanoparticles were explored using transmission electron microscopy (TEM). The band gap measured from Tauc's plot, using UV-Vis spectroscopy, showed reduction in its values upto Zn 4% and the reverse trend was obtained in higher concentrations. The dielectric properties of pure and Zn doped hematite were investigated at room temperature and followed the same trends as that of XRD parameters and band gap. Photocatalytic properties of nanoparticles were performed for hazardous Rose bengal dye and showed effective degradation in the presence of UV light. Hence, Zn^{2+} doped hematite can be considered as an efficient material for the potential applications in the domain of photocatalysis and also higher value of dielectric constant at room temperature makes them applicable in high energy storage devices.

Keywords: α-Fe₂O₃; photocatalytic activity; dielectric properties

---

1. Introduction

In recent years, several dyes have been frequently used in textiles, printing, paper and pharmaceutical industries. The untreated hazardous dyes are discharged into the water, leading to enormous environmental problems, like perturbation of aquatic life and human health. Therefore, the removal of these dyes from water is of the utmost priority for the scientific community. Several approaches have been made to remove the toxic dye molecules from wastewater, such as adsorption, coagulation, membrane separation and ion exchange process. However, these methods fail on a larger scale due to their expensive equipments, slow processes and toxic byproducts [1]. Effective and successful methods to remove dye include photocatalytic activity in which metal oxide semiconductors are used as catalysts due to their large specific surface area, chemical stability and high photocatalytic response [2,3]. It is vitally important to establish the stability and activity of the photocatalyst to propose a photocatalytic system. From the existing transition metal oxide semiconductors, iron oxide has drawn scientific interest due to its outstanding physical and chemical properties. A variety of crystalline phases are exhibited by iron oxides, such as hematite (α-Fe₂O₃), akaganeite (β-Fe₂O₃), maghemite (γ-Fe₂O₃) and magnetite (Fe₃O₄) [4]. Among them, α-Fe₂O₃ exhibits thermodynamical, as well as chemical stability, over a broad pH scale. This compound has drawn significant interest for their potential applications such as photocatalysts, magnetic data storage, gas sensors, lithium-ion batteries, spintronics and ferrofluids [1,5–8]. The atomic arrangement possessed by hematite is similar to that of corundum α-Al_2O_3 structure, in which anions (O^{2-} ions) are stacked in hexagonal close-packed arrangement (framework by the regular alternating layers, in each layer the atoms lie at the vertices of a series of equilateral triangles and the atoms overlie one another in one layer), with cations (Fe^{3+}) occupying 2/3 octahedral coordination geometry [9].

α-Fe_2O_3 is a promising photocatalyst with optical band gap of ~2.6 eV. Also, hematite is one of the few semiconductors having valence band edge position suitable for oxygen evolution and the conduction band edge is more negative than the redox potential of H^+/H_2, thus, requiring an electrical bias to generate hydrogen [10]. However, the catalytic activity of α-Fe_2O_3 nanoparticles remains much lower due to rapid recombination of charge carriers, which reduces the degradation performance [1]. Thus, several methodologies have been made in order to sort out this problem. An effective process is doping of α-Fe_2O_3 with other metal ions, which can overcome their limitations. Doping of various metal ions such as Cr, Ti, Mn, Al, Zn, Ni, Ga, Rh, Zr and Co at Fe site in hematite influence the physical and photocatalytic properties. It is observed that Zr dopant limits the recombination of electron hole pairs in Fe_2O_3 nanorods array that act as a better catalyst for dye degradation [11]. Similarly, Ti-doped Fe_2O_3 enhances the donor density and lowers the electron-hole pair recombination rate that improves the photocatalytic activity [12].

The influence of divalent Zn cation on structural, electrical and optical behavior of hematite has become a field of scientific research. The substitution of Zn^{2+} at Fe^{3+} site causes the charge imbalance in the host lattice [13]. In order to maintain charge neutrality, one or more of the following mechanisms can occur: Transformation of Fe^{3+} to Fe^{2+} state, creation of cation vacancies and filling of oxygen vacancies. The physical properties of hematite are effected by the degree of crystallinity, particle size, doping and pressure [14–17]. A report by Velev et al. [18] showed that Zn^{2+} affects electronic properties of hematite that causes the creation of a hole in the oxygen valence band. The extra hole from Zn^{2+} is situated on the neighboring O sites inducing an acceptor level just below the fermi energy. This hole is relatively delocalized, and hence, provides good hope for high conductivity. The purpose of incorporation of Zn^{2+} ions is to promote the hopping mechanism of electrons by Fe^{3+}-Fe^{2+} pairs and also modifying the optical properties. Based on these factors, we have synthesized Zn-doped Fe_2O_3 nanoparticles, with dilute concentrations, to study their structural, optical, dielectric and photocatalytic properties.

2. Materials and Methods

2.1. Synthesis of Nanoparticles

Pure and Zn doped α-Fe_2O_3 samples were synthesized by sol-gel method using high purity precursors $Fe(NO_3)_3 \cdot 9H_2O$ and $Zn(NO_3)_2 \cdot 6H_2O$ in a stoichiometry ratio with distilled water as solvent (as shown in Figure 1). The solutions were mixed and subjected to vigorous stirring for about 15 min to obtain a clear homogenous solution. The sol mixture was heated on hot plate at 60 °C with constant stirring for 1 h until the gel type solution was obtained. This gel was converted to solid particles by heating at 90 °C for 4 h and then crushed to get nanoparticles. The as obtained powder was annealed at 400 °C for 2 h and then grinded. The sample without doping is represented as pure Fe_2O_3 and Zn doped as Zn 2%, Zn 4% and Zn 6% corresponding to samples α-$Zn_xFe_{2-x}O_3$ where x = 0, 0.02, 0.04 and 0.06.

2.2. Characterizations

Structural study of prepared samples was carried out by Rigaku X-ray Diffractometer (XRD) (installed at DCRUST, Murthal, India) as Cu Kα radiation source with wavelength 1.54 Å. The size and surface morphology of prepared nanoparticles were studied using Thermo Scientific Talos Cryo TEM (installed at AIIMS, New Delhi). Raman spectra were recorded with a STR 500 Confocal Micro Raman Spectrometer (DPSS Laser of wavelength of 532 nm at 12.5 MW power source) (installed at

MNIT, Jaipur, India). Fourier transform infrared spectroscopy (FTIR) was explored by a NICOLET 5700 (present at DCRUST, Murthal, India) with transmittance in the range 400–4000 cm^{-1}. Dielectric measurements were done using Novacontrol broadband (installed at Delhi University) impedance at room temperature to measure complex dielectric permittivity and tangent loss. UV-Vis absorption spectra were recorded at different time intervals to monitor the degradation process using LABINDIA UV 3092 UV-VIS spectrophotometer (present at DCRUST, Murthal, India).

2.3. Photocatalytic Test

Photocatalytic performance of Zn doped α-Fe$_2$O$_3$ samples were investigated by decomposition of Rose bengal (RB) dye using 300 W UV light source having 365 nm wavelength at room temperature which is shown in Figure 1. In this experiment, 0.05 g of catalyst was sprinkled in 50 mL of 5 ppm RB dye solution and vigorously stirred for 45 min in the dark to achieve equilibrium adsorption/desorption at the surface of photocatalyst. Then, the dye solution was continuously stirred throughout the experiment under UV light and 3 mL of solution was collected at regular 15 min time intervals to monitor the degradation process using UV-VIS spectrophotometer.

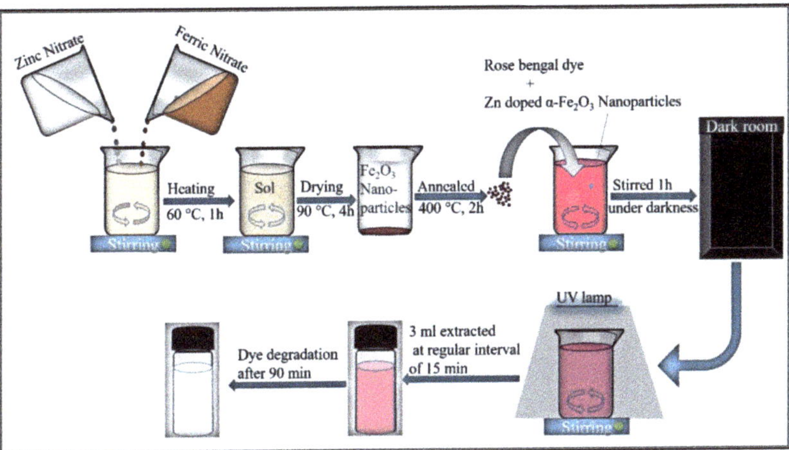

Figure 1. Schematic representation of synthesis and photocatalytic test of pure Fe$_2$O$_3$, Zn 2%, Zn 4% and Zn 6% nanoparticles.

3. Results and Discussion

3.1. XRD Analysis

Structural and phase identification of the materials were confirmed using X-ray diffraction (XRD) as shown in Figure 2. All exhibited XRD peaks of pure and Zn doped hematite assigned to (012), (104), (110), (113), (024), (116), (214) and (300) planes can be easily indexed to rhombohedral with space group *R3c* phase of hematite (JCPDS card no. 84-0311) [19]. No diffraction peaks other than hematite has been observed, indicating that Zn atoms were incorporated in α-Fe$_2$O$_3$ matrices. Thus, crystallinity is altered by dopant atoms without disturbing the rhombohedral structure of hematite. A visual inspection of XRD reveals that (104) diffraction peak is shifting towards lower angle up to 4% of Zn doping and then shifts toward higher angle side for 6% Zn concentration. This shifting of XRD peaks result in the variation of lattice parameters (a and c) as shown in Table 1. It is contemplated that lower doping (≤4% Zn) concentration occupies substitutional sites, whereas, higher doping of Zn occupies partial interstitial sites or segregate on the surface which distorts the host lattice structure. In other words, higher concentrations of Zn^{2+} ions causes non uniform distribution in the host lattice, which plays a dominant role in modifying the various physical properties. Distortion in host matrix is expected due to incorporation of large size Zn^{2+} ions in place of smaller size Fe^{3+} ions which in turn leads to stress (σ) in the system. This can be obtained using the relation [20],

$$\sigma = \frac{226.28\,(c - c_0)}{c_0} \tag{1}$$

where $\frac{(c-c_0)}{c_0}$ represents strain, c_0 and c corresponds to the lattice parameter values from JCPDS card and XRD results, respectively. The obtained negative values of stress indicates the compressive stress in the system. A report by K. Vijayalakshmi et al. stated that compressive stress (negative sign in stress value) may be attributed to zinc interstitials and tensile stress (positive sign in stress values) is associated with oxygen vacancy present in the Mg doped ZnO thin films [20]. The crystallite size (D) of these nanoparticles was calculated from the full-width half maxima (FWHM) of (104) peak using Debye-Scherer formula. It is observed that crystallite size increases up to 4% Zn concentration then decreases for 6% Zn concentration. The enhancement in crystallite size after Zn doping plays an important role in crystal growth and also in crystallization of Fe_2O_3. The enlargement in size is due to the substitution of Fe^{3+} ions with relatively large sized Zn^{2+} ions. The obtained trend in crystallite size for higher Zn doped Fe_2O_3 samples has a similar trend as also discussed in previous reports for Y doped ZnO, Mn-doped CeO_2 and Mg-doped ZnO samples [21–23].

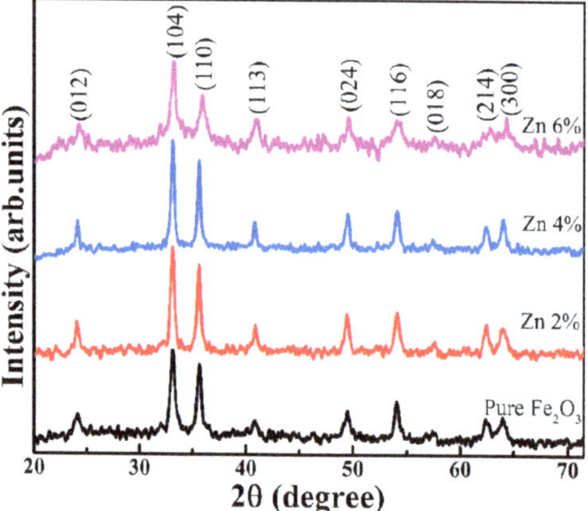

Figure 2. X-ray diffraction pattern of pure Fe_2O_3, Zn 2%, Zn 4% and Zn 6% nanoparticles.

Table 1. Structural parameters of pure Fe_2O_3, Zn 2%, Zn 4% and Zn 6% synthesized nanoparticles.

Samples	Lattice Parameter (Å)		Crystallite Size (nm)	Dislocation Density $(nm)^{-2} \times 10^{-4}$	Stress (GPa)	Particle Size From TEM (nm)
	a–axis	*c*–axis				
Pure Fe_2O_3	5.035	13.229	15	44.44	−8.41	18
Zn 2%	5.041	13.242	18	30.86	−8.20	20
Zn 4%	5.043	13.248	21	22.67	−8.10	23
Zn 6%	5.002	13.221	14	51.02	−8.54	16

Additionally, to obtain more information about the defects present in the synthesized samples, dislocation density (δ) is evaluated from $\delta = \frac{1}{D^2}$. The obtained dislocation density is significantly low for Zn 4% indicating the presence of large number of defects which is helpful in photocatalytic degradation. However, the defects are reduced for Zn 6%. The increase in crystallite size and decrease in dislocation density up to Zn 4% indicates that dopant atoms are entirely included in the lattice. While, in higher Zn dopant concentrations, the decrease in crystallite size and increase in dislocation density infers that dopant atoms occupy interstitial positions in the matrix. This results in a decrease in crystalline order and an increase in dislocation density. The change in dislocation density and stress in synthesized samples confirm the presence of defects in the lattice structure that are responsible for modification in various physical properties.

3.2. TEM Analysis

The surface morphology and particle size of Zn doped Fe_2O_3 nanoparticles were examined by transmission electron microscopy (TEM) measurements. It can be clearly seen from Figure 3a that Fe_2O_3 nanoparticles are almost spherical in shape. The estimated average particle size from TEM lies between 16 nm to 23 nm for pure, as well as Zn doped Fe_2O_3 nanoparticles. The particle size increases up to 4% of Zn concentration and then decreases which is consistent with XRD measurements and the values are given in Table 1. Figure 3a shows agglomerated nanoparticles of Fe_2O_3. The agglomeration is found to be decreasing with Zn content in Fe_2O_3 lattice (Figure 3b–d).

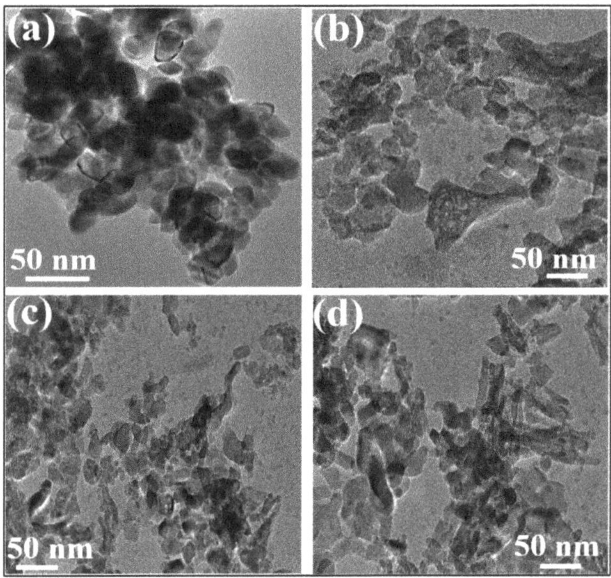

Figure 3. TEM images of (**a**) pure Fe_2O_3, (**b**) Zn 2%, (**c**) Zn 4% and (**d**) Zn 6% nanoparticles.

3.3. Raman Analysis

Raman spectroscopy is a fast and non-destructive tool for identifying the vibrational phonon modes that access the clear identification of compounds. Hence, the structural properties of as synthesized samples are further studied using Raman spectroscopy. The optical vibrational modes can be assumed as lattice waves arising due to an out of phase movement of atoms inside the crystal lattice. As these waves can interact with applied external electric field so, it is easy to excite them through conventional spectroscopic techniques. For a particular vibrational mode to be Raman active, it should be accompanied by change in polarizability. Whereas, changes in the dipole moment are required for vibrations to be infrared active. Vibrational modes of α-Fe_2O_3 at the first Brillouin zone center are represented by [24]:

$$\Gamma = 2A_{1g} + 2A_{1u} + 3A_{2g} + 2A_{2u} + 5E_g + 4E_u. \tag{2}$$

Among these, the acoustic modes (A_{1u} and A_{2g}) are optically silent, due to an in-phase movement of atoms inside the crystal lattice, and cannot be identified by these techniques, as they propagate with the speed of sound of a much lower frequency. The six antisymmetric modes ($2A_{2u}$ and $4E_u$) are infrared active vibrations and seven symmetrical ($2A_{1g}$ and $5E_g$) modes are Raman active vibrations. As the rhombohedral crystal structure of α-Fe_2O_3 features an inversion center, no modes are both infrared and Raman active.

Raman spectra of Zn doped α-Fe_2O_3 in the range 200–800 cm^{-1} at room temperature is shown in Figure 4. The assignment of Raman active modes are consistent with the group theory predicted for the space group *R3c* of hematite. Five phonon modes (2 A_{1g} and 3 E_g) of hematite corresponding to transverse optical (TO) modes are detected by group theory at $A_{1g}(1)$ ~215 cm^{-1}, $E_g(1)$ ~280 cm^{-1}, $E_g(2)$ ~398 cm^{-1}, $A_{1g}(2)$ ~492 cm^{-1}, $E_g(3)$ ~544 cm^{-1} respectively, which are well in agreement with existing literature, thereby confirming the rhombohedral structure of synthesized samples [9]. The expected Raman spectra, corresponding to E_g modes at ~245 cm^{-1} and ~412 cm^{-1} is missing in the present case due to crystalline disorder or broadening of peaks. A_{1g} symmetry can be viewed as the movement of Fe atoms along the crystallographic *c-axis* of the unit cell, while E_g symmetry involves the symmetric breathing mode of O atoms correlated to each iron cation (Fe) in the plane perpendicular to the *c-axis* of the unit cell. It is observed from Figure 4 that peaks shift towards higher wavenumber till 4% Zn doping and then shift towards lower wave number on further doping. This shifting in Raman modes is governed by the change in host lattice strain with the addition of foreign atoms. The observed variation in Raman spectra correlates well with the XRD results of variation in lattice parameter and stress values. Apart from these symmetrical phonon modes, it is observed that there is an additional feature illustrating IR (infrared) - active longitudinal optical (LO) E_u mode at ~597 cm^{-1} which is forbidden in Raman scattering, but is activated by surface defects or disorder in hematite crystalline lattice [25]. The intensity of this mode is maximum for Zn 4% sample. These defects attribute to oxygen vacancies and modify the electronic structure that, in turn, enhances the photocatalytic activity.

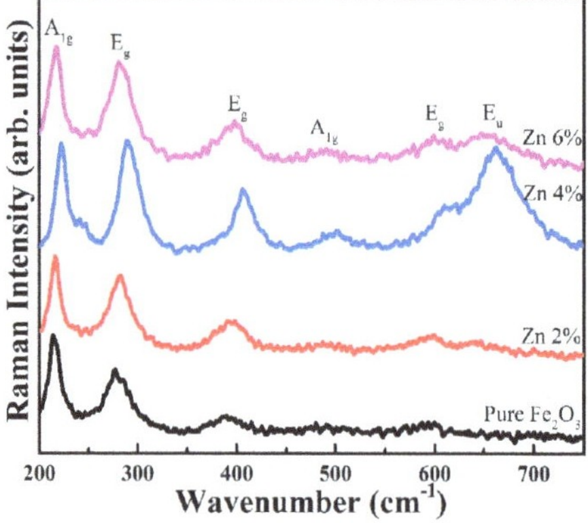

Figure 4. Raman spectrum of pure Fe_2O_3, Zn 2%, Zn 4% and Zn 6% nanoparticles.

3.4. FTIR Analysis

Fourier transform infrared (FTIR) spectroscopy is a powerful method to get the information related to chemical bonds adsorbed on the surface of the material. FTIR spectrum of pure and Zn doped α-Fe_2O_3 nanoparticles was recorded in the range 400–4000 cm^{-1} given in Figure 5a. Different bands in FTIR spectra arises due to various functional groups. As discussed above in Raman studies, group theory analysis predicts six infrared active modes corresponding to α-Fe_2O_3 lattice, out of which two infrared active A_{2u} modes are associated with the vibrations polarized parallel to crystallographic *c-axis*, while other four active E_u modes are polarized perpendicular to crystallographic *c-axis* [26]. The spectra features two prominent peaks at ~463 cm^{-1} and ~551 cm^{-1} are assigned to E_u and A_{2u} + E_u (overlapping of A_{2u} and E_u) phonon modes, respectively. These sharp and strong intensity bands at ~463 cm^{-1} and ~551 cm^{-1} indicate the metal oxygen (Fe–O) vibrations in rhombohedral lattice of hematite [27]. Also, these peaks confirm the existence of α-Fe_2O_3 and are consistent with XRD data. In addition, peak observed at 1095 cm^{-1} is attributed to the presence of adsorbed CO_2 and peak centered at ~1633 cm^{-1} is assigned to O-H bending of water [28,29]. Further, the band at ~2929 cm^{-1} is attributed to CH symmetric stretching vibrations and very broad peak observed at ~3413 cm^{-1} corresponds to the presence of hydroxyl group [30].

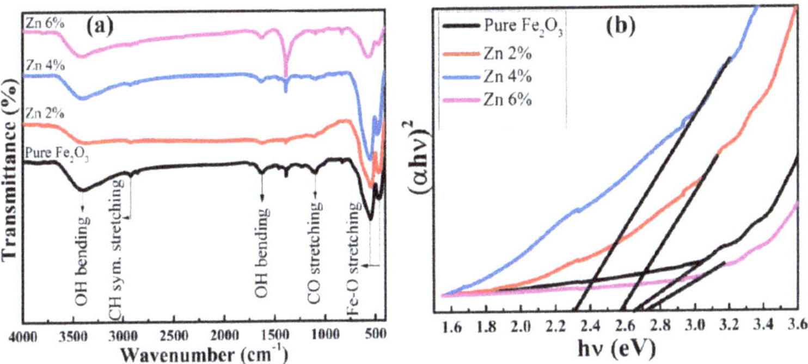

Figure 5. (**a**) FTIR transmittance (%) spectra and (**b**) Tauc plot of pure Fe_2O_3, Zn 2%, Zn 4% and Zn 6% nanoparticles.

It has been demonstrated that band positions in FTIR spectra are sensitive to lattice parameters, particle size and the presence of impurities. As discussed above, bands at ~463 cm^{-1} and ~551 cm^{-1} are related with Fe–O stretching vibrations and these bands are shifting toward lower wavenumber side up to Zn 4% and then shift to higher wavenumber for Zn 6% due to variation in cation-oxygen bond length [31]. Also, it is well-established that bond length is inversely proportional to wavenumber or frequency. The shifting in these bands are analogous to the change in lattice parameter values analyzed through XRD measurements and reveals the strengthening of metal oxygen bond with the change in Zn content in the host matrix. Moreover, intensity of peaks increases up to Zn 4% doping and decreases for Zn 6% which is in accordance with crystallinity of XRD pattern.

3.5. UV–Vis Analysis

UV-Vis measurements in the absorption mode were carried out to reveal the electronic structure and size effect of as prepared nanoparticles. The optical band gap energy for synthesized nanoparticles has been calculated using the Tauc's relation: $(\alpha h\nu)^{n} = A(h\nu - E_g)$, where, α is the absorption coefficient, $h\nu$ is incident photon energy of light, A is a constant, E_g denotes the band gap energy and n is constant that depends on the nature of optical transition (n = 2 and 0.5 for direct and indirect transition respectively) [32]. Figure 5b shows the plot of $(\alpha h\nu)^2$ versus $h\nu$ for Zn doped α-Fe_2O_3 nanoparticles that exhibits a direct band gap with n = 2. The charge transfer in α-Fe_2O_3 takes place between occupied O^{2-} $2p$ state to empty Fe^{3+} $3d$ upper state that is responsible for direct band gap transition in Fe_2O_3. It is found that pure Fe_2O_3 nanoparticles has band gap of 2.66 eV which is higher than the reported band gap of 2.1 eV for pure Fe_2O_3 [1]. This indicates existence of Fe^{3+} in lower spin state that results in higher value of band gap for pure Fe_2O_3. Moreover, the obtained results show reduced band gap from 2.66 eV for pure Fe_2O_3 to 2.31 eV for Zn 4% and then increases for higher Zn concentration. This decrease in band gap may be ascribed to an increase in structural disorder or defects with increase in Zn doping up to 4% concentration. In addition, this decrease in band gap may also be due to partial hybridization between Zn t_{2g} and O 2p states to empty Fe t^{*}_{2g} 3d orbitals. A report by Mashiko et al. explained the decrease in band gap on the basis of decrease in residual in-plane strain [33]. Based on the above considerations, the sequence of band gap for synthesized samples is Zn 6% > Pure Fe_2O_3 > Zn 2% > Zn 4% which agrees well with experimental data and measured band gap values are given in Table 2.

Table 2. The calculated band gap, valence band and conduction band positions corresponding to pure Fe_2O_3, Zn 2%, Zn 4% and Zn 6% nanoparticles.

Samples	Band Gap (eV)	Valence Band Position (eV)	Conduction Band Position (eV)
Pure Fe_2O_3	2.66	2.70	0.04
Zn 2%	2.58	2.66	0.08
Zn 4%	2.31	2.52	0.21
Zn 6%	2.72	2.73	0.01

Band edge positions, bandgap as well as the overall band structure of semiconductors play an important role in photocatalytic applications. The energy position of the band edge level can be controlled by the electronegativity of the dopants, as well as by the quantum confinement effects. The valence band and conduction band edge potential of a semiconductor can be deduced from the relation [34,35],

$$E_{VB} = \chi - E_e + 0.5\,E_g \tag{3}$$

$$E_{CB} = E_{VB} - E_g \tag{4}$$

where, E_{VB} and E_{CB} are the valence band and conduction band edge potential, respectively, χ is the absolute electronegativity of a semiconductor oxide and its value for Fe_2O_3 is 5.87 eV, E_e represents the energy of free electrons, which is about 4.5 eV on hydrogen scale. The calculated valence and conduction band edge position for synthesized samples are given in Table 2.

3.5.1. Photocatalytic Activity

The photocatalytic performance of Zn doped α-Fe_2O_3 (catalyst) was investigated by recording the time-dependent degradation of RB dye (as a contaminant). Figure 6a–d showing the change in absorption spectra over time for RB dye solution with catalysts under the presence of UV light irradiation. Figure 6a demonstrates that pure α-Fe_2O_3 shows poor performance, compared to Zn doped (2% and 4%) samples as the high recombination rate between electrons and holes, and which cannot be easily separate out due to the short hole diffusion length in case of pure α-Fe_2O_3. However, the addition of Zn as dopant is a useful tactic to introduce localized electronic band structure which improves the charge separation efficiency. The appearance of no new absorption peak during whole process indicates the degradation in presence of proposed photocatalyst. The characteristic absorption peak intensity of RB dye gradually decreases with increasing exposure time from 0 min to 90 min. The intense absorption peak of RB dye around 562 nm decreases much faster in the presence of catalyst (Zn 4%) compared to other synthesized samples. The photodegradation activity increases with Zn dopant concentration of α-Fe_2O_3 in the following order: Zn 6% < pure Fe_2O_3 < Zn 2% < Zn 4% as shown in Figure 6a–d. It is well-established that synthesized samples in nano-region exhibit unique surface chemical reactivity for photocatalytic activity. There are several factors that influence the photocatalytic activity, such as type of dopant, recombination of electron hole pairs and band gap of semiconductors. Researchers have claimed that Cu^{2+} doping in α-Fe_2O_3 creates a trap state (separate band) which controls the electron hole recombination in photocatalytic process [1]. In the present study, Zn^{2+} forms a trap state in the band gap of α-Fe_2O_3, i.e., a separate band between conduction band and valence band. The trap state induces defect state/impurity level, which entraps the charge carriers, as soon as they have been generated by UV light illumination, and inhibits the recombination so that charge carriers can be used for the redox process. The band gap decreases in Zn doping (up to 4% concentration), resulting in further surface defects (as clearly seen in Raman spectra), as well as delaying the recombination of charge carriers also which yields better catalyst for the degradation of RB dye.

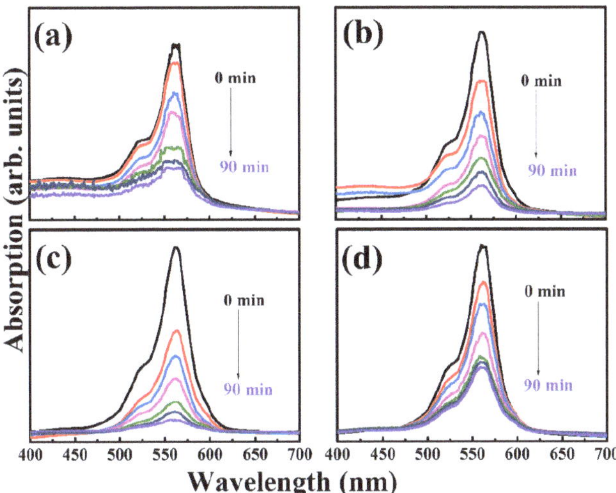

Figure 6. Time-dependent UV–Vis absorption spectra for RB dye in the presence of Catalyst: (**a**) pure Fe_2O_3, (**b**) Zn 2%, (**c**) Zn 4% and (**d**) Zn 6%.

Photocatalytic activity generally includes the partial/complete degradation of organic waste dyes with the assistance of active species existing on the surface of the catalyst. When the catalyst is exposed to UV light, the photogenerated electrons (e^-) are excited from top of valence band to the bottom of conduction band, leaving behind the holes in valence band. This lead to positive holes and negative electrons on the catalyst surface. The photogenerated holes interact with adsorbed water present on the surface of catalyst to generate reactive hydroxyl free radical ($^{·}OH$), while O_2 acts as an electron acceptor to form a superoxide ($O_2^{·-}$) anion radical which on protonation yields $HOO^·$ in the presence of water [36]. Further, the $O_2^{·-}$ can act as an oxidizing agent or as an additional source of $OH^·$ radicals. These hydroxyl radicals are, thus, more efficient for degradation of RB dye into some non-toxic organic compounds, such as CO_2 and H_2O, as shown in Figure 7. The oxidative (using holes) and reductive (using electrons) pathway, followed by the degradation process, are summarized as follows [33,37]:

$$Zn - (\alpha - Fe_2O_3) + h\nu \rightarrow Zn - (\alpha - Fe_2O_3) + e_{CB}^- + h_{VB}^+ \tag{5}$$

$$e_{CB}^- + O_2 \rightarrow O_2^- \tag{6}$$

$$O_2^- + H_2O \rightarrow HOO^· + OH^- \tag{7}$$

$$h_{VB}^+ + OH^- \rightarrow OH^· \tag{8}$$

$$OH^· + O_2^- + RB\ dye\ intermediates \rightarrow CO_2 + H_2O \tag{9}$$

This is in accordance with significant activity of samples which is attributed to the effective inhibition of (e^-/h^+) recombination and migrates to the photocatalyst surface to generate highly reactive free radicals that in turn oxidize RB dye (Figure 7).

Figure 8 displaying the experimental and linear plot of $-\ln(c/c_0)$ versus time (t) for RB dye with different Zn concentration in hematite. It suggests that photodegradation of RB molecules by catalyst follows the pseudo-first-order kinetics [38]:

$$-\ln(C/C_0) = kt \tag{10}$$

$$t_{1/2} = \ln 2/k \tag{11}$$

where, C_0 is the initial concentration of pollutant (RB dye) when the UV light is turned on, while C is the real-time concentration of pollutant under UV light irradiation, and k is the apparent rate constant of pseudo-first-order equation, t is the irradiation time. The half-life time ($t_{1/2}$) is defined as the time required to degrade 50% of initial RB dye concentration. The slope of the plot $-\ln(C/C_0)$ with irradiation time provides the estimated apparent rate constant as given in Table 3. The observed degradation rate constant of RB dye in the presence of a catalyst Zn 4% is 0.02277 min^{-1}, which is significantly larger than other synthesized samples.

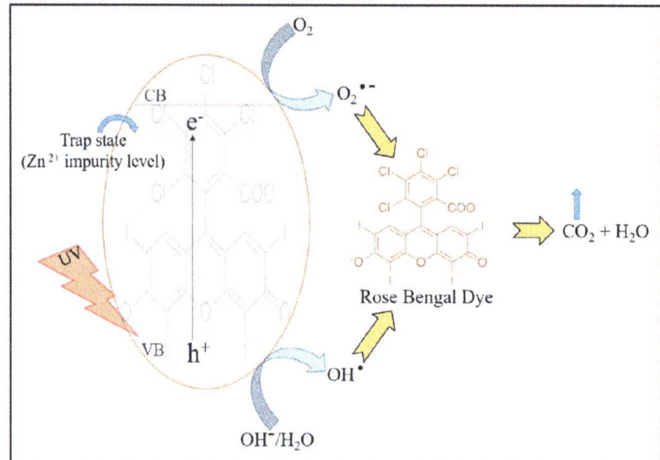

Figure 7. Proposed photocatalytic mechanism in α-Fe$_2$O$_3$ for degrading RB dye.

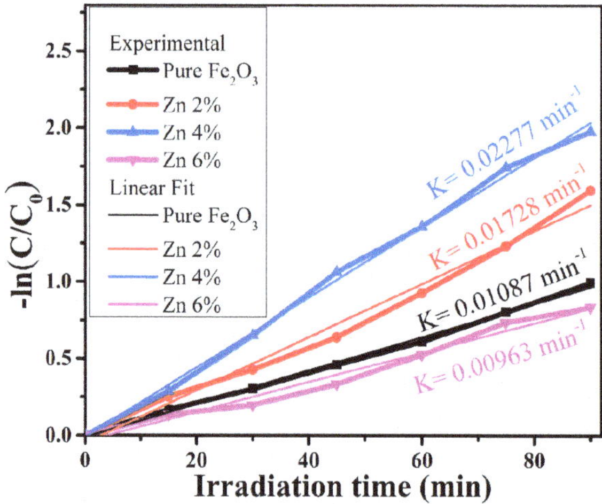

Figure 8. Experimental and linear plot of $-\ln(C/C_0)$ versus irradiation time for pure Fe$_2$O$_3$, Zn 2%, Zn 4% and Zn 6% nanoparticles.

Table 3. Calculated photodegradation parameters of pure Fe$_2$O$_3$, Zn 2%, Zn 4% and Zn 6% nanoparticles.

Samples	Rate Constant (k) (min)$^{-1}$	% Degradation (in 90 min)	R^2	$t_{1/2}$ (min)	t_{90} (min)
Pure Fe$_2$O$_3$	0.01087	63	0.9962	63.8	211.8
Zn 2%	0.01728	80	0.98381	40.1	133.3
Zn 4%	0.02277	87	0.99605	30.4	101.1
Zn 6%	0.00963	57	0.97569	72.0	239.1

The percentage degradation of RB dye, using pure hematite as a catalyst, is 63% after UV irradiation for 90 min. Degradation % increases with an increase in Zn content up to Zn 4% and reached 87% as shown in Figure 9a. Further increase in Zn content decrease the degradation efficiency towards RB dye. Notably, the degradation rate of Zn 6% is even less than that of pure Fe_2O_3, due to the fact that Zn ions occupy interstitials site in the host matrix for this concentration responsible for the enhanced recombination rate between electrons and holes.

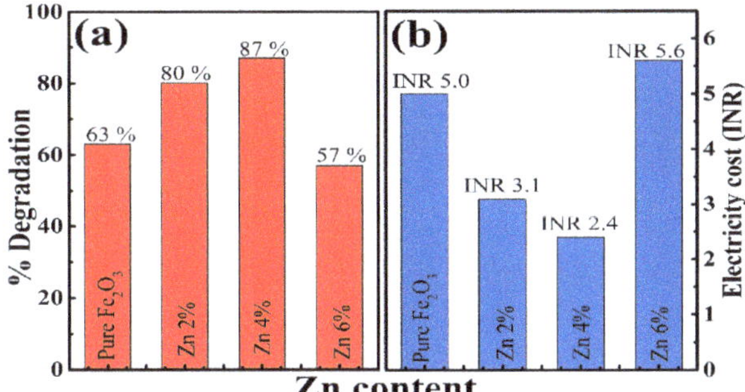

Figure 9. Bar diagram of (**a**) % degradation and (**b**) electricity cost in Indian rupees for degradation of RB dye with pure Fe_2O_3, Zn 2%, Zn 4% and Zn 6% nanoparticles.

3.5.2. Electricity Cost

Cost evaluation is one of the most important factors in waste water treatment. As saving energy (electricity) benefits the world at large scale. The main reason behind saving electricity is that burning of fossil fuels in plants causes several environmental issues, such as global warming and the greenhouse effect, which directly affect human life. Our present study aims to reduce energy to mitigate the effects of greenhouse gases. The power consumption can be estimated using the following relation [39],

$$t_{90} = \ln 10/k \tag{12}$$

$$E_c = \frac{P \times t_{90} \times 4.68}{1000 \times 60} \tag{13}$$

where, t_{90} signifies the time taken by any dye to be degraded 90% of its initial concentration, E_C is electricity cost, P is power consumed (in Watt) of UV light source. Power consumers consuming a maximum 500 units of electricity per month pay INR 4.68 per unit in our locality, as shown in Figure 9b. The electricity cost is also found to be minimum for 4% Zn doped sample which has maximum % degradation.

3.6. Dielectric Properties

Materials under the oscillating electric field impart dielectric behavior which can be expressed as a complex form consisting of real (ε') and imaginary (ε'') components that can be represented as $\varepsilon* = \varepsilon' + i\varepsilon''$. The real component ($\varepsilon'$) of dielectric constant depicts the energy storage and imaginary component signifies the dissipated energy in the material. Various external parameters like microstructure, frequency of applied electric field, sintering temperature, type of cation substitution, etc. affect the dielectric properties. Both components of the dielectric constant can be evaluated using the following relation:

$$\varepsilon'' = \varepsilon' \times tan\delta \tag{14}$$

Both real, and imaginary, components of dielectric constant have strong frequency dependence at room temperature in Zn doped hematite, and is demonstrated in Figure 10. The dielectric constant decreases with an increase in frequency, which agrees well with previous studies [40]. The strong decrease in dielectric constant with rise in frequency can be understood on the basis of Maxwell Wagner model and Koop's phenomenological theory, which explains that ferrites are formed by highly conducting grains, embedded in the insulating matrix, i.e., grain boundaries [3]. High dielectric constant value at lower frequencies is contributed by grain boundaries.As the frequency increases, grains start to predominate over grain boundaries, which reduces the dielectric constant. The dispersion in dielectric constant with frequency can also be understood in terms of space charge polarization, due to the hopping of electrons between ferric and ferrous ions [41]. At low frequencies, hopping of electrons within grains causes the electrons to pile up at grain boundaries resulting in space charge polarization and contributes to higher value of dielectric constant. On the other hand, a reduction in orientation polarizability can be seen with increase in frequency, as the electron exchange between Fe^{2+} and Fe^{3+} ions loses the ability to follow alternative field and lags behind the field. As a result, probability of electrons reaching the grain boundary reduces. Consequently, the dielectric constant decreases and becomes almost constant at higher frequencies.

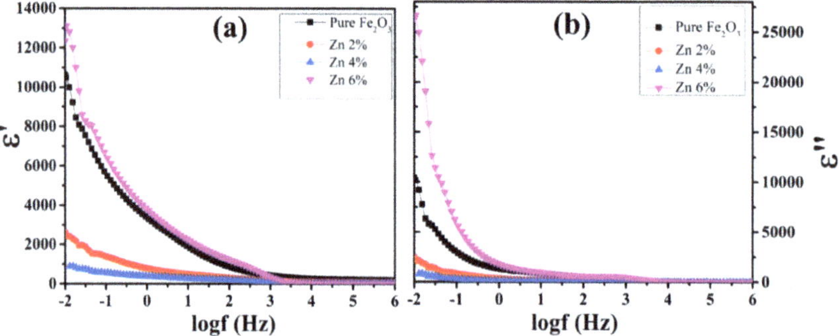

Figure 10. Variation of (**a**) real component (ε') and (**b**) Imaginary component (ε'') of dielectric constant with frequency of pure Fe_2O_3, Zn 2%, Zn 4% and Zn 6% nanoparticles.

Compositional Dependent Dielectric Constant

The behavior of the dielectric constant (ε' and ε'') with Zn concentration is dependent on many factors like hopping mechanism at octahedral site, lattice parameter and crystallite size. It is well known that the polarization and volume of the unit cell are inversely proportional to each other [42]. It is clearly seen from Figure 10 that value of dielectric constant decreases continuously (pure Fe_2O_3 > Zn 2% > Zn 4%) with increase in Zn^{2+} ions up to 4% concentration due to its increased lattice parameter, which yields increase in unit cell volume. Also, this can be justified based on the hopping mechanism. The hopping of ions between Fe^{2+} and Fe^{3+} in octahedral site is responsible for polarization. The decrease in dielectric constant with addition of Zn ions up to 4% arises from a decrease in Fe^{2+} ions at the octahedral site which reduces the electrons (n-type charge carriers) at the cost of increased holes (p-type charge carriers). The mobility of holes is less comparable to electrons and these holes contribute towards polarization resulting in a reduction of dielectric constant. The increase in dielectric constant for Zn 6% is due to decrease in its lattice parameter which reduces the cell volume resulting in large polarization and consequently increase in dielectric constant. Secondly, dielectric constant depends on crystallite size also. It is clearly seen from XRD that crystallite size is increased up to Zn 4% and then decreases for Zn 6%, which is in accordance with variation of dielectric constant with composition.

3.7. Dielectric Loss Tangent

Dissipation of energy is measured with respect to alternating external field which is recorded in terms of dielectric loss. The variation in dielectric loss as a function of frequency for varying Zn concentration in hematite is shown in Figure 11.

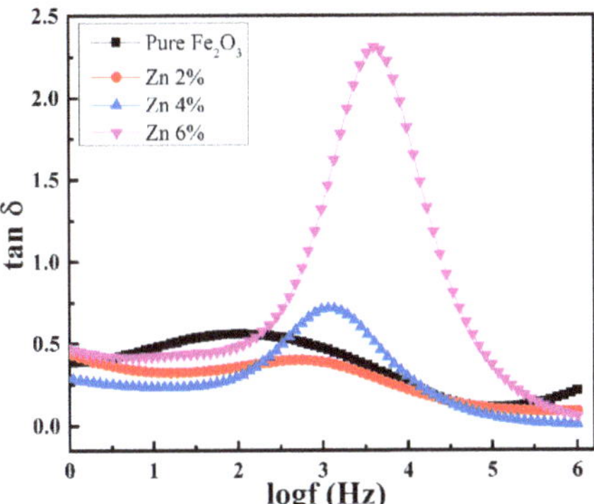

Figure 11. Variation of dielectric loss with frequency of pure Fe_2O_3, Zn 2%, Zn 4% and Zn 6% nanoparticles.

A report by Iwauchi et al. [43] showed that conduction hopping and dielectric behavior are strongly correlated. The dielectric loss is high at lower frequencies, due to the grain boundary behaving as an insulating interface, as the charge carriers undergo space charge polarization [44]. Diamagnetic dopant and organized growth of domains have a major impact on decrease of loss tangent at small frequencies. The loss factor decreases at higher frequencies, due to the mismatch of electrons with applied field frequency, as discussed above. Loss is dependent on various factors, such as ferric and ferrous content, the stoichiometry of material and heterogeneous domain wall geometry. It can be observed that loss tangent has a relaxation peak for α-$Zn_xFe_{2-x}O_3$, which is consistent with earlier reports [45,46]. According to Rezlesque model [47], a peak in dielectric loss is expected when the hopping frequency of electrons between Fe^{2+} and Fe^{3+} states is in resonance with the external applied electric field's frequency. The maxima in dielectric loss can be expressed as a relation $\omega\tau = 1$, where ω is the angular frequency of field and τ is relaxation time for hopping mechanism. The increase in peak height, as well as shifting in peak position with doping of Zn shows the variation in hopping probability of electrons between Fe^{2+} and Fe^{3+} states, and this is influenced by the number of Fe^{3+} ions in the octahedral site [48].

3.8. AC Conductivity

To study the hopping mechanism, ac conductivity (σ_{ac}) versus logf of Zn doped hematite at room temperature is plotted (Figure 12a). At lower frequencies, conductivity seems to be constant and increases with rise in frequency. The type of polarons involved in hopping mechanism was estimated using log σ_{ac} versus logf as shown in Figure 12b. In large polaron model, ac conductivity decreases, while in the small polaron model, ac conductivity increases with rise in frequency. In the present study, conductivity shows almost linear behavior with increases in frequency that indicates conduction hopping is followed by small polaron mechanism, as evident from Figure 12b. Conductivity is more affected by grain boundaries at lower frequencies, while grains have more impact on conduction at high frequencies [49]. The increase in frequency enhances the hopping of charge carriers between ferric and ferrous ions that leads to increase in conductivity. Low conductivity is observed at lower frequencies which is due to the blocking effects at grain boundaries.

The relation between frequency and ac conductivity can be depicted as [50],

$$\sigma_{ac} = 2\pi f \varepsilon_0 \varepsilon' tan\delta \tag{15}$$

where, f is frequency in Hz. The conductivity decrement with Zn dopant concentration could be described by the microstructures of the material, the probability of hopping and hopping duration of the electrons. This may arises due to the reduction of Fe^{3+} ions in octahedral site and creation of Fe^{3+} vacancies by substitution of Zn^{2+} ions.

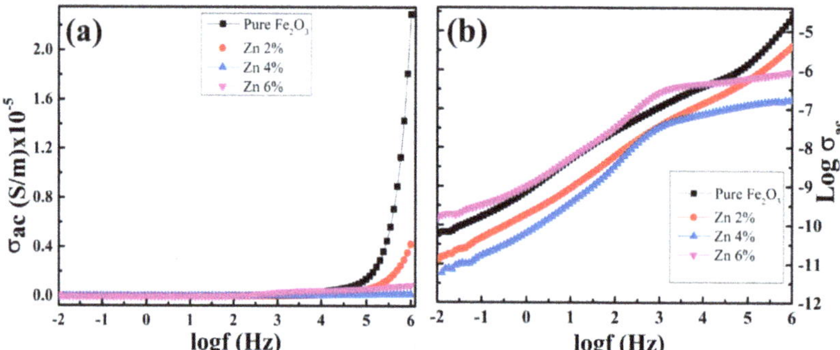

Figure 12. (**a**) Frequency dependency AC conductivity of pure Fe_2O_3, Zn 2%, Zn 4% and Zn 6% nanoparticles and (**b**) Linear plot of log σ_{ac} versus logf.

3.9. Modulus Properties

The electric modulus plays an important role in the study of conduction and relaxation behavior of materials, and also, in detecting the impedance sources like grains, grain boundary conduction effect, electrode polarization and electrical conductivity. The real (M′) and imaginary (M″) components of electric modulus can be obtained using dielectric constant (ε' and ε'') [51]:

$$M^* = \frac{\varepsilon'}{\varepsilon'^2 + \varepsilon''^2} + i\frac{\varepsilon'}{\varepsilon'^2 + \varepsilon''^2} = M' + iM''. \tag{16}$$

The real component (M′) of electric modulus represents the energy given to the system, and the imaginary component (M″) represents the dissipated energy during the conduction process.

The frequency dependence of electric modulus (M′, M″) at room temperature is shown in Figure 13. It is observed that at lower frequencies M′ is very small nearly to be zero and a continuous dispersion with frequency increases having a tendency to saturate at a maximum value for all the samples at higher frequencies due to the relaxation process. These observations implies the lack of restoring force for flow of charge carriers under the action of steady electric field. The small value of M′ at low frequency supports the long range mobility of charge carriers. While, in higher frequencies, M′ increases rapidly with frequency, indicating that the conduction mechanism, which may be due to the short range mobility of charge carriers.

The imaginary part of electrical modulus M″ shows an increasing trend compared to frequency with relaxation peaks for all samples. The frequency region below peak frequency represents the frequency range by which ions drift to long distance, i.e., performing successful hopping from one site to neighboring site. Whereas, the high frequency region above the peak shows that the carriers are confined to their potential wells and can make localized motion inside the well. The occurrence of peak in electrical modulus M″ indicates the transition from long range to short range mobility of charge carriers with rise in frequency. The behavior of the modulus spectrum is indicative of hopping type mechanism for electrical conduction in the system. The broadening of the peaks is the consequence of the distributions of relaxation time arise from the non-Debye type of the material. Further, it is observed that the peaks shift towards the higher frequency side, with Zn doping and the obtained relaxation peaks having resonance peaks, where the oscillating dipoles frequency matches the applied field frequency.

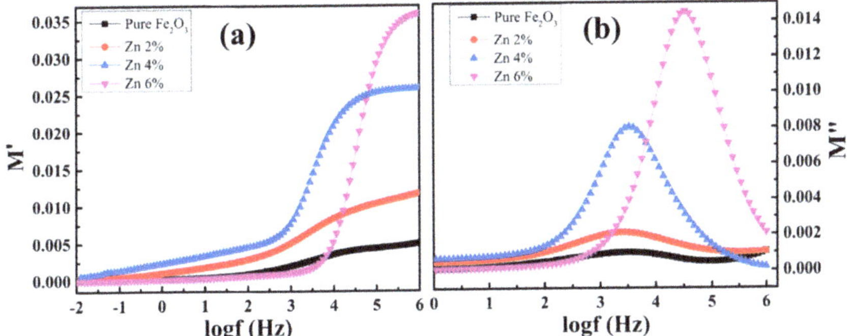

Figure 13. Frequency dependence (**a**) real component (M′) and (**b**) Imaginary component (M″) of electric modulus of pure Fe_2O_3, Zn 2%, Zn 4% and Zn 6% nanoparticles.

4. Conclusions

Zn-doped α-Fe_2O_3 nanoparticles were synthesized by sol-gel method to investigate the effect of Zn doping on structural, optical and dielectric behavior of hematite. XRD reveals the rhombohedral structure of as-prepared samples with average crystallite size lying between 14–21 nm and other various parameters like lattice parameter, strain, stress, dislocation density also has been discussed. Raman spectroscopy confirms defects in the host lattice exist that are consistent with XRD results. FTIR spectra reveals the metal oxygen vibrations and shifting in the spectra with doping ion concentration. The optical band gap obtained from Tauc plot decreases from 2.66 eV for pure Fe_2O_3 to 2.31 eV for 4% Zn doping and then increases for Zn 6%, which is in accordance with structural parameters. Photodegradation analysis shows that 4% Zn doped α–Fe_2O_3 is a good catalyst for degradation of RB dye under the illumination of UV light and almost 87% of RB dye has been degraded in 90 min in the presence of catalyst. In future, Zn doped α-Fe_2O_3 have been used as a very good photocatalyst to remove organic pollutants from waste water. AC conductivity increases with the increase in frequency for all the samples, due to the conduction hopping mechanism, revealing a small polaron hopping mechanism involved in conduction. The dispersion in dielectric constant spectra with frequency can be understood based on Koop and Maxwell-Wagner's models and also on basis of space charge polarization. Higher value of dielectric constant of synthesized samples at room temperature makes them suitable for high energy storage applications.

Author Contributions: Conceptualization, S. & S.C.; investigation, S.; resources, P.K.; writing—original draft preparation, S.; writing—review and editing, A.K. & P.K.; visualization, A.K.; supervision, P.K.; project administration, P.K.; funding acquisition, P.K. All authors have read and agreed to the published version of the manuscript.

Funding: The present work was financially supported by DST, New Delhi for providing research grant under DST-INSPIRE faculty award (No. DST/INSPIRE/04/2015/003149) to Parmod Kumar. Also, Suman acknowledges University Grant Commission, India (UGC-JRF) for providing research fellowship (UGC-Ref. No.: 1320/ (OBC) (CSIR-UGC NET DEC. 2016)).

Conflicts of Interest: The authors declare no conflict of interest.

References

1. Satheesh, R.; Vignesh, K.; Suganthi, A.; Rajarajan, M. Visible light responsive photocatalytic applications of transition metal (M = Cu, Ni, and Co) doped α-Fe_2O_3 nanoparticles. *J. Environ. Chem. Eng.* **2014**, *2*, 1956–1968. [CrossRef]
2. Nguyen, C.C.; Vu, N.N.; Do, T.O. Recent Advances in the Development of Sunlight-Driven Hollow Structure Photocatalysts and their Applications. *J. Mater. Chem. A* **2015**, *3*, 18345–18359. [CrossRef]

3. Chahal, S.; Rani, N.; Kumar, A.; Kumar, P. Electronic Structure and Photocatalytic Activity of Samarium Doped Cerium Oxide Nanoparticles for Hazardous Rose Bengal Dye Degradation. *Vacuum* **2020**, *172*, 109075. [CrossRef]

4. Bagheri, S.; Chandrappa, C.K.G.; Hamid, S.B.A. Generation of Hematite Nanoparticles via Sol-Gel Method. *Res. J. Chem. Sci.* **2013**, *3*, 62–68.

5. Showa, B.; Mukherjee, N.; Mondal, A. α-Fe$_2$O$_3$ nanospheres: Facile synthesis and highly efficient photo-degradation of organic dyes and surface activation by nano-Pt for enhanced methanol sensing. *RSC Adv.* **2016**, *6*, 75347–75358. [CrossRef]

6. Dong, H.; Zhang, H.; Xu, Y.; Zhao, C. Facile synthesis of α-Fe$_2$O$_3$ nanoparticles on porous human hair-derived carbon as improved anode materials for lithium-ion batteries. *J. Power Sources* **2015**, *300*, 104–111. [CrossRef]

7. Saritas, S.; Sakar, B.C.; Kundakci, M.; Yildirim, M. The effect of Mg dopants on magnetic and structural properties of iron oxide and zinc ferrite thin films. *Results Phys.* **2018**, *9*, 416–423. [CrossRef]

8. Abareshi, M.; Sajjadi, S.H.; Zebarjad, S.M.; Goharshadi, E.K. Fabrication, characterization, and measurement of viscosity of α-Fe$_2$O$_3$-glycerol nanofluids. *J. Mol. Liq.* **2011**, *163*, 27–32. [CrossRef]

9. Kumar, P.; Sharma, V.; Singh, J.P.; Kumar, A.; Chahal, S.; Sachdev, K.; Chae, K.H.; Kumar, A.; Asokan, K.; Kanjilal, D. Investigations on Magnetic and Electrical Properties of Zn doped Fe$_2$O$_3$ Nanoparticles and their correlation with local electronic structure. *J. Magn. Magn. Mater.* **2019**, *489*, 165398. [CrossRef]

10. Boudjemaa, A.; Boumazaa, S.; Trari, M.; Bouarab, R.; Bouguelia, A. Physical and photo-electrochemical characterizations of α-Fe$_2$O$_3$. Application for hydrogen production. *INT J. Hydrog. Energ.* **2009**, *34*, 4268–4274. [CrossRef]

11. Shen, S.; Guo, P.; Wheeler, D.A.; Jiang, J.; Lindley, S.A.; Kronawitter, C.X.; Zhang, J.Z.; Guoa, L.; Mao, S.S. Physical and Photoelectrochemical Properties of Zr-doped Hematite Nanorod Arrays. *Nanoscale* **2013**, *5*, 9867. [CrossRef] [PubMed]

12. Wang, G.; Ling, Y.; Wheeler, D.A.; George, K.E.N.; Horsley, K.; Heske, C.; Zhang, J.Z.; Li, Y. Facile Synthesis of Highly Photoactive α-Fe$_2$O$_3$-Based Films for Water Oxidation. *Nano Lett.* **2011**, *11*, 3503–3509. [CrossRef] [PubMed]

13. Yogi, A.; Varshney, D. Magnetic and structural properties of pure and Cr-doped hematite: α-Fe$_{2-x}$Cr$_x$O$_3$ ($0 \leq x \leq 1$). *J. Adv. Ceram.* **2013**, *2*, 360–369. [CrossRef]

14. Dang, M.Z.; Rancourt, D.G.; Dutrizac, J.E.; Lamarche, G.; Provencher, R. Interplay of surface conditions, particle size, stoichiometry, cell parameters, and magnetism in synthetic hematite-like materials. *Hyperfine Interact.* **1998**, *117*, 271–319. [CrossRef]

15. Zeng, S.; Tang, K.; Li, T. Controlled synthesis of α-Fe$_2$O$_3$ nanorods and its size-dependent optical absorption, electrochemical, and magnetic properties. *J. Colloid Interface Sci.* **2007**, *312*, 513–521. [CrossRef]

16. Varshney, D.; Yogi, A. Structural and Electrical conductivity of Mn doped Hematite (α-Fe$_2$O$_3$) phase. *J. Mol. Struct.* **2011**, *995*, 157–162. [CrossRef]

17. Bruzzone, C.L.; Ingalls, R. Mossbauer-effect study of the Morin transition and atomic positions in hematite under pressure. *Phys. Rev. B* **1983**, *28*, 2430–2440. [CrossRef]

18. Velev, J.; Bandyopadhyay, A.; Butler, W.H.; Sarker, S. Electronic and magnetic structure of transition-metal-doped α-hematite. *Phys. Rev. B* **2005**, *71*, 205208. [CrossRef]

19. Lassoued, A.; Dkhil, B.; Gadri, A.; Ammar, S. Control of the shape and size of iron oxide (α-Fe$_2$O$_3$) nanoparticles synthesized through the chemical precipitation method. *Results Phys.* **2017**, *7*, 3007–3015. [CrossRef]

20. Vijayalakshmin, K.; Renitta, A.; Karthick, K. Growth of high-quality ZnO:Mg films on ITO coated glass substrates for enhanced H$_2$ sensing. *Ceram. Int.* **2014**, *40*, 6171–6177. [CrossRef]

21. Kumar, P.; Singh, V.; Sharma, V.; Rana, G.; Malik, H.K.; Asokan, K. Investigation of phase segregation in yttrium-doped zinc oxide. *Ceram. Int.* **2015**, *41*, 6734–6739. [CrossRef]

22. Kumar, P.; Kumar, P.; Kumar, A.; Meena, R.C.; Tomar, R.; Chand, F.; Asokan, K. Structural, morphological, electrical and dielectric properties of Mn-doped CeO$_2$. *J. Alloys Compd.* **2016**, *672*, 543–548. [CrossRef]

23. Kumar, P.; Singh, J.P.; Malik, H.K.; Gautam, S.; Chae, K.H.; Asokan, K. Structural, transport and ferroelectric properties of Zn$_{1-x}$Mg$_x$O samples and their local electronic structure. *Superlattices Microst.* **2014**, *78*, 183–189. [CrossRef]

24. Jubb, A.M.; Allen, H.C. Vibrational Spectroscopic Characterization of Hematite, Maghemite and Magnetite Thin Films Produced by Vapor Deposition. *ACS Appl. Mater. Interfaces* **2010**, *2*, 2804–2812. [CrossRef]

25. Mccarty, K.F.; Boehme, D.R. A Raman Study of the Systems $Fe_{3-x}Cr_xO_4$ and $Fe_{2-x}Cr_xO_3$. *J. Solid State Chem.* **1989**, *79*, 19–27. [CrossRef]

26. Chernyshova, I.V.; Hochella, M.F., Jr.; Madden, A.S. Size-dependent structural transformations of hematite nanoparticles. 1. Phase transition. *Phys. Chem. Chem. Phys.* **2007**, *9*, 1736. [CrossRef] [PubMed]

27. Rani, B.J.; Mageswari, R.; Ravi, G.; Ganesh, V.; Yuvakkumar, R. Design, fabrication, and characterization of hematite (α-Fe_2O_3) Nanostructures. *JOM* **2017**, *69*, 2508–2514. [CrossRef]

28. Wang, W.; Liang, L.; Johs, A.; Gu, B. Thin films of uniform hematite nano-particles: Control of surface hydrophobicity and self-assembly. *J. Mater. Chem.* **2008**, *18*, 5770–5775. [CrossRef]

29. Darezereshki, E. One-step synthesis of hematite (α-Fe_2O_3) nano-particles by direct thermal-decomposition of maghemite. *Mater. Lett.* **2010**, *65*, 642–645. [CrossRef]

30. Lian, J.; Duan, X.; Ma, J.; Peng, P.; Kim, T.; Zheng, W. Hematite (α-Fe_2O_3) with Various Morphologies: Ionic Liquid-Assisted Synthesis, Formation Mechanism and Properties. *ACS Nano* **2009**, *3*, 3749–3761. [CrossRef]

31. Mohammed, K.A.; Al-Rawas, A.D.; Gismelseed, A.M.; Sellai, A.; Widatallah, H.M.; Yousif, A.; Elzain, M.E.; Shongwe, M. Infrared and structural studies of $Mg_{1-x}Zn_xFe_2O_4$ ferrites. *Phys. B* **2012**, *407*, 795–804. [CrossRef]

32. Lassoued, A.; Lassoued, M.S.; Granda, S.G.; Dkhil, B.; Ammar, S.; Gadri, A. Synthesis and characterization of Ni-doped α-Fe_2O_3 nanoparticles through co-precipitation method with enhanced photocatalytic activities. *J. Mater. Sci. Mater. Electron.* **2018**, *29*, 5726–5737. [CrossRef]

33. Mashiko, H.; Oshima, T.; Ohtomo, A. Band-gap narrowing in α-$(Cr_xFe_{1-x})_2O_3$ solid-solution films. *Appl. Phys. Lett.* **2011**, *99*, 241904. [CrossRef]

34. Arul, N.S.; Mangalaraj, D.; Ramachandran, R.; Grace, A.N.; Hana, J.I. Fabrication of CeO_2/Fe_2O_3 composite nanospindles for enhanced visible light driven photocatalyst and supercapacitor electrode. *J. Mater. Chem.* **2015**, *3*, 15248–15258. [CrossRef]

35. Nagaraju, P.; Kumar, Y.V.; Reddy, M.V.R.; Reddy, C.V.; Reddy, D.M. Phase, Preparation, Micro structural characterization and Optical characterization of pure and Gd doped ceria thin films. *J. Sci. Eng. Res.* **2014**, *5*, 2229–5518.

36. Chahal, S.; Rani, N.; Kumar, A.; Kumar, P. UV-irradiated photocatalytic performance of yttrium doped ceria for hazardous Rose Bengal dye. *Appl. Surf. Sci.* **2019**, *493*, 87–93. [CrossRef]

37. Anandan, K.; Rajendran, V. Effects of Mn on the magnetic and optical properties and photocatalytic activities of NiO nanoparticles synthesized via the simple precipitation process. *Mater. Sci. Eng. B Adv.* **2015**, *199*, 48–56. [CrossRef]

38. Bharatha, G.; Ponpandiana, N. Hydroxyapatite nanoparticles on dendritic α-Fe_2O_3 hierarchical architectures for heterogeneous photocatalyst and adsorption of Pb (II) ions from industrial wastewater. *RSC Adv.* **2015**, *5*, 84685–84693. [CrossRef]

39. Chahal, S.; Kumar, A.; Kumar, P. Erbium-doped oxygen deficient cerium oxide: Bi-functional material in the field of spintronics and photocatalysis. *Appl. Nanosci.* **2020**, 1–13. [CrossRef]

40. Şafak-asar, Y.; Asar, T.; Altindal, Ş.; Özçelik, S. Investigation of dielectric relaxation and ac electrical conductivity using impedance spectroscopy method in (AuZn)/TiO_2/p-GaAs(110) schottky barrier diodes. *J. Alloys Compd.* **2015**, *628*, 442–449. [CrossRef]

41. Phor, L.; Kumar, V. Structural, Magnetic and Dielectric properties of Lanthanum substituted $Mn_{0.5}Zn_{0.5}Fe_2O_4$. *Ceram. Int.* **2019**, *17*, 22972–22980. [CrossRef]

42. Rana, G.; Johri, U.C.; Asokan, K. Correlation between structural and dielectric properties of Ni-substituted magnetite nanoparticles. *Europhys. Lett.* **2013**, *103*, 17008. [CrossRef]

43. Iwauchi, K. Dielectric Properties of Fine Particles of Fe_3O_4 and Some Ferrites. *Jpn. J. Appl. Phys.* **1971**, *10*, 1520–1528. [CrossRef]

44. Pervaiz, E.; Gul, I.H. High frequency AC response, DC resistivity and magnetic studies of holmium substituted Ni-ferrite: A novel electromagnetic material. *J. Magn. Magn. Mater.* **2014**, *349*, 27–34. [CrossRef]

45. Dar, M.A.; Batoo, K.M.; Verma, V.; Siddiqui, W.A.; Kotnala, R.K. Synthesis and characterization of nano-sized pure and Al-doped lithium ferrite having high value of dielectric constant. *J. Alloys Compd.* **2010**, *493*, 553–560. [CrossRef]

46. Ramesh, S.; Dhanalakshmi, B.; Sekhar, B.C.; Rao, P.S.V.S.; Rao, B.P. Effect of Mn/Co substitutions on the resistivity and dielectric properties of nickel-zinc ferrites. *Ceram. Int.* **2016**, *42*, 9591–9598. [CrossRef]

47. Rezlescu, N.; Rezlescu, E. Dielectric properties of copper containing ferrites. *Phys. Status Solidi A* **1974**, *23*, 575–582. [CrossRef]

48. Ashwini, L.S.; Sridhar, R.; Bellad, S.S. Dielectric and magnetoelectric properties of LiMg ferrite: Barium titanate composites. *Mater. Chem. Phys.* **2017**, *200*, 136–145. [CrossRef]

49. Gul, I.H.; Ahmed, W.; Maqsood, A. Electrical and magnetic characterization of nanocrystalline Ni-Zn ferrite synthesis by co-precipitation route. *J. Magn. Magn. Mater.* **2008**, *320*, 270–275. [CrossRef]

50. Alia, I.; Islama, M.U.; Ashiqb, M.N.; Iqbala, M.A.; Khana, H.M.; Karamat, N. Effect of Tb-Mn substitution on DC and AC conductivity of Y-type hexagonal ferrite. *J. Alloys Compd.* **2013**, *579*, 576–582. [CrossRef]

51. Bindu, K.; Ajith, K.M.; Nagaraja, H.S. Electrical, dielectric and magnetic properties of Sn-doped hematite (α-$Sn_xFe_{2-x}O_3$) nanoplates synthesized by microwave-assisted method. *J. Alloys Compd.* **2017**, *735*, 847–854. [CrossRef]

crystals

Article

Seed-Mediated Preparation of Ag@Au Nanoparticles for Highly Sensitive Surface-Enhanced Raman Detection of Fentanyl

Yazhou Qin *, Binjie Wang, Yuanzhao Wu, Jiye Wang, Xingsen Zong and Weixuan Yao *

Key Laboratory of Drug Prevention and Control Technology of Zhejiang Province, Zhejiang Police College, 555 Binwen Road, Hangzhou 310053, China; wangbinjie@zjjcxy.cn (B.W.); wuyuanzhao@zjjcxy.cn (Y.W.); wangjiye@zjjcxy.cn (J.W.); zongxingsen@zjjcxy.cn (X.Z.)
* Correspondence: yazhouqin@zju.edu.cn (Y.Q.); yaoweixuan@zjjcxy.cn (W.Y.)

Abstract: Bimetallic nanoparticles have received extensive attention due to their unique physical and chemical properties, including enhanced optical properties, chemical stability, and better catalytic activity. In this article, we have successfully achieved the controllable preparation of Ag@Au nanoparticles via a seed-mediated growth method. By regulating the amount of seeds—silver nanospheres—we realized that Ag@Au nanoparticles gradually changed from spherical to a sea-urchin-like structure. The structure and composition of the prepared nanoparticles were characterized via scanning electron microscopy (SEM), transmission electron microscopy (TEM), energy-dispersive X-ray spectroscopy (EDS), and high-angle circular dark field scanning transmission electron microscopy (HAADF-STEM). In addition, we use the prepared Ag@Au nanoparticles as a substrate material for highly sensitive surface-enhanced Raman spectroscopy (SERS). Using 4-aminothiophenol (4-ATP) as the test molecule, we explored the SERS enhancement effects of Ag@Au nanoparticles with different structures. Furthermore, we used Ag@Au nanoparticles for SERS to detect the drug fentanyl, and realized the label-free detection of fentanyl, with the lowest detection concentration reaching 10^{-7} M. This research not only provides a method for preparing bimetallic Ag@Au nanoparticles with different structures, but also provides a reference for the application of Ag@Au nanoparticles in the field of detection technology.

Keywords: Ag@Au nanoparticle; core–shell structure; sea-urchin-like structure; SERS; fentanyl

Citation: Qin, Y.; Wang, B.; Wu, Y.; Wang, J.; Zong, X.; Yao, W. Seed-Mediated Preparation of Ag@Au Nanoparticles for Highly Sensitive Surface-Enhanced Raman Detection of Fentanyl. *Crystals* **2021**, *11*, 769. https://doi.org/10.3390/cryst11070769

Academic Editors: Raghvendra Singh Yadav, Anju and Kottakkaran Sooppy Nisar

Received: 8 June 2021
Accepted: 26 June 2021
Published: 1 July 2021

Publisher's Note: MDPI stays neutral with regard to jurisdictional claims in published maps and institutional affiliations.

1. Introduction

Noble metal nanocrystals have received extensive attention due to their unique localized surface plasmon resonance characteristics. A large number of studies have reported the application of noble metal nanocrystals in the fields of sensing, catalysis, and detection [1–5]. The characteristics of metal nanomaterials depend largely on their size, structure, and composition. In the past few decades, much research has been devoted to the controllable preparation of noble metal nanocrystals with uniform morphology and size. Various preparation methods have been developed one after another, including the solvothermal method [6–8], electrochemical method [9,10], and photochemical method [11]. Many metal nanocrystals with various regular morphologies have also been prepared. For example, Mirkin et al. realized the transformation of silver nanocrystals from spherical to icosahedral [12]. Yan et al. successfully prepared branched gold nanocrystals with tunable local surface plasmon resonance (LSPR) characteristics, and used the prepared gold nanocrystals as a substrate to achieve quantitative detection of the heme concentration in the cytosol of human red blood cells [13]. By controlling the kinetics involved in the growth of preformed Rh cube seeds, Xia et al. achieved the synthesis of Rh nanocrystals, including cubes, cubic octahedrons, and octahedrons [14]. In addition to polyhedral nanocrystals with regular morphologies, researchers have found that nanomaterials with sharp protrusions and pores often exhibit specific properties. For example, core–satellite structures and sea-urchin-like structures with sharp protrusions promote electromagnetic field enhancement by generating multiple "hot spots". This phenomenon has also aroused the interest of a large number of researchers.

In addition, researchers have found that alloy nanomaterials with different elemental compositions have the advantages of various nanomaterials, and show better performance than single-element nanomaterials in various fields. The preparation and performance of bimetallic nanoparticles is now an exciting research area, because they provide a new way to change the properties of the particles by mixing two metals in one particle. Especially for two kinds of metal nanoparticles with different advantages, we can realize the combination of their advantages through the controllable preparation of bimetallic nanoparticles, and avoid their respective disadvantages. For example, for two metal nanomaterials—gold and silver—studies have shown that silver nanomaterials have a larger excitation window— from blue to near-infrared—than gold nanoparticles (from red to near-infrared) [15]. In addition, it can be said that silver is a more effective optical material than gold. Due to its greater scattering contribution, which is related to the real part of its dielectric constant, the surface-enhanced Raman scattering (SERS) signal generated by silver is more than 100 times higher than that of similar gold nanostructures [16]. However, the size and shape of gold nanoparticles are easier to control, and have good chemical resistance—that is, higher biocompatibility [17]. Therefore, silver nanomaterials are used more in in vitro research, while gold nanoparticles are used more in in vivo research [18,19].

In order to make full use of the advantages of gold and silver nanoparticles, and avoid their disadvantages, we urgently need to develop an Ag@Au nanomaterial that can coat the surface of silver nanoparticles with a gold layer, or form an alloy of gold and silver on the surface of silver nanoparticles. In general, alloy nanoparticles are obtained via the simultaneous reduction of two metal precursors (Ag^+ and Au^{3+}). Conversely, the controlled growth of the metal on the preformed colloidal surface produces good results if the inner metal (the metal forming the colloid) has a greater surface free energy than the metal on its surface. However, because the surface free energy of gold is greater than that of silver (1.128 and 0.923 Jm^{-2}, respectively), it is easier to grow silver on the surface of gold nanoparticles. In fact, a number of studies have also reported that Au@Ag nanoparticles were obtained by growing silver layers on gold seeds [20,21]. For example, Xia et al. recently achieved regulation of the plasma performance of Au@Ag nanoparticles by adjusting the thickness of the silver layer grown on the surface of gold nanospheres and gold nanorods [22]. Cheng et al. adjusted the concentration of the surfactant cetyltrimethylammonium chloride (CTAC) to precisely control the position of the Ag coating on the convex gold nanoarrows. Three different nanostructures were obtained at low, medium, and high CTAC concentrations—namely, anisotropic coating, intermediate coating, and conformal coating, respectively [23]. Conversely, depositing gold onto silver nanoparticles by reducing gold precursors yields a solid solution of the two metals [24]. In general, the gold precursor will undergo a galvanic replacement reaction with the silver nanoparticles, so that the internal silver atoms are first replaced by silver ions, and then the silver ions are reduced to silver atoms by the reducing agent in the solution and deposited on the surface, so that the hollow structure is finally obtained [25–28]. As far as we know, only a few studies have reported the method of growing a gold layer on the surface of silver nanoparticles to prepare Ag@Au core–shell-structured nanoparticles. For example, Xia et al. successfully prepared Ag@Au nuclear sheath nanowires by depositing Au atoms on the surface of pre-synthesized Ag nanowires, which greatly improved their stability under different corrosive environmental conditions [29]. Kim et al. used sodium sulfite to selectively bind Au cations, and thereby reduce the reduction potential of Au ($E° = 0.111$ V). The gold(I) sulfite complex is relatively benign to the Ag nanowire surface, so no oxidative etching will occur; on the contrary, the Au coating is promoted, and finally Ag@Au nanowires are formed [30]. Kim et al. achieved gold-spiked coating of silver particles through controlled Ostwald ripening of small gold nanoparticles on the surface of larger silver particles [31].

In this article, in order to fully combine the advantages of gold and silver nanoparticles and avoid their disadvantages, we have developed a method for preparing Ag@Au nanoparticles. Ag@Au nanoparticles have both the excellent surface plasmon resonance characteristics of silver nanoparticles and the stability of gold. First, Au^{3+} is combined with Br^- in CTAB to form $[AuBr_4]^-$, and then the addition of ascorbic acid reduces Au^{3+} to Au^+. Finally, an Au(I)–GSH complex is formed by combining Au^+ with glutathione modified on the surface of silver nanoparticles, and then Au^+ is reduced by glutathione in situ to grow gold on the surface of silver nanoparticles. By simply adjusting the amount of silver seeds added, we realized the transformation of the Ag@Au nanoparticle structure from spherical to a sea-urchin-like structure.

2. Materials and Methods

2.1. Materials

The hydrochloroauric acid trihydrate ($HAuCl_4 \cdot 3H_2O$, 99.9% trace metals basis), hexadecyltrimethylammonium bromide (CTAB, AR, 99%), glutathione (Mw = 307.32, $\geq$98%), 4-aminothiophenol (4-ATP, GC, $\geq$98%), nitric acid (HNO_3, AR), and hydrochloric acid (HCl, AR) were purchased from Aladdin. Silver nanospheres (diameter: 80 nm; purity: >99% concentration: 0.1 mg/mL, XFJ63) were purchased from Nanjing/Jiangsu XFNANO Materials Tec Co., Let. Fentanyl hydrochloride standard product was purchased from the Shanghai Institute of Criminal Science and Technology (Shanghai, China). All reagents were used directly, without further processing. All of the water used in the experiment was ultrapure water (18.2 M·cm) purified using a Milli-Q Lab System (Nihon Millipore Ltd., Hangzhou, China). The flasks and glass slides used in the experiment were first washed with aqua regia ($HCl:HNO_3$ = 3:1) for 30 min, then washed with water (18.2 MΩ·cm) and absolute ethanol twice, and finally dried for use.

2.2. Synthesis of Ag@Au Nanoparticles

First, we mixed the silver nanoball seeds with 100 μL of 1 mM glutathione solution in a 1.5 mL centrifuge tube at room temperature; using an IKA shaker, we then vortexed it for 1 min, and incubated it for 1 h to make glutathione modified into silver nanospheres by the Ag–S bond. At the same time, we added 4.0 mL ultrapure water to a 20 mL glass flask, followed by 200 μL of 10 mM $HAuCl_4 \cdot 3H_2O$ solution and 0.8 mL of 0.1 M CTAB solution. The mixed solution was stirred at 500 rpm for 30 min at room temperature using an IKA magnetic stirring device, mixed well, and then 475 μL of 0.1 M ascorbic acid solution was added, stirring was continued at room temperature for 5 min, and the mixed solution was used as a growth solution. Finally, the silver nanospheres modified with glutathione were added to the growth solution, and the reaction was carried out at room temperature under magnetic stirring for 2 h. The resulting product was centrifuged at 5000 rpm for 10 min; the supernatant was removed, and washed twice with water and ethanol, respectively. In order to explore the effect of the amount of silver nanoball seeds added on the produced Ag@Au NPs, we added 40 μL, 80 μL, 160 μL, and 500 μL of silver nanosphere seeds to the reaction, without changing any of the other reaction conditions.

2.3. Characterization and SERS Test

The morphology and structure of the prepared nanoparticles were passed through a field emission scanning electron microscope (SEM, 3.0 kV, JEOL, JSM-6700F, Japan), transmission electron microscope (TEM, 100.0 kV, JEOL-2100f, Japan), high-resolution TEM (HRTEM, 200 kV, JEOL-2100f, Japan), and selected-area electron diffraction (SAED, JEOL-2100f, Japan) for characterization. The SERS test was conducted using a Raman spectrometer (Thermo Fisher DXR2xi, America) under 633 nm laser excitation; the laser power was 6 mW, and the acquisition time was 10 s. For surface-enhanced Raman spectroscopy testing, we first added 10 μL of Ag@Au nanoparticles to the centrifuge tube, and then added 50 μL of 4-ATP solutions of different concentrations, mixing well. The mixed solution was stored at room temperature for 30 min, and then 5 μL was dropped on a glass slide, dried at room temperature, and used for surface-enhanced Raman spectroscopy (SERS) testing. The SERS test used a laser Raman spectrometer (Thermo Fisher DXR2xi, America) equipped with a microscope (50× objective lens) and a CCD detector; the laser source was 633 nm, the acquisition time was 10 s, and the laser power was 6 mW. We used the supporting software to analyze the Raman peak spectrum. The SERS test for the drug fentanyl followed the same process.

3. Results and Discussion

Characterization of Ag@Au NPs

We studied the morphology and size of Ag@Au nanoparticles via SEM and TEM. Figure 1a is the SEM image of Ag@Au nanoparticles prepared when the amount of silver seed added is 40 µL, and Figure 1b is the corresponding TEM image. As we can see, the surface of the prepared Ag@Au nanoparticles is distributed with dense, thorn-like structures. From the HRTEM image (Figure 1c) of the gold thorns grown on the surface, it can be seen that the lattice spacing is 0.24 nm, indicating that the exposed crystal faces of the gold thorns are mainly composed of {111} planes [4]. Figure 1d shows the corresponding HAADF map. We performed EDS characterization of the prepared Ag@Au nanoparticles, as shown in Figure 1e–g. Figure 1e,f are the distribution diagrams of silver and gold, respectively. Figure 1g shows the overlapping distribution of the two elements. From the EDS results, it can be seen that the atomic ratio of gold to silver is 68:32.

Figure 1. (**a**) SEM, (**b**) TEM, (**c**) HRTEM, and (**d**) HAADF images of as-prepared Ag@Au NPs, when the amount of Ag seed was 40 uL. (**e–g**) The corresponding EDS images.

Figure 2 shows the characterization results of Ag@Au nanoparticles obtained when the amount of silver seed added is 500 µL. Figure 2a,b show the SEM and TEM images of Ag@Au nanoparticles, respectively. From the figure, we can clearly see that the Ag@Au nanoparticles obtained still maintain a spherical structure, and have a uniform size distribution of about 95 nm. From its HRTEM image (Figure 2c), we can see that the lattice spacing of gold nanoparticles distributed on the surface is 0.24 nm, indicating that it is mainly composed of {111} planes. In addition, from the HAADF map (Figure 2d), it can be seen that only a small part of the Ag@Au nanoparticles are gray in the central area, indicating that the Ag@Au nanoparticles have formed a partially hollow structure, while the main part is still a solid structure. Furthermore, we collected the distribution of Au and Ag elements in Ag@Au nanoparticles via EDX, as shown in Figure 2e–g. From the figure, we can see that gold and silver form an alloy structure distributed over the entire surface.

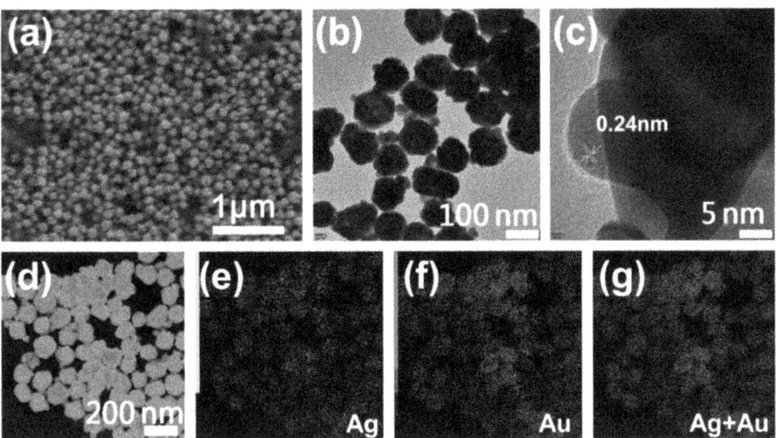

Figure 2. (a) SEM, (b) TEM, (c) HRTEM, and (d) HAADF images of as-prepared Ag@Au nanoparticles, when the amount of Ag seed was 500 µL. (e–g) The corresponding EDS images.

Furthermore, we explored the effects of the amount of silver seeds on the Ag@Au nanoparticles. As shown in Figure 3, when there was no silver seed, we prepared a mesoporous gold structure (Figure 3a). When adding 40 µL of silver seeds, we prepared Ag@Au nanoparticles with dense, thorn-like structures on their surface (Figure 3b). When the amount of silver seeds was doubled (80 µL), the thorn-like structures on the surface of the Ag@Au nanoparticles were reduced (Figure 3c). When the amount of silver seeds was further increased to 160 µL, the thorn-like structures on the surface were further reduced (Figure 3d). When the amount of silver seed was increased to 500 µL, we achieved a spherical structure with shape retention (Figure 3e). According to the EDS characterization results of the nanoparticles prepared under different concentrations of silver seeds, we performed a statistical analysis of the proportions of gold and silver atoms, as shown in Figure 3f. The EDS diagrams of Ag@Au nanoparticles prepared under different silver seed conditions are shown in Figure A1. Analyzing the EDS results, when the amount of silver seeds added was 40, 80, 160, and 500 µL, the proportion of tightness in the prepared Ag@Au nanoparticles was 0.68, 0.44, 0.38, and 0.20, respectively. The corresponding proportions of silver were 0.32, 0.56, 0.62, and 0.80, respectively.

Based on the above experimental results, we can make the following inferences on the formation mechanism of Ag@Au nanoparticles: Figure 3g is a schematic diagram of the growth of Ag@Au nanoparticles. First, Au^{3+} ions in $HAuCl_4 \cdot 3H_2O$ solution combine with Br^- ions in CTAB to form $[AuBr_4]^-$, and then ascorbic acid reduces $[AuBr_4]^-$ to get Au^+. This process is consistent with previous reports [32,33]. When silver seeds modified with glutathione were added, Au^+ reduced glutathione adsorbed on the surface of silver seeds in situ to form Au^0. It can be seen from Figure 2D that Au^+ inevitably had a galvanic replacement reaction with silver seeds (forming few hollow structures). However, the majority of Ag@Au nanoparticles are still solid, which also indicates that the glutathione on the surface of the silver seeds largely inhibits the galvanic replacement reaction. In addition, the reduction of glutathione also causes the gold ions to be reduced to gold atoms and deposited on the surface of the silver seeds, which also prevents the occurrence of the galvanic replacement reaction. Thus, when the amount of silver seeds is low, there are enough gold atoms to grow along the direction of glutathione, and eventually form sea-urchin-like Ag@Au nanoparticles. When the amount of silver seeds is greater, the Ag@Au spherical shell structure with the original shape will be formed. The above results show that we have successfully prepared Ag@Au nanoparticles. Compared with the previously reported Au@Ag core–shell structure and the Ag@Au hollow structure prepared by the displacement reaction, the Ag@Au nanoparticles we prepared have a solid structure. A gold thorn structure grows on the surface of the seed, which realizes the combination of the advantages of gold and silver nanoparticles.

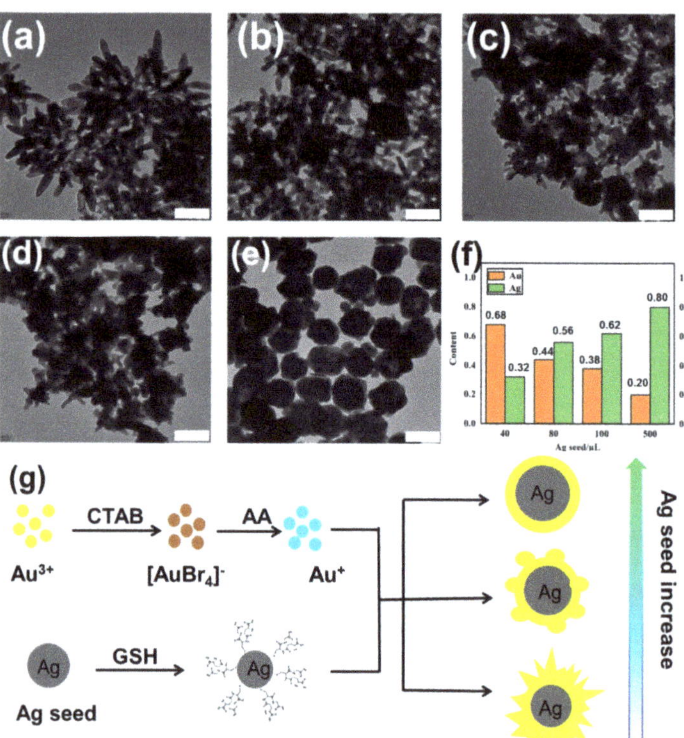

Figure 3. TEM images of as-prepared Ag@Au nanoparticles, when the amount of Ag seed added was (**a**) 0 µL, (**b**) 40 µL, (**c**) 80 µL, (**d**) 160 µL, and (**e**) 500 µL. (**f**) A statistical graph of the atomic ratio of Ag and Au in Ag@Au nanoparticles prepared with different amounts of silver seeds, according to the EDS results. (**g**) Schematic diagram of the growth process of Ag@Au nanoparticles. Scale bar: 100 nm.

Furthermore, we used the prepared Ag@Au nanoparticles as the substrate material for surface-enhanced Raman spectroscopy (SERS) for highly sensitive detection of the drug fentanyl. First, we explored the SERS enhancement effect of Ag@Au nanoparticles with different structures. Figure 4 shows the SERS spectra obtained with 10^{-4} M *p*-aminothiophenol (4-ATP) as the test molecule and Ag@Au nanoparticles with different structures as the base material. We can see that the order of SERS enhancement effects, from high to low, is spherical Ag@Au nanoparticles, sea-urchin-shaped Ag@Au nanoparticles, and mesoporous gold nanoparticles. Therefore, in the follow-up test, we use spherical Ag@Au nanoparticles as the enhancement reagent. The Raman characteristic peaks of 4-ATP molecules are located at 1076 cm^{-1} and 1585 cm^{-1} [34], and we see obvious SERS peaks at 1140 cm^{-1}, 1389 cm^{-1}, and 1432 cm^{-1}, which can be attributed to βCH + νCN, νNN + νCN, and νNN + βCH of *p,p′*-dimercaptoazobenzene (DMAB), respectively [4]. This indicates that the prepared nanomaterials have catalytic activity, so that 4-ATP is partially converted to DMAB by catalytic oxidation.

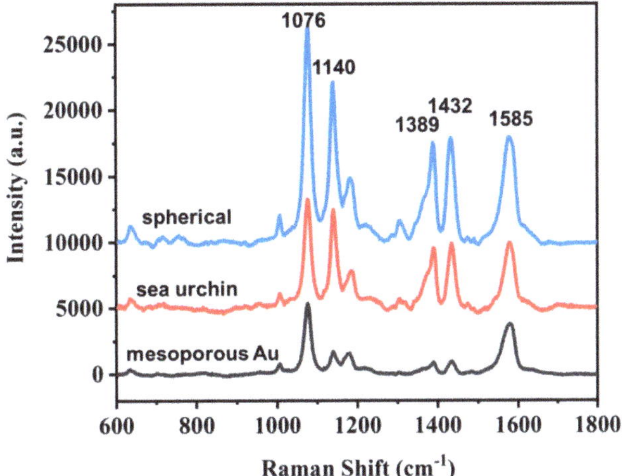

Figure 4. SERS spectra of 10^{-4} M 4-ATP molecules with different substrate materials.

We used spherical Ag@Au nanoparticles to detect fentanyl. Figure 5a shows the Raman spectrum and SERS spectrum of the fentanyl standard substance. As shown in Figure 5a (black line), the Raman peaks of fentanyl solid powder are located at 621 cm^{-1}, 746 cm^{-1}, 831 cm^{-1}, 1002 cm^{-1}, 1030 cm^{-1}, 1201 cm^{-1}, 1447 cm^{-1}, and 1600 cm^{-1}, which is consistent with previous reports [35]. However, the peak at 621 cm^{-1} in the SERS spectrum (red line) disappeared. This is because, for a group, the Raman peak of tensile vibration usually appears in a higher frequency range, and is less affected by the external environment. The Raman peak of deformation vibration is usually located in a lower frequency range, and is sensitive to environmental changes [36]. The peak at 621 cm^{-1} is the bending vibration of C–C–C [35], which is easily affected by the SERS detection environment, and disappears. We assigned the SERS peak of fentanyl, as shown in Table A1. Furthermore we performed SERS tests on different concentrations of fentanyl, and the results are shown in Figure 5b. We can see that the lowest detection concentration is 10^{-7} M.

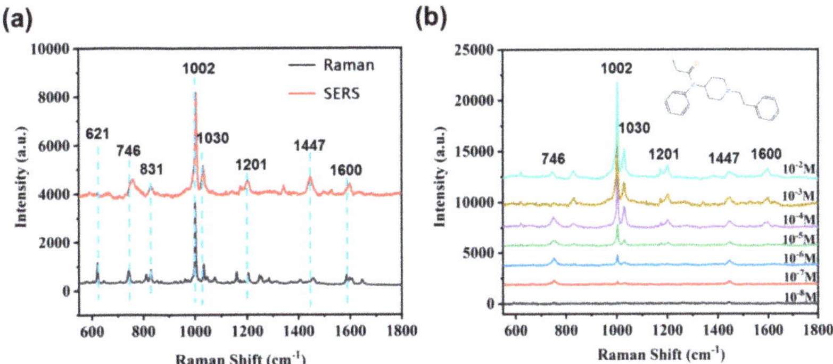

Figure 5. (**a**) Normal and surface-enhanced Raman spectra for fentanyl (10^{-4} M). (**b**) SERS spectra of fentanyl at different concentrations.

4. Conclusions

In this paper, we prepared Ag@Au nanoparticles via a seed-mediated growth method. By adjusting the amount of silver seeds, we realized the transformation of Ag@Au nanoparticles from a spherical to a sea-urchin-like structure, and proposed the possible growth mechanism. We further explored the SERS enhancement performance of the prepared nanoparticles, using 4-ATP as the test molecule. The results show that the order of SERS enhancement effects, from high to low, is spherical Ag@Au nanoparticles, sea-urchin-shaped Ag@Au nanoparticles, and mesoporous gold nanoparticles. Finally, with Ag@Au spherical nanoparticles as the enhancement material, we achieved the highly sensitive SERS detection of fentanyl, with the lowest detection concentration reaching 10^{-7}.

Author Contributions: Methodology, Y.Q.; validation, W.Y., B.W. and Y.Q.; resources, W.Y., J.W. and X.Z.; data curation, J.W.; writing—original draft preparation, Y.Q.; writing–review and editing, Y.W., B.W. and Y.Q.; supervision, X.Z.; project administration, J.W. and W.Y.; funding acquisition, W.Y. All authors have read and agreed to the published version of the manuscript.

Funding: This research was funded by "the National Key Research and Development Program of Zhejiang Province, grant number 2021C03135", "the National Key Research and Development Program of China Public Safety Risk Prevention and Control, grant number 2018YFC0807201", Zhejiang Police College scientific research project, grant number 2019XJY001 and "Zhejiang Provincial Natural Science Foundation of China under grant number, LGF20C090001" The APC was funded by the National Key Research and Development Program of Zhejiang Province.

Institutional Review Board Statement: The Institutional Review Committee has informed and agreed to the publication of this article.

Informed Consent Statement: All authors know and agree to the publication of this article.

Data Availability Statement: The data presented in this study are available within the article.

Acknowledgments: We would like to express our heartfelt thanks to Mao Zhanpeng for his assistance in the electron microscope test and the assistance provided by XFNANO with the experimental materials.

Conflicts of Interest: The authors declare no conflict of interest.

Appendix A

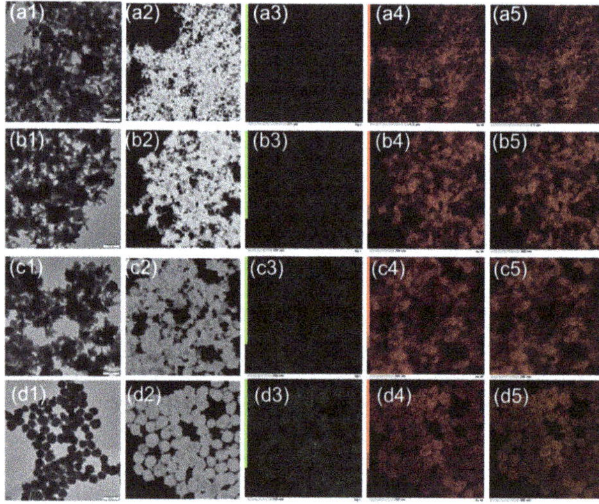

Figure A1. TEM, HAADF, and EDS characterization of the prepared nanoparticles with different amounts of Ag seeds: (**a1–a5**) 40 µL, (**b1–b5**) 80 µL, (**c1–c5**) 160 µL, and (**d1–d5**) 500 µL.

Table A1. Characteristic vibrations of fentanyl.

Fentanyl	
746	δ(C–C–C), δ(N–CH$_3$)
831	ν(N–C–C–C)
1002	δ(CH) Ar
1030	ν(C=C), δ(CH$_2$) twisting in aliphatic ring
1201	δ(C–C) benzyl stretch
1447	δ(CH$_2$) scissoring
1600	ν(C=C) Ar

References

1. Ding, S.Y.; Yi, J.; Li, J.F.; Ren, B.; Wu, D.Y.; Panneerselvam, R.; Tian, Z.Q. Nanostructure-based plasmon-enhanced Raman spectroscopy for surface analysis of materials. *Nat. Rev. Mater.* **2016**, *1*, 16021–16036. [CrossRef]
2. Li, J.F.; Zhang, Y.J.; Ding, S.Y.; Panneerselvam, R.; Tian, Z.Q. Core–Shell Nanoparticle-Enhanced Raman Spectroscopy. *Chem. Rev.* **2017**, *117*, 5002–5069. [CrossRef]
3. Dong, D.; Yap, L.W.; Smilgies, D.M.; Si, K.J.; Shi, Q.; Cheng, W. Two-dimensional gold trisoctahedron nanoparticle superlattice sheets: Self-assembly, characterization and immunosensing applications. *Nanoscale* **2018**, *10*, 5065–5071. [CrossRef]
4. Qin, Y.; Lu, Y.; Pan, W.; Yu, D.D.; Zhou, J.G. One-pot synthesis of hollow hydrangea Au nanoparticles as a dual catalyst with SERS activity for in situ monitoring of a reduction reaction. *RSC Adv.* **2019**, *9*, 10314–10319. [CrossRef]
5. Ha, M.J.; Kim, J.H.; You, M.; Li, Q.; Fan, C.H.; Nam, J.M. Multicomponent Plasmonic Nanoparticles: From Heterostructured Nanoparticles to Colloidal Composite Nanostructures. *Chem. Rev.* **2019**, *119*, 12208–12278. [CrossRef]
6. Qin, Y.; Pan, W.F.; Yu, D.D.; Lu, Y.X.; Wu, W.H.; Zhou, J.G. Stepwise evolution of Au micro/nanocrystals from an octahedron into a truncated ditetragonal prism. *Chem. Commun.* **2018**, *54*, 3411–3414. [CrossRef]
7. Qin, Y.Z.; Lu, Y.X.; Yu, D.D.; Zhou, J.G. Controllable synthesis of Au nanocrystals with systematic shape evolution from an octahedron to a truncated ditetragonal prism and rhombic dodecahedron. *CrystEngComm* **2019**, *21*, 5602–5609. [CrossRef]
8. Litti, L.; Reguera, J.; Abajo, F. Manipulating chemistry through nanoparticle Morphology. *Nanoscale Horiz.* **2020**, *5*, 102–108. [CrossRef]
9. Tian, N.; Zhou, Z.Y.; Yu, N.F.; Wang, L.Y.; Sun, S.G. Direct electrodeposition of tetrahexahedral Pd nanocrystals with high-index facets and high catalytic activity for ethanol electrooxidation. *J. Am. Chem. Soc.* **2010**, *132*, 7580–7581. [CrossRef]
10. Du, J.H.; Sheng, T.; Xiao, C.; Tian, N.; Xiao, J.; Xie, A.; Liu, S.; Zhou, Z.; Sun, S.G. Shape transformation of {hk0}-faceted Pt nanocrystals from a tetrahexahedron into a truncated ditetragonal prism. *Chem. Commun.* **2017**, *22*, 3236–3238. [CrossRef] [PubMed]
11. Guillerm, G.R.; Pablo, D.N.; Antonio, R.; Alejandro, P.; Gloria, T.; Jesús, G.; Luis, B.; Pablo, L.; Luis, G.M.; Mauricio, A.P.; et al. Femtosecond laser reshaping yields gold nanorods with ultranarrow surface plasmon resonances. *Science* **2017**, *358*, 640–644.
12. Langille, M.R.; Zhang, J.; Personick, M.L.; Li, S.; Mirkin, C.A. Stepwise evolution of spherical seeds into 20-fold twinned icosahedra. *Science* **2012**, *337*, 954–957. [CrossRef] [PubMed]
13. Cai, Z.; Hu, Y.; Sun, Y.; Gu, Q.; Wu, P.; Cai, C.; Yan, Z. Plasmonic SERS Biosensor Based on Multibranched Gold Nanoparticles Embedded in Polydimethylsiloxane for Quantification of Hematin in Human Erythrocytes. *Anal. Chem.* **2021**, *93*, 1025–1032. [CrossRef] [PubMed]
14. Zhao, M.; Chen, Z.; Shi, Y.; Hood, Z.D.; Lyu, Z.; Xie, M.; Chi, M.; Xia, Y. Kinetically Controlled Synthesis of Rhodium Nanocrystals with Different Shapes and a Comparison Study of Their Thermal and Catalytic Properties. *J. Am. Chem. Soc.* **2021**, *143*, 6293–6302. [CrossRef] [PubMed]
15. Zhao, J.; Pinchuk, A.O.; McMahon, J.M.; Li, S.; Ausman, L.K.; Atkinson, A.L.; Schatz, G.C. Methods for Describing the Electromagnetic Properties of Silver and Gold Nanoparticles. *Acc. Chem. Res.* **2008**, *41*, 1710–1720. [CrossRef]
16. Abajo, F.J. Colloquium: Light scattering by particle and hole arrays. *Rev. Mod. Phys.* **2007**, *79*, 1267–1290. [CrossRef]
17. Murphy, C.J.; Gole, A.M.; Stone, J.W.; Sisco, P.N.; Alkilany, A.M.; Goldsmith, E.C.; Baxter, S.C. Gold Nanoparticles in Biology: Beyond Toxicity to Cellular Imaging. *Acc. Chem. Res.* **2008**, *41*, 1721–1730. [CrossRef] [PubMed]
18. Schlücker, S. Surface-enhanced Raman spectroscopy: Concepts and chemical applications. *Angew. Chem. Int. Ed.* **2014**, *53*, 4756–4795. [CrossRef]
19. Sperling, R.A.; Gil, P.R.; Zhang, F.; Zanella, M.; Parak, W.J. Biological applications of gold nanoparticles. *Chem. Soc. Rev.* **2008**, *37*, 1896–1908. [CrossRef]
20. Guisbiers, G.; Mendoza-Cruz, R.; Bazán-Díaz, L.; Elázquez-Salazar, J.J.; Mendoza-Perez, V.R.; Torres, J.A.; Lopez, J.L.; Carrizales, J.M.; Whetten, R.L.; Yacamán, M.J. Response to "Comment on 'Electrum, the Gold–Silver Alloy, from the Bulk Scale to the Nanoscale: Synthesis, Properties, and Segregation Rules'". *ACS Nano* **2016**, *10*, 188–198. [CrossRef] [PubMed]

21. Cardinal, M.F.; González, B.R.; Puebla, R.A.; Juste, J.P.; Marzán, L.M. Modulation of Localized Surface Plasmons and SERS Response in Gold Dumbbells through Silver Coating. *J. Phys. Chem. C* **2010**, *114*, 10417–10423. [CrossRef]
22. Gao, Z.; Shao, S.; Gao, W.; Tang, D.; Tang, D.; Zou, S.; Kim, M.J.; Xia, X. Morphology-Invariant Metallic Nanoparticles with Tunable Plasmonic Properties. *ACS Nano* **2021**, *15*, 2428–2438. [CrossRef]
23. Dong, D.; Shi, Q.; Sikdar, D.; Zhao, Y.; Liu, Y.; Fu, R.; Premaratne, M.; Cheng, W. Site-specific Ag coating on concave Au nanoarrows by controlling the surfactant concentration. *Nanoscale Horiz.* **2019**, *4*, 940–946. [CrossRef]
24. Liao, H.; Fisher, A.; Xu, Z.J. Surface Segregation in Bimetallic Nanoparticles: A Critical Issue in Electrocatalyst Engineering. *Small* **2015**, *11*, 3221–3246. [CrossRef]
25. Sun, Y.; Mayers, B.T.; Xia, Y. Template-Engaged Replacement Reaction: A One-Step Approach to the Large-Scale Synthesis of Metal Nanostructures with Hollow Interiors. *Nano Lett.* **2002**, *2*, 481–485. [CrossRef]
26. Sun, Y.; Xia, Y. Shape-Controlled Synthesis of Gold and Silver Nanoparticles. *Science* **2002**, *298*, 2176–2179. [CrossRef]
27. Xuan, H.; Nguyen, D.D.; Thi, T.H.P.; Nguyen, V.T.; Nguyen, X.C.; Vu, V.T. Tunable LSPR of silver/gold bimetallic nanoframes and their SERS activity for methyl red detection. *RSC Adv.* **2021**, *11*, 14596–14606.
28. Ahn, J.; Wang, D.; Ding, Y.; Zhang, J.; Qin, D. Site-Selective Carving and Co-Deposition: Transformation of Ag Nanocubes into Concave Nanocrystals Encased by Au–Ag Alloy Frames. *ACS Nano* **2017**, *12*, 298–307. [CrossRef] [PubMed]
29. Yang, M.; Hood, Z.D.; Yang, X.; Chi, M.; Xia, Y. Facile synthesis of Ag@Au core–sheath nanowires with greatly improved stability against oxidation. *Chem. Commun.* **2017**, *53*, 1965–1968. [CrossRef]
30. Choi, S.; Han, S.I.; Jung, D.; Hwang, H.J.; Lim, C.; Bae, S.; Park, O.K.; Tschabrunn, C.M.; Lee, M.; Bae, S.Y.; et al. Highly conductive, stretchable and biocompatible Ag–Au core–sheath nanowire composite for wearable and implantable bioelectronics. *Nat. Nanotech.* **2018**, *13*, 1748–3387. [CrossRef]
31. Irene, C.; Ramon, A.A.; Nicolas, P.P. Gold-spiked coating of silver particles through cold nanowelding. *Nanoscale* **2021**, *13*, 4530–4536.
32. Lee, H.; Ahn, H.Y.; Mun, J.; Lee, Y.Y.; Kim, M.; Cho, N.H.; Chang, K.; Kim, W.S.; Rho, J.; Nam, K. Amino-acid-and peptide-directed synthesis of chiral plasmonic gold nanoparticles. *Nature* **2018**, *556*, 360–365. [CrossRef] [PubMed]
33. Qin, Y.Z.; Wu, Y.; Wang, B.; Wang, J.; Zong, X.; Yao, W. Controllable preparation of sea urchin-like Au NPs as a SERS substrate for highly sensitive detection of the toxic atropine. *RSC Adv.* **2021**, *11*, 19813–19818. [CrossRef]
34. Huang, Y.F.; Zhu, H.P.; Liu, G.K.; Wu, D.Y.; Ren, B.; Tian, Z.Q. When the signal is not from the original molecule to be detected: Chemical transformation of para-aminothiophenol on Ag during the SERS measurement. *J. Am. Chem. Soc.* **2010**, *132*, 9244–9246. [CrossRef] [PubMed]
35. Haddad, A.; Greencand, O.; Lombardi, J.R. Detection of fentanyl in binary mixtures with cocaine by use of surface-enhanced Raman spectroscopy. *Spectrosc. Lett.* **2019**, *8*, 462–472. [CrossRef]
36. Matas, M.D.; Edwards, H.G.; Lawson, E.E.; Shields, L.; York, P. FT-Raman spectroscopic investigation of a pseudopolymorphic transition in caffeine hydrate. *J. Mol. Struct.* **1998**, *440*, 97–104. [CrossRef]

Review

A Review on Metamaterials for Device Applications

N. Suresh Kumar [1], K. Chandra Babu Naidu [2,*], Prasun Banerjee [3], T. Anil Babu [2] and B. Venkata Shiva Reddy [2,4]

1   Department of Physics, JNTUA College of Engineering, Anantapuramu 515002, India;
    sureshmsc6.physics@jntua.ac.in
2   Department of Physics, Gandhi Institute of Technology and Management (GITAM) Deemed to be University,
    Bangalore 562163, India; atadiboy@gitam.edu (T.A.B.); 321962704001@gitam.in (B.V.S.R.)
3   Multiferroic and Magnetic Materials Research Laboratory (MMMRL), Gandhi Institute of Technology and
    Management (GITAM) Deemed to be University, Bangalore 562163, India; pbanerje@gitam.edu
4   Department of Physics, The National College, Bagepalli 561207, India
*   Correspondence: ckadiyal@gitam.edu

Abstract: Metamaterials are the major type of artificially engineered materials which exhibit naturally unobtainable properties according to how their microarchitectures are engineered. Owing to their unique and controllable effective properties, including electric permittivity and magnetic permeability, the metamaterials play a vital role in the development of meta-devices. Therefore, the recent research has mainly focused on shifting towards achieving tunable, switchable, nonlinear, and sensing functionalities. In this review, we summarize the recent progress in terahertz, microwave electromagnetic, and photonic metamaterials, and their applications. The review also encompasses the role of metamaterials in the advancement of microwave sensors, photonic devices, antennas, energy harvesting, and superconducting quantum interference devices (SQUIDs).

Keywords: metamaterials; electronic materials; electromagnetic; Fano resonances; SQUID

Citation: Suresh Kumar, N.; Naidu, K.C.B.; Banerjee, P.; Anil Babu, T.; Venkata Shiva Reddy, B. A Review on Metamaterials for Device Applications. *Crystals* **2021**, *11*, 518. https://doi.org/10.3390/cryst11050518

Academic Editors: Yuri Kivshar, Raghvendra Singh Yadav, Anju Deswal and Kottakkaran Sooppy Nisar

Received: 12 March 2021
Accepted: 30 April 2021
Published: 7 May 2021

Publisher's Note: MDPI stays neutral with regard to jurisdictional claims in published maps and institutional affiliations.

1. Introduction

In the modern society, we are surrounded by an amazing variety of materials, such as polymers, ceramics, composites, high strength alloys, superalloys, and so on. Naturally, existing materials possess some properties. However, one can change or modify the naturally occurring properties by altering the microstructure of the materials. The resultant materials are also called as the engineered materials. Recently, the engineered materials have gained much attention by researchers. The engineered materials exhibit better performance, such as large cyclic life, broad temperature ranges, lighter in weight, etc., than conventional materials. There exist numerous engineered materials such as metal-matrix composites, polymer-based composites, piezoelectric materials, magnetostrictive materials, metamaterials, etc. Owing to their large lifetime and high performance, these materials are significantly used almost in all fields of science and engineering. In this article, we have focused on metamaterials and their applications.

Metamaterials

In Greek, the word meta means "beyond". The metamaterials are new class of engineered materials which exhibit unusual electromagnetic properties that do not occur in natural materials. In general, natural materials like diamond, glass, etc., have positive refractive index, magnetic permeability, and electrical permittivity. Whereas these new engineered materials show negative index of refraction, negative electrical permittivity, and magnetic susceptibility. The metamaterials are also called as left-handed (LH) materials or backward wave (BW) media or negative index materials (NIM) or double negative (DNG) media. In addition, metamaterials have some special properties such as perfect lensing [1], classical electromagnetically induced transparency [2–5], cloaking capability [6], high frequency magnetism [7], dynamic modulation of Terahertz (THz) radiation [8], reverse

Doppler effect, and reverse Cerenkov effect [9]. These unique properties of metamaterials enable them to fabricate functional devices with switching and tuning capabilities [10–14]. Further, based on the permittivity and the permeability, the metamaterials are categorized as mu-negative material (MNG), epsilon negative material (ENG), double positive material (DPS), and double negative material (DNG) [15]. The MNG and ENG are also called as single negative materials. Whereas the double negative and double positive materials can be engineered at particular frequency band. In the year 1968, Victor Veselago, who is a Russian physicist, first theoretically proposed the metamaterials [16]. He theoretically explored the electrodynamics of the materials with negative values of the magnetic permeability (μ) and relative permittivity (ϵ). Moreover, the propagation of electromagnetic waves is discussed in those materials. Nevertheless Smith et al. practically demonstrated for first time a structure which exhibits negative refraction in the microwave region [15]. In addition, J.B. Pendry et al. [17] fabricated the first metamaterial by two interpenetrating subsystems. Particularly, by using an array of thin metallic (copper) wires and rings, they fabricated split ring resonators (SRRs), which provide negative values of permittivity and permeability [17–19]. In metamaterials, the split rings play the role of atoms in natural materials and act as electrically small resonant particles which contribute negative permeability. Herein, copper wire contributes negative permittivity. Combined array exhibits negative permeability and permittivity. Figure 1 gives the pictorial representation of a metamaterial [20] (Figure taken from ref. [20]). Later, Ziolkowski [21] reported another metamaterial which consists of substrate entrenched with capacitively loaded strips (CLSs) and square SRRs. In this material, capacitively loaded strips gives a strong response to electric fields and gives negative ϵ and SRRs interacts with magnetic fields and provides negative μ. Afterwards, numerous researchers designed and fabricated different metamaterials by using different methods like shadow mask/etching, clean room etching, etc., in many frequency bands. In addition, based on the presence of SRRs, the metamaterials are available in various forms. That means the metamaterials consisting SRRs are available in one, two, and three dimensions [22–25]. Further, those metamaterials without SRRs (for example, fishnet structure) are available in two, quasi-two, and three dimensions [26–29]. Hence, owing to their unique properties and structures, the metamaterials find their applications in various devices such as sensors [30–32], superlens [33], antennas [34,35], superconductors [36], absorbers [37–42], energy harvesting [43,44], etc. In this article, we have concentrated on recently developed metamaterials and their applications.

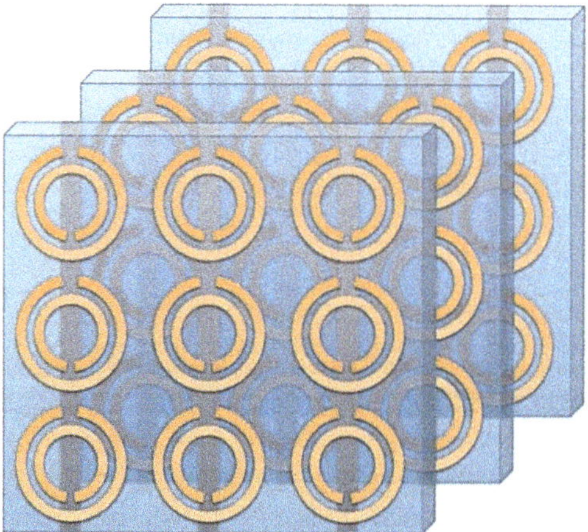

Figure 1. Pictorial representation of a metamaterial [20].

2. Applications of Metamaterials

As we know, the metamaterials are engineered metamaterials which exhibit exceptional properties that cannot be found naturally. In metamaterials, the resonant nature of the atoms leads to the enhancement in light matter interaction that affords dramatic changes in the light properties [45]. Especially, chiral metamaterials give strong response to light, this chiroptical response surpasses the natural materials. Owing to these special properties, the chiral meta devices have potential applications in polarization sensitive nano optical devices [46,47]. For instance, 3D nanoarchitecture generally exhibit chiroptical response [48,49], which are widely used for fast polarization switching [50]. Compared to bulk materials, the metamaterials or metasurfaces show domination in controlling the amplitude, phase, and polarization of light. In recent years, researchers have concentrated on achieving strong chiroptical responses in metasurfaces by breaking the rotational symmetries at unit cell level [51–55]. In this regard, 2D patterned layer of chiral metamirrors present a decent light matter interaction [53,54]. In addition, the metal/dielectric/metal sandwich structures display improved chiroptical responses. Recently, Kang et al. [56] demonstrated an ultrathin nonlinear chiral meta mirror. The proposed metamirror consists an array of SRRs made up of amorphous silicon (α-Si). They reported that the metamirror exhibit fast optical polarization switching (picosecond) of near infrared light at pump energies of picojoule per resonator. This can be achieved by strong chiroptic response of the atoms in the proposed mirror. Hence the chiral metamirrors are promising candidates for power efficient and high-speed polarization state modulators in optical information processing.

Further, it is important to guide the waves for reliable transportation of information through a physical channel. But, due to attenuation and back-scattering, the transportation of energy is sensitive to sharp turns and defects in high frequency systems such as signal processing [57,58]. However, in matter systems topological phenomenon is considered for unidirectional and attenuation free energy transportation [59]. Moreover, the topological metamaterials can transport the energy effectively, which is called phononic energy transportation. Till now, the mechanical metamaterials are being used in large scale systems such as gyroscopic lattices [60,61], arrays of pendulums [62,63], structured plates [64], and so on. For utilizing the mechanical materials in high frequency energy transportation, the metamaterial topological systems must be scaled on to the chip level [65,66]. In this view, Cha et al. [67] proposed topological nanoelectromechanical metamaterials. The proposed metamaterials consist of 2D array of free-standing SiN nanomembranes which can operate at high frequencies of 10–20 MHz. They also demonstrated the presence of edge states and frequency dispersion. Finally, the proposed on-chip topological metamaterials are suitable for high frequency signal processing applications.

Furthermore, Cho et al. [68] proposed the concept of virtualized metamaterials to overcome the shortcomings of traditional metamaterials. Through engineered resonant modes with structured atoms the metamaterials permit to attain constitutive parameters beyond their natural range. Generally, the tunability of constitutive parameters in real time applications is a fundamental challenge in traditional metamaterials. This can overcome by the proposed concept of virtualized metamaterials. In addition, they reported that the replacement of the physical structure with designed fast signal processing Kernel circuit. The circuit allows exhibiting defined frequency dispersion through which one can control mass density and bulk modulus of the metamaterials. In addition, the proposed concept is helpful in designing topological, non-Hermitian, and non-reciprocal systems by tuning the frequency, dispersion, and amplitude for fast signal processing applications.

Afterwards, the fluorescent resonant energy transfer (FRET) [69,70] between the donor and acceptor enhances the efficiency of solar cells [71], organic light emitting diodes [72], photosynthesis [73], and so on. Nevertheless, in case of direct dipole–dipole interaction, the resonance energy transfer occurs only at very short distances of 10 nm, beyond this distance efficiency of the energy transfer decreases significantly. This is one of the major limitations of transportation of energy in solar cells. Continuous research efforts are going on to enhance the efficiency of the energy transportation by using hyperbolic metamaterials [74,75], plasmonic nanostructures [76,77], and planar silver films [78]. Especially, the optical topological transition (OTT) in a metamaterial theoretically considered the ideal model for the enhancing energy transfer over long range orders [79]. In this regard, recently Deshmukh et al. [80] demonstrated the long range (about 160 nm) direct energy transfer between the donor CdSe/ZnS core-shell quantum dots and acceptor Cy3 organic dye molecules by using OTT in a metamaterial. They reported that the metamaterial consists the alternating layers of germanium (Ge) deposited silver (Ag) and alumina (Al_2O_3). The OTT in the metamaterial alters the density of states between donor and acceptor, which leads to the long-range energy transfer with 32% transfer efficiency. Further the experimental values are in good agreement with theoretical values. Finally, they concluded that the OTT in metamaterials enhances the efficiency of energy transportation and controls the transfer process. Owing to this, the OTT in metamaterials have potential applications in numerous fields such as organic solar cells, quantum entanglement, etc.

In addition, the hyperbolic metamaterials are also gained considerable attention. A hyperbolic metamaterial is a special type of anisotropic metamaterial whose isofrequency contour (IFC) takes the form of an open hyperboloid because the principal components of its electric or magnetic tensor have opposite signs [81–84]. The unusual nature of IFC enables the hyperbolic metamaterials to be used for controlling the electromagnetic waves in new ways. Through controlling the shape of the hyperbolic dispersion, one can flexibly control the propagation of light in hyperbolic metamaterial, resulting abnormal scattering [85–88] splitting [89–91], all angle negative refraction [92–96], etc. These hyperbolic metamaterials are also called indefinite media [97,98] or polaritonic crystal in which the coupled states of matter and light give rise to a larger bulk density of electromagnetic states [99,100]. Owing to this, hyperbolic metamaterial possesses strong enhancement of spontaneous emission [101–105] and Cherenkov emission with low energy electrons [106–109]. These hyperbolic metamaterials can be used in multifunctional platform for sensing, quantum engineering, waveguiding, super resolution imaging, and so on [110,111]. At first, the hyperbolic nature of metamaterial is first observed in microwave region. The natural as well as engineered material exhibits the hyperbolic dispersion. In infrared and visible region, some natural materials such as SiC, graphite, Bi_2Se_3, etc., exhibit the hyperbolic dispersion due to excitation of phonon polaritons [112–118]. Similarly, the engineered structures such as multilayer fishnets [119], uniaxial metasurfaces [120,121], and metal-dielectric structures [122,123] have subwavelength unit cells that display the hyperbolic dispersion. The dispersion manipulation in hyperbolic metamaterials can be used in many applications like sub-diffraction imaging [124], sub-wavelength modes [125], thermal emission engineering [126–128], high sensitivity sensors, hyper lens, etc. Therefore, the hyperbolic metamaterials with unusual properties will play a significant role in the advancement of novel optical devices in future. Recently, Schoche et al. [129] reported the behaviour of tunable hyperbolic metamaterials which are derived from self-assembled carbon nanotubes. To obtain the hyperbolic metamaterials from carbon nanotubes, they employed the Muller matrix ellipsometry over the broad spectral range from mid-IR to UV region to specify the dielectric tensor function on high dense films consisting single wall carbon nanotubes. Optically these films are anisotropic and act as metamaterials with effective medium response. Further, an oscillator model developed from the proposed metamaterials exhibit broad range of hyperbolic dispersion compared to film consisting unordered carbon nanotubes. Hence, the aligned carbon nanotubes form the metamaterial and play a prominent role in accomplishing tunable hyperbolic nature.

In addition, with the help of face-to-face SRRs and a central bar Hu et al. [130] demonstrated the design of tunable terahertz metamaterial (TTM) for switching applications. They reported that, the free spectrum ranges (FSR) of TTM varies bi-directionally, i.e., broadened, and shortened with respect to variation of gap between SSR and central bar. In TE mode, FSR is broadened by 0.14 THz (from 0.65 to 0.79 THz), whereas in TM mode it is shortened by 0.19 THz (from 0.30 to 0.19 THz). Moreover, TTM shows the dependence of polarization when it is subjected to external electromagnetic radiation. Moreover, in both single band and dual band proposed TTM acts like switch at TE mode, whereas at TM mode, the resonance of TTM insensitive to the displacement of central bar. Owing to this the proposed TTM can be utilized as switch in terahertz frequency range.

Li and Cheng [131] proposed a temperature tunable metamaterial absorber (MMA) by using two stacked square shaped strontium titanate resonator structures (STOs) and copper substrate. From the analysis of simulation studies, they reported that, at specific frequencies, i.e., 0.114 THz and 0.181 THz in terahertz region the reflectance of the proposed MMA is about 2.1% and 0.2% at room temperature. Moreover, it has maximum absorption peaks 97.9 and 99.8% at the same frequencies. In addition, owing to symmetry of unit cell structure STO based MMA shows polarization insensitive nature to both transverse electric and transverse magnetic modes. Further, the distribution studies of power flow and electromagnetic fields disclosed that, the excitations of fundamental dipole modes in the MMA are the origin of the high level of absorption. In addition, the absorbance of the proposed MMA can be altered by varying the external temperature and, by varying its structural parameters. Owing to the excellent characteristics, the proposed STO-based MMA may have potential prospects in thermal imaging, temperature sensing, thermal emission, etc. In addition, scientists designed strontium titanate (STO) based single band tunable metamaterial absorber [131]. They reported that, with respect to structural parameters, absorption peak is increased and resonant frequency is reduced. This means the absorption peaks are enhanced from 90 to 99.4%, whereas the resonant frequency is shifted from 6.2 to 5.8 THz. As the temperature is increased from 300 to 380 K the absorption peak shows 2% decrement. While the resonant frequency shifted from 6.2 to 7.1 THz. So, the proposed metamaterial can be tuned by temperature. Moreover, the temperature tunable metamaterial exhibits the excellent absorption property.

Next, using four L-shaped anisotropic metamaterials Lu et al. [132] demonstrated an ultrathin reflective linear polarization converter. The proposed design can convert the linearly polarized radiation to cross polarized radiation with the conversion ratio of 98% from 4.2 to 5.2 THz. In addition, the reflection coefficient of cross polarization is greatly influenced by the deflection angles relative to y-axis [133]. So, the proposed metamaterial-based device can be useful for the manipulation of polarization in optical instruments. In recent years, planar metamaterials are studied extensively owing to their potential applications in design of lasing spacers, flat optical components, nonlinear devices, etc. In these metamaterials high Q-factor (quality factor) is essen26GH26tial for enabling strong light-matter coupling [134,135]. In this regard, the Fano resonances support the narrow linewidths with high Q-factor values [136–138]. The Fano resonance in metasurfaces is originated from increased electric field in the subwavelength capacitive gaps of symmetry broken resonators. Earlier the Fano resonance is modulated by changing the coupling distance in the unit cell or changing the asymmetry parameters of the unit cell [139–141]. But, in this technique the Fano resonance works in passive mode. Therefore, the researchers searched for an alternative design for active control of metasurfaces. They suggested that active control is possible by controlling the conductivity of the Fano structure. Until now, very less work has been done on the Fano structures [142–145]. In this connection, Ma et al. [146] demonstrated an active switching of extremely large Q-factor Fano resonance by using VO_2 (vanadium oxide)-implanted THz asymmetric double C-shaped metamaterial structures. They reported that the double Fano resonances are highly sensitive to the temperatures and, the Fano resonances can be switched at low thermal pumping of 68 °C. The exited Fano resonances in the metadevices can be completely vanished by cooling. However, the tuning of Fano resonances in cooling and heating are different. To understand the internal mechanism of the temperature tuned Fano resonances, at respective resonant frequencies they simulated the distributions of surface current and electric field for high (70 °C) and low (45 °C) temperatures. In addition, at 1.16 THz, the resonant linewidth of the proposed device is 0.015 THz, and the Q-factor is 98, which is very high compared to traditional metamaterials. Hence, the proposed temperature tunable Fano resonant devices are promising candidates for designing advanced high-performance photonic devices like ultrasensitive temperature sensors. Later, Li et al. [147] reported an active Fano metasurfaces with the help of graphene-based terahertz asymmetric split-ring resonant structure (Gr-TASR) on silicon (Si) substrate. The active control in the proposed design is possible through controlling the conductivity of the graphene by a combination of CW (continuous wave) illumination and bias voltage. Further, they experimentally observed that, the clear modulation of Fano resonance is achieved by optical illumination with 0.7 Wcm^{-2} power density at very low voltage of about -0.8 V. This may be due to shorting effect between the capacitive split gaps of Gr-TASR. This is an effective approach for controlling the metamaterials actively with both electrical and optical fields which can play a prominent role in designing of variety of futuristic graphene-silicon based terahertz photonic devices.

Further, Lin et al. [148] demonstrated the design of a tunable terahertz (THz) resonator by using AFSM (asymmetrical F-shaped metamaterial) which is composed of Au-layer fabricated on SOI (silicon-on-insulator) substrate. Figure 2 shows the schematic representation of the proposed F-shaped THz resonator (Figure 1 of Ref. [148]). Without changing the other parameters, they designed three F-shaped resonators of lengths 60 μm, 65 μm, and 70 μm. When the proposed resonator is exposed to electromagnetic radiation, it shows the switch function for single band resonance at TM mode and dual band resonance at TE mode. Owing to this one can use the proposed device as a THz switch in TE-mode and as a THz filter in TM-mode. In addition, the tuning resonances are in between 0.2 and 0.4 THz in TE-mode and the resonator having length of 60 μm exhibits largest Q-factor of 40. Moreover, at TM-mode, all the proposed resonators show same Q-factor of 20. All the outcomes are evinced that the proposed AFSM device is suitable for designing environmental sensor. They reported that, the proposed device is subjected to surrounding environmental factors with different refractive indices for further enhancement of the flexibility and efficiency. Therefore, this new AFSM device provides a way to enhance the sensitivity of metamaterials in THz region for polarizer, sensor, and switching applications.

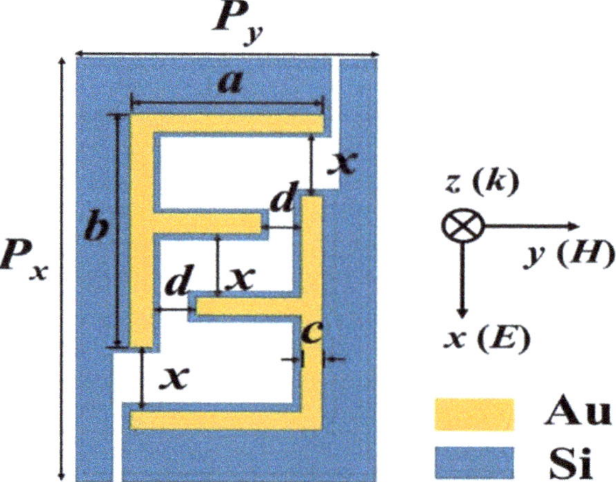

Figure 2. Schematic representation of the F-shaped THz resonator (Figure 1 of Ref. [148]).

3. Metamaterials in Photonic Devices

Over past few years, researchers have been searching for highly integrated light source. The photon transparency in metallic photonic crystals which were doped with nanoparticles have been studied by Mahi R Singh; his study used new optical devices like switches and optical transistors [149]. There is a high demand for the development of practical optical photonic devices and its application systems for controlling absorption/transmission/reflection of electromagnetic waves. It was observed that by adjusting the graphene fermi level a tunable metamaterial attains electronically reconfigurable terahertz reflection, absorption, and transmission in an effective manner [150]. In this view, various designs for free electron radiation emitters have seen light, which includes light wells [151,152], graphene based plasmonic, and dielectric based undulators [153,154], metallic nano gratings [155], etc. Along with the advancement of narrow linewidth miniaturized lasers [156] and highly integrated electron accelerators [157,158], the progress in nanoscale undulators paves the way toward the realization of X-ray light sources and on-chip extreme ultraviolet (EUV) sources. These advanced sources have promising applications in various fields such as natural sciences, engineering, and medicine [159,160]. Owing to strong confinement, low losses and dynamic tunability, the graphene plasmons are much suitable for manipulation of light matter interaction [161–165]. In addition, highly confined plasmons interact effectively with the matter. Nevertheless, small transverse extent of polaritonic field of such highly confined plasmons limits the interaction. This type of limitation exists in the light matter interaction in micro/nanoscale photonic wavelengths. Such limitations can be conquered by introducing metamaterials. Amongst, various kinds of materials graphene exhibit outstanding optical properties are highly appealing for optoelectronics and energy conversion applications [166–173]. However, the low optical absorption (2.3%) and ultrathin nature (3.4 Å) of monolayer graphene over a broadband wavelength limits its ability to provide sufficient optical modulation that restricts the performance in optical applications [174,175]. However, the graphene based-metamaterials consisting alternating dielectric layers and graphene can enhance the optical modulation which can be useful in advanced photonic devices [176–178]. Recently, Pizzi et al. [179] reported the enhancement of electron–plasmon interaction area by graphene metamaterials. They found that the output intensity is scaleup by a factor of 580 with respect to single graphene layer. For example, for 5 MeV electrons, a single layer metamaterial having 50 μm length and 50 layers, and a beam of current 1.7 μA can generate 1.5×10^7 photons. This is due to the ability of the graphene multilayer structures to support MRPs (multilayer resonant plasmons) [180–186]. Further, the layered conducting structures can also be generated the visible Cerenkov radiation [187]. In addition, they reported that through allowing large amount of electron beams, the graphene multilayers produces the significant improvement in output intensities. Based on the conductivity of the graphene, there will be optimum layers in the metamaterial that enhances the intensity of the output. Further, by varying the resonating modes at different laser frequencies, the proposed metamaterials can generate multiple X-ray harmonics, which are used in time-resolved X-ray spectroscopy for ultrafast imaging of chemical reactions and electronic state transitions [188,189]. In addition, Yang et al. [190] suggested a low-cost and transfer free, solution-phase technique for the fabrication of multilayer graphene-based metamaterials. This consists of alternating monolayer graphene oxide/graphene and dielectric layers. They reported that the optical properties of the prepared graphene-based metamaterials can be tuned dynamically by controllable laser mediated conversion. Graphene-enabled active metamaterials may provide new platform for dynamical manipulation of light matter interactions [191]. Figure 3 shows the schematic representation of dynamic process for in situ phototunable graphene based metamaterial (Figure 1 of Ref. [190]). In addition, the laser patterning leads to functional photonic devices like ultrathin flat lenses embedded in the lab-on-chip device. In general, these ultrathin flat lenses maintain consistency and shows subwavelength focusing resolution at ambient environment without any observable degradation compared with the original lens. Therefore, these graphene-based metamaterials provide a new insight for widespread applications in on-chip integrated photonic devices.

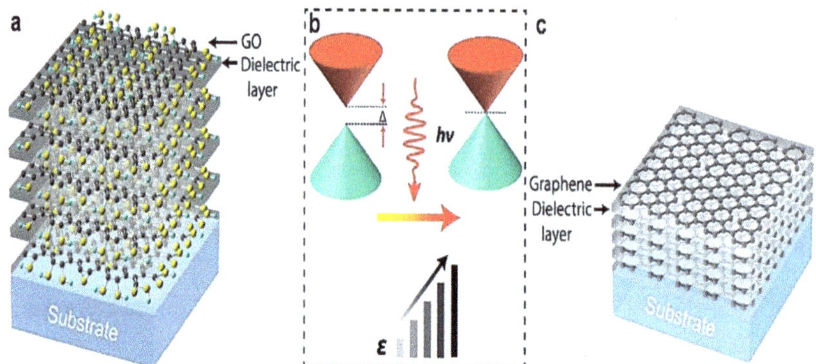

Figure 3. The representation of dynamic process for (**a**) initial, (**b**) schematic and (**c**) final state of in situ phototunable graphene based metamaterial (Figure 1 of Ref. [190]).

4. Metamaterials in Microwave Sensors

Nowadays, the characterization and quantification of liquids have become crucial in different fields such as biomedical engineering, agriculture, pharmaceutical, etc. [192–195]. In general, the characteristics of various liquids can be analyzed with the help of their polar nature and specific electrical properties. Moreover, the performance of the microwave devices is greatly influenced by the electrical properties of the liquids. Further, the interaction of the electromagnetic radiation with the polar liquid materials led to variation of the direction of the polarization for different molecules. Especially, microwave sensors employ that kind of interaction to manipulate the dielectric properties of the liquids for characterization. Due to simple procedure, non-invasiveness and quick response, the electromagnetic approach-based sensors offer several advantages than the normal ones. Further, owing to their unique properties the electromagnetic metamaterials gained much attention in the advancement of the electromagnetic devices over last two decades. Recently, the designs based on electromagnetic metamaterials have been employed for microfluidic sensing applications [196,197]. These sensors exhibit high sensitivity and strong interaction between the analytes and electric field. However, the requirement of large sample volume is one of the shortcomings of the microfluidic sensors. This can be mitigated by using unit cell resonation structures in microfluidic sensors [198,199]. Still, achieving the significant sensitivity in microlevel is a major problem. To overcome this, numerous researchers are putting their continuous efforts in designing the miniaturized sensors with significant sensitivity and selectivity for consistent characterization of a liquid. In continuing this, Xu et al. [200] demonstrated a lightweight, low cost, portable, biocompatible, and flexible metamaterial (metaflex) based photonic device for the biological and chemical sensing and high sensitivity strain applications. The device can be operated invisible as well as IR regions. The proposed device consists U-shaped SRRs of 30 nm thick gold (Au) or silver (Ag) which are deposited on poly(ethylene-naphthalate) substrate with the help of electron beam lithography [201]. In addition, the U-shaped SRRs metamaterials display an electric resonance of 542 nm and magnetic resonance of 756 nm. Both magnetic and electric resonant modes give highly sensitive response to surrounding dielectric media, bending strain, and surface chemical environment. Owing to coupling of electric and magnetic fields, the proposed metamaterial-based photonic device shows greater response to nonspecific bovine serum albumin protein binding with a shift of magnetic resonance of 4.5 nm. It also shows superior response for self-assembled monolayer of 2-naphthalenethiol with a shift of magnetic resonance of 65 nm. These outcomes suggest that the proposed device is a prominent candidate for chemical and biological sensing. Microwave metamaterials are electromagnetic devices which are synthesized to control microwave fields. These materials can be used as future healthcare systems that can overcome technical restrictions after they are interfaced with human body. John S. Ho et al. [202] studied the working principles and applications of microwave metamaterials for biomedical sensing. Further, Kayal et al. [203] demonstrated a compact microwave sensor with the help of mu negative (MNG) metamaterial for liquid characterization. They reported that the prepared sensor exhibits high sensitivity along with noticeable compactness. The square spiral metamaterial (MNG) plays a significant role in accomplishing this sensitivity in small cross-sectional area as well as notable compactness. In addition, the sensing behaviour of the prepared device is confirmed through least square technique followed by development of two nonlinear equations for calibration purpose. These nonlinear equations (Equations (9) and (10) of Ref. [203]) are useful for finding the permittivity of the unknown samples. Hence, the compactness and high sensitivity of the prepared sensor make it a prominent candidate for liquid sensing applications.

In addition, the electromagnetic waves in tera Hertz (THz) region shows sensitive responses to intra and intermolecular vibration modes and low photon energy (few meV) [204,205]. Due to this the THz electromagnetic waves display potential applications in bio-sensing, microscopy, and spectroscopy. In THz region, with the help of tens of metamaterials and graphene, Xu et al. [206] introduced a platform for bio molecular sensing. Further, by using graphene assisted nano metamaterials, Lee et al. [207] demonstrated a label free sensing technique for discrimination of single-stranded deoxyribonucleic acids (ssDNAs) in THz region. The combination of unusual properties of metamaterial and electrooptical properties of graphene provides biomolecule sensing property even using THz photons with very low energy. Additionally, they reported that the enhancement of THz field at resonance frequency causes the rise in absorption cross section of the graphene sheet which in turn provides ultrahigh sensitivity. The sensing mechanism includes the direct transfer of graphene onto a nano-slot metamaterial and tightly binding the targeted DNA molecules without modifying the structure. Here, the nano-slot metamaterials enhance the THz transmittance which is proportional to the absorption cross-section of DNA adsorbed graphene layer. So, greater number of DNA molecules can be observed through the strongly focused THz electromagnetic waves. Further, adsorbed molecules change the intrinsic electrical properties of the graphene which can easily be detected. This mechanism is allowed to sense of different biomolecules. Specifically, considering suitable receptor to capture DNA molecules followed by rapid primary screening, finally applied for sequencing the DNA. Hence, the graphene assisted THz metamaterial sensing platform is suitable for biological sensing applications also understanding the electro-optic behaviour of 2D materials.

The sensitivity of the microwave sensor can be enhanced by coupling the transmission lines with metamaterial based open loop resonators [208]. However, during coupling the resonator with transmission line there exist shift in resonant frequency which strongly affects attain high sensitivity. To overcome this, Abdolrazzaghi et al. [209] proposed a novel metamaterial based planar microwave sensor which can be operated at 2.5 GHz. They prepared the above proposed sensor by coupling negative refractive indexed metamaterials with transmission lines which exhibits greatly improved resonant properties [210]. Afterwards, they developed the signal flow analysis to estimate the transmission response of the prepared sensor. In comparison with the microstrip or conventional sensor the proposed sensor shows very high sensitivity along with large complex permittivity. They reported that the proposed sensor displays superior properties, particularly in water host medium and high permittivity materials. The concentration measurements of the methanol or ethanol in water medium reveals the outstanding performance of the proposed sensor over conventional sensor. Hence, the proposed metamaterial based planar microwave sensor is useful for characterization of high permittivity materials, highly sensitive concentration measurement of methanol or ethanol in water and also biomolecule detection.

5. Metamaterials in Antennas

In modern era, the communication system is shifting from wired to wireless. In this regard, there is a need for antennas for wireless transmission of signals. Hence, the demand for antennas which are having large operating bandwidth and high gain has increased. The researchers are searching for new kind antennas which satisfy the demand of modern communication. In this concern, DRA (dielectric resonator antenna) attracted much attention owing to its novel features such as high radiation efficiency and broad bandwidth [211–213]. Initially, Long and his coworkers started the experimental investigations on DRAs, followed by many researchers, now the DRAs are available in different shapes like rectangular, gammadion cross, cylindrical, quadruple, spherical, hemispherical, and so on [214–217]. Amongst, rectangular DRA exhibits good attractive properties. However, limited bandwidth and manufacturing cost are the major drawbacks of DRAs. To reduce the fabrication cost, cost effective additive manufacturing (like 3D printing) can be used in the place of the costly conventional manufacturing. In addition to cost effectiveness 3D printing allows to develop complex structures with specific features of antennas for demanding applications. With the help of SRR, biodegradable PLA and cost-effective 3D printing technique Kumar et al. [218], recently demonstrated the design and development of star shaped dielectric resonator antenna (SDRA). They reported that a special type of polymer called PLA is fabricated by additive manufacturing technique using renewable energy sources. In addition, SDRA shows high bandwidth of 37% that can be attained by modifying the shape of rectangular DRA. Further, the circular shaped SRR can serve as a metamaterial which lead to enhancement in gain, the maximum gain attained by the antenna is of 82.7%. Moreover, within in the operating band the proposed antenna exhibits the average efficiency of 80.51% and is circularly polarized at 5.8 GHz. Therefore, the outcomes are evidenced that the proposed antenna fulfils the WLAN bandwidth requirements (5.15–5.35 GHz and 5.725–5.825 GHz). This is useful in communication, C-band (aeronautical and meteorological radio navigation and satellite navigation systems.

In addition, the metamaterials are widely used in reduce the size of the antennas to attain multiband frequency response [219]. Negative order and zeroth order resonances are useful to design miniaturized antennas which are very useful in wireless vehicular communication systems [220–223]. Mehdipour et al. [224] demonstrated the design and applications of monopole antennas that are loaded by complementary split ring resonators (CSRR) and zeroth order resonator units. The proposed antennas can be operated in three tunable frequency bands. They reported that, the miniaturization of the proposed antenna is achieved by loading the zeroth order resonators. In addition, they observed the good agreement between the simulated and experimental results. Hence, the multi-band tunable response and miniaturized structure of the proposed antenna evinced that it is a prominent candidate for vehicular communication system. Further, Elwi [225] introduced a novel cylindrical antenna with miniaturized structure for multi-input and multi output systems. The proposed antenna consists a cylindrical shaped Kodak photo paper substrate (with height $\lambda_o/4.5$ and diameter $\lambda_o/4.5$, where λ_o is the wavelength of the free space at 2.25 GHz) on which four omega shaped monopoles (the separation between each monopole is of $\lambda_o/29$) are folded. In the frequency range 2 to 3 GHz there exists the maximum coupling between the monopoles. The reduction in coupling is done by mounting the SRRs between the monopoles. With the help of Ink-jet deposition process the silver nanoparticles are printed on the substrate. Finally, it is observed that the proposed array shows a gain of 2.5 dB with wide radiation patterns which is suitable for the applications in the systems having multi-inputs and multi outputs.

6. Metamaterials in Energy Harvesting

Metamaterials play a vital role in energy harvesting. As we know that acoustic energy is inexhaustible and present in the ambient environment, together with common voice, rustle of falling leaves, sound near jet plane, operating sound of large grinding machine and so on [226,227]. However, if the energy density is low, most of the acoustic energy is dissipated into thermal energy during its propagation. In recent years, the researchers are focused on energy harvesting technology to convert energy into electrical energy to power up low power electronic devices [228,229]. To scavenge and confine the acoustic energy distinct energy harvesters have been proposed [230]. The advancement of acoustic energy harvesters has begun from classic Helmholtz resonators [231], photonic crystal resonators [232], and quarter-wave resonators [233] to local acoustic metamaterials resonators [234] for large energy focusing on short structural dimensions. Wireless energy harvesting (WEH) from electromagnetic fields is flattering an emerging technology. Recent advances in broadband rectennas for wireless power transfer (WPT) and ambient RF energy harvesting was studied by C. Song [235]. The review of wireless and battery-free platforms for collection of bio signals, biosensors and bioelectronics was done by Tucker Stuart et al. [236]. They summarize present methods to realize such device architectures and deliberates their building blocks. On-site and external energy harvesting in underground wireless was studied by Raza, U [237]. These energy harvesting methods lead to design of a competent wireless underground communication system to power underground nodes for extended field operation. Additionally, key energy harvesting tools are offered that use available energy sources in the fields like vibration, solar, and wind. In this concern, the Electromagnetic (EM)- and Magnetic Induction (MI)-based approaches are important for underground wireless communication system. Recently, WPT and energy harvesting: current status and future prospects was reviewed by J. Huang et al. [238]. A meta-material based on a cubic high-dielectric resonator (CHDR) for coupled WPT (wireless power transfer) system was studied by R. Das et al. [239], they have observed that the proposed CHDR system providing more than 90% power transfer efficiency at a distance of 0.1λ. Amongst, the photonic crystals are suitable harvesters for high frequency acoustic environment, because the Bragg scattering controls the scattering characteristics of the waves in photonic crystals and the wavelength of the harvested sound is of the order of magnitude (periodic parameters) of the scatterer.

The acoustic metamaterials alter the propagation of the wave depends on the local resonance bandgap principle. In which at a specific frequency, the propagation of the incident wave is suppressed due to interaction of the resonant modes of the structural units with the travelling wave [240]. In addition, the acoustic metamaterials exhibit special wave propagation characteristics in wave vector space, spectral space, and phase space [241,242]. In acoustic energy harvesters, the acoustic energy is focused onto piezoelectric crystal and its density is improved significantly to reach the usable magnitude. The piezoelectric material converts the incident energy into electrical energy by inverse piezoelectric effect. The generated electrical energy can be used in low power consuming devices and to power up the wireless sensor network. Different types of acoustic metamaterials have been developed for harvesting of acoustic energy. With the help of lead zirconate titanate transducer and double layered acoustic metamaterial Wang et al. [243] demonstrated the acoustic energy harvester which exhibits the maximum power of 73.1 nW at the incident wave frequency of 318 Hz and pressure of 2 Pa. In addition, using spring-mass resonators Oudich et al. [244] developed acoustic metamaterial thin-plate for harvesting of acoustic energy. The designed harvester shows large out power of 18.1 μW at 519 Hz and 2 Pa. Recently, Ma et al. [245] designed a two-dimensional local resonant acoustic metamaterial for energy harvesting applications. They reported that the designed energy harvester shows voltage enhancement of 950% compared to bare plate energy harvester. In addition, at resonant conditions and the sound pressure of 20 Pa the proposed harvester shows maximum voltage of 291 mV, average power of 28 μW and power density of 1.24 mW cm^{-3}. Further, the outdoor studies revealed that the acoustic material-based harvester shows 18 times better open circuit voltage compared to the bare plate harvester. Hence, the outstanding properties of the proposed harvester evinced that it could serve as a promising acoustic energy harvester with improved performance.

In addition, the metamaterials can also absorb and harvest electromagnetic signals. Nowadays, the investigations on the metamaterial-based absorbers have increased for electromagnetic signal harvesting and absorbing applications in microwave region. Landy et al. [246] in 2008 designed metamaterial-based absorber consisting of two separated resonators to harvest electromagnetic waves. In addition, for high frequency regions of GHz and THz, Dincer et al. [247] introduced a metamaterial-based absorber with the help of square resonator. In addition, Cheng et al. [248] investigated the polarization insensitive metamaterial-based absorber for harvesting electromagnetic energy at various frequency bands. Yagitani et al. [249] proposed and designed an electrical circuit model and mushroom-like electromagnetic band gap [250] (EBG) structures for obtaining the 2D image of distribution of RF power. Finally, Alkurt et al. [251] demonstrated a metamaterial-based absorber for energy harvesting and imaging applications. The proposed harvester is also called as 2 × 2 patch array antenna and microwave image detector. The unit cell of the proposed harvester is shown in Figure 4 (Figure 1 of the Ref. [251]). They reported that, in the proposed absorber, first the absorbed energy is converted into DC signal with the help of Schottky diodes. Subsequently, the obtained DC signals generate image of the absorbed power. Afterwards, they fabricated 2 × 2 patch array antenna and energy harvester. The outcomes of the numerical and experimental measurements are matched with one another. Hence, all these results evidenced that the proposed metamaterial-based absorber can be promising candidate for energy harvesting, imaging, crack detection, and so on.

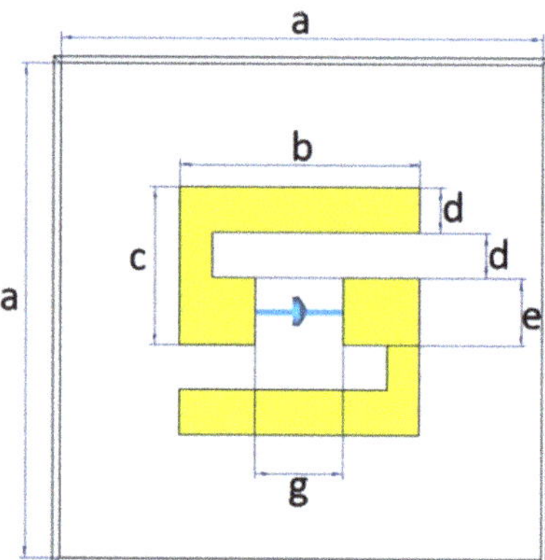

Figure 4. Unit cell of the metamaterial-based harvester (Figure 1 of Ref. [251]).

7. Metamaterials in SQUIDs

The investigations on the metamaterials comprising SQUIDs (Superconducting quantum interference devices) are going on intensively from the last decade [252,253]. These are the artificial materials which exhibit special properties such as negative magnetic permeability [254], dynamic multi-stability and switching [252], broad band tunability [255], etc. As we know that simple version of SQUIDs contains superconducting ring interrupted by Josephson junction [256]. The two-dimensional arrays of SQUIDs in different lattice geometries establish nonlinear metasurfaces. This can project theoretically in both quantum [257] and classical regimes [258,259]. By implementing this various designs and structures have been investigated in one and two dimensions [260–263]. With the help of microwave transmission measurements recently reported the degree of spatiotemporal coherence of SQUID metamaterials [264]. Further, these SQUID metamaterials provide novel testbed for exploring complex spatiotemporal dynamics. Interestingly, it is proved that SQUID metamaterials support spatially inhomogeneous states like chimera states. These chimera states gained much attention for both experimental and theoretical viewpoints [265,266]. In various discrete systems it was identified that different dynamic states such as solitary state chimeras [267], spiral wave chimeras [268], and imperfect chimeras [269]. In addition, the chimera states in SQUID metasurfaces controls the speed of the propagating electromagnetic waves [270]. Lazarides et al. [271] demonstrated the generation of chimera states by SQUID metasurfaces which are subjected to dc-flux gradient and driven by ac-flux. Figure 5 shows the schematic representation of SQUID metamaterial (Figure from Ref. [271]). They reported that the flux gradient of dc-field and amplitude of the ac-flux controls desynchronized clusters of chimera state along with their location and size. Further, the chimera states are distinguished from non-chimera states by ac driving flux. Finally, the proposed SQUID oscillator is an example for system with inertia and driving which plays a prominent role in the advancement of metamaterials science. In addition, J. Hizanidis et al. [272] presented 2D SQUID lattice with nearest neighbour interactions which can form states with Turning-like pattern. In general, single SQUID undergoes complex bifurcations at low coupling limit, where as in 2D SQUID the Turning like pattern arises near synchronization to desynchronization transformation region. Further, owing to extreme multi-stability of single SQUID, 2D chimera states have formed at near resonant regime. The formed chimera states can be tuned by dc-flux also control their position and multiplicity. Finally, recent reports revealed that the imaging of the chimera states in SQUID metamaterials is prominent candidates for verifying the theoretical findings [273].

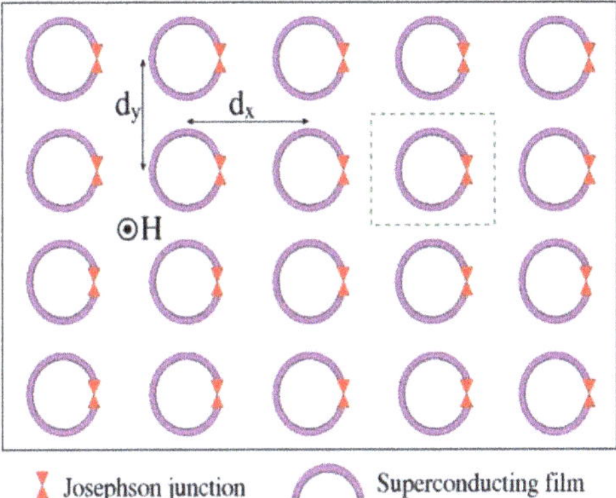

Figure 5. Schematic representation of SQUID metamaterial (Figure from Ref. [271]).

8. Conclusions

Metamaterials are the advanced engineered materials which offer the naturally unobtainable properties. Due to their unique behaviour the metamaterials gained considerable attention in various fields including sensors, energy harvesting, photonics, and so on. Graphene based metamaterials efficiently can control absorption/transmission/reflection of electromagnetic waves. The graphene assisted THz metamaterial sensing platform is suitable for biological sensing applications. Even though, massive research is going on to search new kind of tunable metamaterials with improved performance, significant challenges remain. The star shaped dielectric resonator antenna (SDRA) achieves the WLAN bandwidth requirements (5.15–5.35 GHz and 5.725–5.825 GHz). This is useful in communication, like aeronautical and meteorological radio navigation and satellite navigation systems (c-band). However, in comparison with the normal materials the metamaterials exhibit superior properties. The electrical energy generated by energy harvesting metamaterial can be used in low power consuming devices and to power up the wireless sensor network. Lead zirconate titanate transducer and double layered acoustic energy harvester exhibits the maximum power of 73.1 nW at the incident wave frequency of 318 Hz and pressure of 2 Pa. Metamaterial-based absorbers have increased for electromagnetic signal harvesting and absorbing applications in microwave region. The image of chimera states of SQUID metamaterials is prominent applicants for verifying the theoretical findings. Finally, with the rapid advancement of relevant science and technology in recent years, metamaterials have been developed in the direction of a standard technology platform to accomplish basic tunability and advanced application of metamaterial devices.

Author Contributions: Conceptualization, N.S.K. and K.C.B.N.; methodology, N.S.K.; validation, N.S.K., K.C.B.N., P.B. and T.A.B.; resources, B.V.S.R.; data curation, T.A.B.; writing—original draft preparation, N.S.K.; writing—review and editing, K.C.B.N.; visualization, B.V.S.R.; supervision, N.S.K.; project administration, K.C.B.N.; funding acquisition, P.B. All authors have read and agreed to the published version of the manuscript.

Funding: One of us P. Banerjee received UGC, India start-up grant no F.30-457/2018 (BSR).

Institutional Review Board Statement: Not applicable for studies not involving humans or animals.

Informed Consent Statement: Not applicable for studies not involving humans or animals.

Data Availability Statement: The data will be made immediately available based on the request.

Acknowledgments: The authors express thankfulness to P. Sreeramulu, (English), GITAM, Bangalore for providing English language editing services to this manuscript.

Conflicts of Interest: The authors declare that we have no conflicts of interest.

References

1. Pendry, J.B. Negative refraction makes a perfect lens. *Phys. Rev. Lett.* **2000**, *85*, 3966. [CrossRef] [PubMed]
2. Papasimakis, N.; Fedotov, V.A.; Zheludev, N.I.; Prosvirnin, S.L. Metamaterial analog of electromagnetically induced transparency. *Phys. Rev. Lett.* **2008**, *101*, 253903. [CrossRef] [PubMed]
3. Kurter, C.; Tassin, P.; Zhang, L.; Koschny, T.; Zhuravel, A.P.; Ustinov, A.V.; Anlage, S.M.; Soukoulis, C.M. Classical analogue of electromagnetically induced transparency with a metal-superconductor hybrid metamaterial. *Phys. Rev. Lett.* **2011**, *107*, 043901. [CrossRef] [PubMed]
4. Jin, B.B.; Wu, J.B.; Zhang, C.H.; Jia, X.Q.; Jia, T.; Kang, L.; Chen, J.; Wu, P.H. Enhanced slow light in superconducting electromagnetically induced transparency metamaterials. *Supercond Sci. Technol.* **2013**, *26*, 074004. [CrossRef]
5. Zhang, C.; Wu, J.; Jin, B.; Jia, X.; Kang, L.; Xu, W.; Wang, H.; Chen, J.; Tonouchi, M.; Wu, P. Tunable electromagnetically induced transparency from a superconducting terahertz metamaterial. *Appl. Phys. Lett.* **2017**, *110*, 241105. [CrossRef]
6. Schurig, D.; Mock, J.J.; Justice, B.J.; Cummer, S.A.; Pendry, J.B.; Starr, A.F.; Smith, D.R. Metamaterial electromagnetic cloak at microwave frequencies. *Science* **2006**, *314*, 977–980. [CrossRef]
7. Linden, S.; Enkrich, C.; Dolling, G.; Klein, M.W.; Zhou, J.; Koschny, T.; Soukoulis, C.M.; Burger, S.; Schmidt, F.; Wegener, M. Photonic Metamaterials: Magnetism at Optical Frequencies. *IEEE J. Sel. Top. Quantum Electron.* **2006**, *12*, 1097–1105. [CrossRef]
8. Li, C.; Wu, J.; Jiang, S.; Su, R.; Zhang, C.; Jiang, C.; Zhou, G.; Jin, B.; Kang, L.; Xu, W.; et al. Electrical dynamic modulation of THz radiation based on superconducting metamaterials. *Appl. Phys. Lett.* **2017**, *111*, 092601. [CrossRef]
9. Shelby, R.A.; Smith, D.R.; Schultz, S. Experimental verification of a negative index of refraction. *Science* **2001**, *292*, 77–79. [CrossRef]
10. Zheludev, N.I. The road ahead for metamaterials. *Science* **2010**, *328*, 582–583. [CrossRef]
11. Zheludev, N.I. A Roadmap for Metamaterials. *Opt. Photonics News* **2011**, *22*, 31–35. [CrossRef]
12. Tong, X.C. *Science*; Springer International Publishing AG: New York, NY, USA, 2018; Volume 262.
13. Engheta, N.; Ziolkowski, R. *Metamaterials: Physics and Engineering Explorations*; John Wiley & Sons: New York, NY, USA, 2006; ISBN 9780471784180.
14. Capolino, F. *Theory and Phenomena of Metamaterials: Metamaterials Handbook*; CRC Press: Boca Raton, FL, USA, 2017; ISBN1 1420054260. ISBN2 9781420054262.
15. Smith, D.R.; Padilla, W.J.; Vier, D.C.; Nemat-Nasser, S.C.; Schultz, S. Composite Medium with Simultaneously Negative Permeability and Permittivity. *Phys. Rev. Lett.* **2000**, *84*, 4184–4187. [CrossRef] [PubMed]
16. Veselago, V.G. The Electrodynamics of Substances with Simultaneously Negative Values of ε and μ. *Usp. Fiz. Nauk.* **1967**, *92*, 517–526. [CrossRef]
17. Pendry, J.B.; Holden, A.J.; Robbins, D.J.; Stewart, W.J. Magnetism from Conductors, and Enhanced Non-linear Phenomena. *IEEE Trans. Microw. Theory Tech.* **1999**, *47*, 2075–2084. [CrossRef]
18. Lalas, A.X.; Kantartzis, N.V.; Tsiboukis, T.D. Metamaterial-based wireless power transfer through interdigitated SRRs. *COMPEL:Int. J. Comput. Math. Electr. Electron. Eng.* **2016**, *35*, 1338–1345. [CrossRef]
19. Pendry, J.B.; Holden, A.J.; Stewart, W.J.; Youngs, I. Extremely Low Frequency Plasmons in Metallic Mesostructures. *Phys. Rev. Lett.* **1996**, *76*, 4773–4776. [CrossRef] [PubMed]
20. Manufacturing Disruption. Available online: https://manufacturingdisruption.com/2014/12/31/metamaterials-ultimate-composites/ (accessed on 4 May 2021).
21. Ziolkowski, R.W. Design, Fabrication, and Testing of Double Negative Metamaterials. *IEEE Trans. Antennas Propag.* **2003**, *51*, 1516–1529. [CrossRef]

22. Shamonina, E.; Solymar, L. Magneto-inductive waves supported by metamaterial elements: Components for a one-dimensional waveguide. *J. Phys. D Appl. Phys.* **2004**, *37*, 362–367. [CrossRef]
23. Butz, S.; Jung, P.; Filippenko, L.V.; Koshelets, V.P.; Ustinov, A.V. A one-dimensional tunable magnetic metamaterial. *Opt. Express* **2013**, *21*, 22540–22548. [CrossRef]
24. Zagoskin, A.M. Superconducting quantum metamaterials in 3D: Possible realizations. *J. Opt.* **2012**, *14*, 114011. [CrossRef]
25. Mawatari, Y.; Navau, C.; Sanchez, A. Two-dimensional arrays of superconducting strips as dc magnetic metamaterials. *Phys. Rev. B* **2012**, *85*, 134524. [CrossRef]
26. Kafesaki, M.; Tsiapa, I.; Katsarakis, N.; Koschny, T.; Soukoulis, C.M.; Economou, E.N. Left-handed meta-materials: The fishnet structure and its variations. *Phys. Rev. B* **2007**, *75*, 235114. [CrossRef]
27. Wuestner, S.; Pusch, A.; Tsakmakidis, K.L.; Hamm, J.M.; Hess, O. Overcoming losses with gain in a negative refractive index metamaterial. *Phys. Rev. Lett.* **2010**, 127401. [CrossRef]
28. Liu, N.; Guo, H.; Fu, L.; Kaiser, S.; Schweizer, H.; Giessen, H. Three-dimensional photonic metamaterials at optical frequencies. *Nat. Mater.* **2008**, *7*, 31–37. [CrossRef] [PubMed]
29. Valentine, J.; Zhang, S.; Zentgraf, T.; Ulin-Avila, E.; Genov, D.A.; Bartal, G.; Zhang, X. Three-dimensional optical metamaterial with a negative refractive index. *Nature* **2008**, *455*, 376–379. [CrossRef]
30. Altintas, O.; Aksoy, M.; Akgol, O.; Unal, E.; Karaaslan, M.; Sabah, C. Fluid, Strain and Rotation Sensing Applications by Using Metamaterial Based Sensor. *J. Electrochem. Soc.* **2017**, *164*, B567–B573. [CrossRef]
31. Abdulkarim, Y.I.; Deng, L.; Altintas, O.; Unal, E.; Karaaslan Physica, M. Low-Dimens, E. Metamaterial absorber sensor design by incorporating swastika shaped resonator to determination of the liquid chemicals depending on electrical characteristics. *Syst. Nanostruct.* **2019**, *114*, 113593.
32. Bakir, M.; Dalgaç, Ş.; Karaaslan, M.; Karada, F.; Akgol, O.; Unal, E.; Depçi, T.; Sabah, C. A comprehensive study on fuel adulteration sensing by using triple ring resonator type metamaterial. *J. Electrochem. Soc.* **2019**, *166*, B1044–B1052. [CrossRef]
33. Fang, N.; Zhang, X. Rapid growth of evanescent wave by a silver superlens. *Appl. Phys. Lett.* **2003**, *82*, 161–163. [CrossRef]
34. Zhu, J.; Eleftheriades, G.V. Dual-band metamaterial-inspired small monopole antenna for WiFi applications. *Electron. Lett.* **2009**, *45*, 1104–1106. [CrossRef]
35. Erentok, A.; Ziolkowski, R.W. Metamaterial-Inspired Efficient Electrically Small Antenna. *IEEE Trans. Antennas Propag.* **2008**, *56*, 691–707. [CrossRef]
36. Aydin, K.; Bulu, I.; Ozbay, E. Subwavelength resolution with a negative-index metamaterial superlens. *Appl. Phys. Lett.* **2007**, *90*, 254102. [CrossRef]
37. Hu, T.; Landy, N.I.; Bingham, C.M.; Zhang, X.; Averitt, R.D.; Padilla, W.J. A metamaterial absorber for the terahertz regime: Design, fabrication and characterization. *Opt. Express* **2008**, *16*, 7181–7188.
38. Dincer, F.; Karaaslan, M.; Sabah, C. Design and analysis of perfect metamaterial absorber in GHz and THz Frequencies. *J. Electromagn. Waves Appl.* **2015**, *29*, 2492–2500. [CrossRef]
39. Li, M.; Yang, H.L.; Hou, X.W.; Tian, Y.; Hou, D.Y. Perfect Metamaterial Absorber with Dual Bands. *Prog. Electromagn. Res.* **2010**, *108*, 37–49. [CrossRef]
40. Ma, Y.; Chen, Q.; Grant, J.; Saha, S.C.; Khalid, A.; Cumming, D.R. A terahertz polarization insensitive dual band metamaterial absorber. *Opt. Lett.* **2011**, *36*, 945–947. [CrossRef]
41. Watts, C.M.; Liu, X.; Padilla, W.J. Metamaterial electromagnetic wave absorbers. *Adv. Mater.* **2012**, *24*, OP98–OP120. [CrossRef]
42. Akgol, O.; Altintas, O.; Dalkilinc, E.E.; Unal, E.; Karaaslan, M.; Sabah, C. Metamaterial absorber-based multisensor applications using a meander-line resonator. *Opt. Eng.* **2017**, *56*, 087104. [CrossRef]
43. Bagmanci, M.; Karaaslan, M.; Unal, E.; Özaktürk, M.; Akgol, O.; Karadag, F.; Bhadauria, A.; Bakir, M. Wide band fractal-based perfect energy absorber and power harvester. *Int. J. RF Microw. Comput. Aided Eng.* **2019**, *29*, e21597. [CrossRef]
44. Mulla, B.; Sabah, C. Multiband metamaterial absorber design based on plasmonic resonances for solar energy harvesting. *Plasmonics* **2016**, *11*, 1313–1321. [CrossRef]
45. Zheludev, N.I.; Kivshar, Y.S. From metamaterials to metadevices. *Nat. Mater.* **2012**, *11*, 917–924. [CrossRef]
46. Wang, Z.; Cheng, F.; Winsor, T.; Liu, Y. Optical chiral metamaterials: A review of the fundamentals, fabrication methods and applications. *Nanotechnology* **2016**, *27*, 412001. [CrossRef]
47. Qiu, M.; Zhang, L.; Tang, Z.; Jin, W.; Qiu, C.-W.; Lei, D.Y. 3D metaphotonic nanostructures with intrinsic chirality. *Adv. Funct. Mater.* **2018**, *28*, 1803147. [CrossRef]
48. Gansel, J.K.; Thiel, M.; Rill, M.S.; Decker, M.; Bade, K.; Saile, V.; von Freymann, G.; Linden, S.; Wegener, M. Gold helix photonic metamaterial as broadband circular polarizer. *Gold Helix Sci.* **2009**, *325*, 1513–1515. [CrossRef]
49. Gao, W.; Leung, H.M.; Li, Y.; Chen, H.; Tam, W.Y.J. Circular dichroism in double-layer metallic crossed-gratings. *Optics* **2011**, *13*, 115101.
50. Yin, X.; Schäferling, M.; Michel, A.-K.U.; Tittl, A.; Wuttig, M.; Taubner, T.; Giessen, H. Active Chiral Plasmonics. *Nano Lett.* **2015**, *15*, 4255–4260. [CrossRef]
51. Wu, C.; Arju, N.; Kelp, G.; Fan, J.A.; Dominguez, J.; Gonzales, E.; Tutuc, E.; Brener, I.; Shvets, G. Spectrally selective chiral silicon metasurfaces based on infrared Fano resonances. *Nat. Commun.* **2014**, *5*, 3892. [CrossRef] [PubMed]
52. Wang, Z.; Jia, H.; Yao, K.; Cai, W.; Chen, H.; Liu, Y. Circular dichroism metamirrors with near-perfect extinction. *ACS Photonics* **2016**, *3*, 2096–2101. [CrossRef]

53. Li, W.; Coppens, Z.J.; Besteiro, L.V.; Wang, W.; Govorov, A.O.; Valentine, J. Circularly polarized light detection with hot electrons in chiral plasmonic metamaterials. *Nat. Commun.* **2015**, *6*, 8379. [CrossRef] [PubMed]
54. Kang, L.; Rodrigues, S.P.; Taghinejad, M.; Lan, S.; Lee, K.-T.; Liu, Y.; Werner, D.H.; Urbas, A.; Cai, W. Preserving spin states upon reflection: Linear and nonlinear responses of a chiral meta-mirror. *Nano Lett.* **2017**, *17*, 7102–7109. [CrossRef] [PubMed]
55. Taghvaee, H.; Abadal, S.; Pitilakis, A.; Tsilipakos, O.; Tasolamprou, A.; Liaskos, C.K.; Kafesaki, M.; Kantartzis, N.V.; Cabellos-Aparicio, A.; Alarcón, E. Scalability Analysis of Programmable Metasurfaces for Beam Steering. *IEEE Access* **2020**, *8*, 105320–105334. [CrossRef]
56. Kang, L.; Wang, C.-Y.; Guo, X.; Ni, X.; Liu, Z.; Werner, D.H. Nonlinear chiral meta-mirrors: Enabling technology for ultrafast switching of light polarization. *Nano Lett.* **2020**, *20*, 2047–2055. [CrossRef] [PubMed]
57. Olsson, R.H.; El-Kady, I. Microfabricated phononic crystal devices and applications. *Meas. Sci. Technol.* **2009**, *20*, 012002. [CrossRef]
58. Baboly, M.G.; Soliman, Y.M.F.; Reinke, M.; Leseman, Z.C.; El-Kady, I. Demonstration of acoustic waveguiding and tight bending in phononic crystals. *Appl. Phys. Lett.* **2016**, *109*, 183504. [CrossRef]
59. Hasan, M.Z.; Kane, C.L. Colloquium: Topological insulators. *Rev. Mod. Phys.* **2010**, *82*, 3045–3067. [CrossRef]
60. Nash, L.M.; Kleckner, D.; Read, A.; Vitelli, V.; Turner, A.M.; Irvine, W.T.M. Topological mechanics of gyroscopic metamaterials. *Proc. Natl. Acad. Sci. USA* **2015**, *112*, 14495–14500. [CrossRef]
61. Mitchell, N.P.; Nash, L.M.; Hexner, D.; Turner, A.M.; Irvine, W.T.M. Amorphous topological insulators constructed from random point sets. *Nat. Phys.* **2018**, *14*, 380–385. [CrossRef]
62. Süsstrunk, R.; Huber, S.D. Observation of phononic helical edge states in a mechanical topological insulator. *Science* **2015**, *349*, 47–50. [CrossRef] [PubMed]
63. Serra-Garcia, M.; Peri, V.; Süsstrunk, R.; Bilal, O.R.; Larsen, T.; Villanueva, L.G.; Huber, S.D. Observation of a phononic quadrupole topological insulator. *Nature* **2018**, *555*, 342–345. [CrossRef]
64. Yu, S.; He, C.; Wang, Z.; Liu, F.-K.; Sun, X.-C.; Li, Z.; Lu, H.-Z.; Lu, M.-H.; Liu, X.-P.; Chen, Y.-F. Elastic pseudospin transport for integratable topological phononic circuits. *Nat. Commun.* **2018**, *9*, 3072. [CrossRef]
65. Peano, V.; Brendel, C.; Schmidt, M.; Marquardt, F. Topological phases of sound and light. *Phys. Rev.* **2015**, *X5*, 031011. [CrossRef]
66. Brendel, C.; Peano, V.; Painter, O.J.; Marquardt, F. Snowflake phononic topological insulator at the nanoscale. *Phys. Rev.* **2018**, *B97*, 020102. [CrossRef]
67. Cha, J.; Kim, K.W.; Daraio, C. Experimental realization of on-chip topological nanoelectromechanical metamaterials. *Nature* **2018**, *564*, 229–233. [CrossRef]
68. Cho, C.; Wen, X.; Park, N.; Li, J. Digitally virtualized atoms for acoustic metamaterials. *Nat. Commun.* **2020**, *11*, 251. [CrossRef] [PubMed]
69. Forster, T.Z. Experimentelle und theoretische Untersuchung des zwischenmolekularen Übergangs von Elektronenanregungsenergie. *Naturforsch. A* **1949**, *4*, 321. [CrossRef]
70. Scholes, G.D. Long-range resonance energy transfer in molecular systems. *Annu. Rev. Phys. Chem.* **2003**, *54*, 57–87. [CrossRef] [PubMed]
71. Hardin, B.E.; Hoke, E.T.; Armstrong, P.B.; Yum, J.-H.; Comte, P.; Torres, T.; Frechet, J.M.; Nazeeruddin, M.K.; Gratzel, M.; McGehee, M.D. Increased light harvesting in dye-sensitized solar cells with energy relay dyes. *Nat. Photonics* **2009**, *3*, 406–411. [CrossRef]
72. Baldo, M.A.; O'brien, D.; You, Y.; Shoustikov, A.; Sibley, S.; Thompson, M.; Forrest, S. Highly efficient phosphorescent emission from organic electroluminescent devices. *Nature* **1998**, *395*, 151–154. [CrossRef]
73. van Grondelle, R.; Dekker, J.P.; Gillbro, T.; Sundstrom, V. Energy transfer and trapping in photosynthesis. *Biochim. Biophys. Acta Bioenerg.* **1994**, *1187*, 1–65. [CrossRef]
74. Tumkur, T.U.; Kitur, J.K.; Bonner, C.E.; Poddubny, A.N.; Narimanov, E.E.; Noginov, M.A. Control of Förster energy transfer in the vicinity of metallic surfaces and hyperbolic metamaterials. *Faraday Discuss.* **2015**, *178*, 395–412. [CrossRef]
75. Cortes, C.L.; Jacob, Z. Super-Coulombic atom–atom interactions in hyperbolic media. *Nat. Commun.* **2017**, *8*, 14144. [CrossRef]
76. Ren, J.; Wu, T.; Yang, B.; Zhang, X. Simultaneously giant enhancement of Förster resonance energy transfer rate and efficiency based on plasmonic excitations. *Phys. Rev. B Condens. Matter Mater. Phys.* **2016**, *94*, 125416. [CrossRef]
77. Martín-Cano, D.; Martín-Moreno, L.; García-Vidal, F.J.; Moreno, E. Resonance energy transfer and superradiance mediated by plasmonic nanowaveguides. *Nano Lett.* **2010**, *10*, 3129–3134. [CrossRef] [PubMed]
78. Bouchet, D.; Cao, D.; Carminati, R.; De Wilde, Y.; Krachmalnic off, V. Long-range plasmon-assisted energy transfer between fluorescent emitters. *Phys. Rev. Lett.* **2016**, *116*, 037401. [CrossRef]
79. Biehs, S.-A.; Menon, V.M.; Agarwal, G.S. Long-range dipole-dipole interaction and anomalous Förster energy transfer across a hyperbolic metamaterial. *Phys. Rev. B Condens. Matter Mater. Phys.* **2016**, *93*, 245439. [CrossRef]
80. Deshmukh, R.; Biehs, S.-A.; Khwaja, E.; Galfsky, T.; Agarwal, G.S.; Menon, V.M. Long-range resonant energy transfer using optical topological transitions in metamaterials. *ACS Photonics* **2018**, *5*, 2737–2741. [CrossRef]
81. Kruk, S.S.; Wong, Z.J.; Pshenay-Severin, E.; O'Brien, K.; Neshev, D.N.; Kivshar, Y.S.; Zhang, X. Magnetic hyperbolic optical metamaterials. *Nat. Commun.* **2016**, *7*, 11329. [CrossRef]
82. Mirmoosa, M.S.; Kosulnikov, S.Y.; Simovski, C.R. Magnetic hyperbolic metamaterial of high-index nanowires. *Phys. Rev. B* **2016**, *94*, 075138. [CrossRef]

83. Papadaki, G.T.; Fleischma, D.; Davoyan, A.; Yeh, A.; Atwater, H.A. Optical magnetism in planar metamaterial heterostructures. *Nat. Commun.* **2018**, *9*, 296. [CrossRef] [PubMed]

84. Yang, Y.H.; Qin, P.F.; Zheng, B.; Shen, L.; Wang, H.P.; Wang, Z.J.; Li, E.P.; Singh, R.; Chen, H.S. Hyperbolic metamaterials: From dispersion manipulation to applications. *Adv. Sci.* **2018**, *5*, 1801495. [CrossRef]

85. Iorsh, I.V.; Poddubny, A.N.; Ginzburg, P.; Belov, P.A.; Kivshar, Y.S. Compton-like polariton scattering in hyperbolic metamaterials. *Phys. Rev. Lett.* **2015**, *114*, 185501. [CrossRef]

86. Shen, H.; Lu, D.; VanSaders, B.; Kan, J.J.; Xu, H.X.; Fullerton, E.E.; Liu, Z.W. Anomalously Weak Scattering in Metal-Semiconductor Multilayer Hyperbolic Metamaterials. *Phys. Rev.* **2015**, *X5*, 021021. [CrossRef]

87. Qian, C.; Lin, X.; Yang, Y.; Gao, F.; Shen, Y.C.; Lopez, J.; Kaminer, I.; Zhang, B.L.; Li, E.P.; Soljacic, M.; et al. Multifrequency superscattering from subwavelength hyperbolic structures. *ACS Photonics* **2018**, *5*, 1506. [CrossRef]

88. Rituraj; Catrysse, P.B.; Fan, S.H. Scattering of electromagnetic waves by cylinder inside uniaxial hyperbolic medium. *Opt. Express* **2019**, *27*, 3991.

89. Memarian, M.; Eleftheriades, G.V. Light concentration using hetero-junctions of anisotropic low permittivity metamaterials. *Light Sci. Appl.* **2013**, *2*, e114. [CrossRef]

90. Guo, Z.W.; Jiang, H.T.; Zhu, K.J.; Sun, Y.; Li, Y.H.; Chen, H. Focusing and super-resolution with partial cloaking based on linear-crossing metamaterials. *Phys. Rev. Appl.* **2018**, *10*, 064048. [CrossRef]

91. Yang, Y.T.; Jia, Z.Y.; Xu, T.; Luo, J.; Lai, Y.; Hang, Z.H. Beam splitting and unidirectional cloaking using anisotropic zero-index photonic crystals. *Appl. Phys. Lett.* **2019**, *114*, 161905. [CrossRef]

92. Naik, G.V.; Liu, J.; Kildishev, A.V.; Shalaev, V.M.; Boltasseva, A. Demonstration of Al: ZnO as a plasmonic component for near-infrared metamaterials. *Proc. Natl. Acad. Sci. USA* **2012**, *109*, 8834. [CrossRef] [PubMed]

93. Xu, T.; Agrawal, A.; Abashin, M.; Chau, K.J.; Lezec, H.J. All-angle negative refraction and active flat lensing of ultraviolet light. *Nature* **2013**, *497*, 470. [CrossRef]

94. Argyropoulos, C.; Estakhri, N.M.; Monticone, F.; Alù, A. Negative refraction, gain and nonlinear effects in hyperbolic metamaterials. *Opt. Express* **2013**, *21*, 15037. [CrossRef]

95. High, A.A.; Devlin, R.C.; Dibos, A.; Polking, M.; Wild, D.S.; Perczel, J.; de Leon, N.P.; Lukin, M.D.; Park, H. Visible-frequency hyperbolic metasurface. *Nature* **2015**, *522*, 192. [CrossRef] [PubMed]

96. Sheng, C.; Liu, H.; Chen, H.Y.; Zhu, S.N. Definite photon deflections of topological defects in metasurfaces and symmetry-breaking phase transitions with material loss. *Nat. Commun.* **2018**, *9*, 4271. [CrossRef]

97. Smith, D.R.; Schurig, D. Electromagnetic wave propagation in media with indefinite permittivity and permeability tensors. *Phys. Rev. Lett.* **2003**, *90*, 077405. [CrossRef]

98. Smith, D.R.; Schurig, D.R.; Mock, J.J.; Kolinko, P.; Rye, P. Partial focusing of radiation by a slab of indefinite media. *Appl. Phys. Lett.* **2004**, *84*, 2244. [CrossRef]

99. Noginov, M.A.; Li, H.; Barnakov, Y.A.; Dryden, D.; Nataraj, G.; Zhu, G.; Bonner, C.E.; Mayy, M.; Jacob, Z. ZEE Narimanov, Controlling spontaneous emission with metamaterials. *Opt. Lett.* **2010**, *35*, 1863–1865. [CrossRef]

100. Jacob, Z.; Kim, J.; Naik, G.; Boltasseva, A.; Narimanov, E.; Shalaev, V. Engineering photonic density of states using metamaterials. *Appl. Phys. B Lasers Opt.* **2010**, *100*, 215–218. [CrossRef]

101. Smolyaninov, I.I. Giant Unruh effect in hyperbolic metamaterial waveguides. *Opt. Lett.* **2019**, *44*, 2224–2227. [CrossRef]

102. Tumkur, T.; Zhu, G.; Black, P.; Barnakov, Y.A.; Bonner, C.E.; Noginov, M.A. Control of spontaneous emission in a volume of functionalized hyperbolic metamaterial. *Appl. Phys. Lett.* **2011**, *99*, 151115. [CrossRef]

103. Krishnamoorthy, H.N.; Jacob, Z.; Narimanov, E.; Kretzschmar, I.; Menon, V.M. Topological transitions in metamaterials. *Science* **2012**, *336*, 205. [CrossRef] [PubMed]

104. Galfsky, T.; Krishnamoorthy, H.N.S.; Newman, W.; Narimanov, E.E.; Jacob, Z.; Menon, V.M. Active hyperbolic metamaterials: Enhanced spontaneous emission and light extraction. *Optica* **2015**, *2*, 62. [CrossRef]

105. Feng, K.J.; Sivco, D.L.; Hoffman, A.J. Engineering optical emission in sub-diffraction hyperbolic metamaterial resonators. *Opt. Express* **2018**, *26*, 4382. [CrossRef]

106. Fernandes, D.E.; Maslovski, S.I.; Silveirinha, M.G. Cherenkov emission in a nanowire material. *Phys. Rev. B* **2012**, *85*, 155107. [CrossRef]

107. Liu, F.; Xiao, L.; Ye, Y.; Wang, M.X.; Cui, K.Y.; Feng, X.; Zhang, W.; Huang, Y.D. Integrated Cherenkov radiation emitter eliminating the electron velocity threshold. *Nat. Photonics* **2017**, *11*, 289. [CrossRef]

108. Silveirinha, M. A low-energy Cherenkov glow. *Nat. Photonics* **2017**, *11*, 269. [CrossRef]

109. Tao, V.; Wu, L.; Zheng, G.X.; Yu, S.H. Cherenkov polaritonic radiation in a natural hyperbolic material. *Carbon* **2019**, *150*, 136. [CrossRef]

110. Belov, P.A.; Simovski, C.R.; Ikonen, P. Canalization of subwavelength images by electromagnetic crystals. *Phys. Rev. B* **2005**, *71*, 193105. [CrossRef]

111. Poddubny, A.; Iorsh, I.; Belov, P.; Kivshar, Y. Hyperbolic metamaterials. *Nat. Photon.* **2013**, *7*, 948–957. [CrossRef]

112. Folland, T.G.; Fali, A.; White, S.T.; Matson, J.R.; Liu, S.; Aghamiri, N.A.; Edgar, J.H.; Haglund, R.F., Jr.; Abate, Y.; Caldwell, J.D. Reconfigurable infrared hyperbolic metasurfaces using phase change materials. *Nat. Commun.* **2018**, *9*, 4371. [CrossRef] [PubMed]

113. Yoxall, E.; Schnell, M.; Nikitin, A.Y.; Txoperena, O.; Woessner, A.; Lundeberg, M.B.; Casanova, F.; Hueso, L.E.; Koppens, F.H.L.; Hillenbrand, R. Direct observation of ultraslow hyperbolic polariton propagation with negative phase velocity. *Nat. Photonics* **2015**, *9*, 674. [CrossRef]

114. Caldwell, J.D.; Aharonovich, I.; Cassabois, G.; Edgar, J.H.; Gil, B.; Basov, D.N. Photonics with hexagonal boron nitride. *Nat. Rev. Mater.* **2019**, *4*, 552. [CrossRef]

115. Ambrosio, A.; Jauregui, L.A.; Dai, S.; Chaudhary, K.; Tamagnone, M.; Fogler, M.M.; Basov, D.N.; Capasso, F.; Kim, P.; Wilson, W.L. Mechanical detection and imaging of hyperbolic phonon polaritons in hexagonal boron nitride. *ACS Nano* **2017**, *11*, 8741. [CrossRef] [PubMed]

116. Alfaro-Mozaz, F.J.; Alonso-González, P.; Vélez, S.; Dolado, I.; Autore, M.; Mastel, S.; Casanova, F.; Hueso, L.E.; Li, P.; Nikitin, A.Y.; et al. Nanoimaging of resonating hyperbolic polaritons in linear boron nitride antennas. *Nat. Commun.* **2017**, *8*, 15624.

117. Lin, X.; Yang, Y.; Rivera, N.; López, J.J.; Shen, Y.; Kaminerb, I.; Chen, H.; Zhang, B.; Joannopoulos, J.D.; Soljacic, M. All-angle negative refraction of highly squeezed plasmon and phonon polaritons in graphene–boron nitride heterostructures. *Proc. Natl. Acad. Sci. USA* **2017**, *114*, 6717. [CrossRef] [PubMed]

118. Ma, W.; Alonso-González, P.; Li, S.; Nikitin, A.Y.; Yuan, J.; Martín-Sánchez, J.; Taboada-Gutiérrez, J.; Amenabar, I.; Li, P.; Vélez, S.; et al. In-plane anisotropic and ultra-low-loss polaritons in a natural van der Waals crystal. *Nature* **2018**, *562*, 557. [CrossRef]

119. Kruk, S.S.; Powell, D.A.; Minovich, A.; Neshev, D.N.; Kivshar, Y.S. Spatial dispersion of multilayer fishnet metamaterials. *Opt. Express* **2012**, *20*, 15100. [CrossRef]

120. Gomez-Diaz, J.S.; Tymchenko, M.; Alù, A. Hyperbolic plasmons and topological transitions over uniaxial metasurfaces. *Phys. Rev. Lett.* **2015**, *114*, 233901. [CrossRef]

121. Gomez-Diaz, J.S.; Alù, A. Flatland optics with hyperbolic metasurfaces. *ACS Photonics* **2016**, *3*, 2211. [CrossRef]

122. Wood, B.; Pendry, J.B.; Tsai, D.P. Directed subwavelength imaging using a layered metal-dielectric system. *Phys. Rev. B* **2006**, *74*, 115116. [CrossRef]

123. Avrutsky, I.; Salakhutdinov, I.; Elser, J.; Podolskiy, V. Highly confined optical modes in nanoscale metal-dielectric multilayers. *Phys. Rev. B* **2007**, *75*, 242402. [CrossRef]

124. Liu, Z.; Lee, H.; Xiong, Y.; Sun, C.; Zhang, C. Far-field optical hyperlens magnifying sub-diffraction-limited objects. *Science* **2007**, *315*, 1686. [CrossRef]

125. Iorsh, I.; Poddubny, A.; Orlov, A.; Belov, P.; Kivshar, Y.S. Spontaneous emission enhancement in metal–dielectric metamaterials. *Phys. Lett. A.* **2012**, *376*, 185–187. [CrossRef]

126. Guo, Y.; Cortes, C.L.; Molesky, S.; Jacob, S. Broadband super-Planckian thermal emission from hyperbolic metamaterials. *Appl. Phys. Lett.* **2012**, *101*, 131106. [CrossRef]

127. Biehs, S.-A.; Tschikin, M.; Ben-Abdallah, P. Hyperbolic metamaterials as an analog of a blackbody in the near field. *Phys. Rev. Lett.* **2012**, *109*, 104301. [CrossRef] [PubMed]

128. Nefedov, I.S.; Simovski, C.R. Giant radiation heat transfer through micron gaps. *Phys. Rev. B* **2011**, *84*, 195459. [CrossRef]

129. Schoche, S.; Ho, P.-H.; Roberts, J.A.; Yu, S.J.; Fan, J.A.; Falk, A.L. Mid-IR and UV-Vis-NIR Mueller matrix ellipsometry characterization of tunable hyperbolic metamaterials based on self-assembled carbon nanotubes. *J. Vac. Sci. Technol. B* **2020**, *38*, 014015. [CrossRef]

130. Hu, X.; Zheng, D.; Lin, Y.-S. Actively tunable terahertz metamaterial with single-band and dual-band switching characteristic. *Appl. Phys. A* **2020**, *126*, 1–9. [CrossRef]

131. Li, W.; Cheng, Y. Dual-band tunable terahertz perfect metamaterial absorber based on strontium titanate (STO) resonator structure. *Opt. Commun.* **2020**, *462*, 125265. [CrossRef]

132. Lu, T.; Qiu, P.; Lian, J.; Zhang, D.; Zhuang, S. Ultrathin and broadband highly efficient terahertz reflective polarization converter based on four L-shaped metamaterials. Ultrathin and broadband highly efficient terahertz reflective polarization converter based on four L-shaped metamaterials. *Opt. Mater.* **2019**, *95*, 109230. [CrossRef]

133. Li, Z.; Aydin, K.; Ozbay, E. Determination of effective constitutive parameters of bianisotropic metamaterials from reflection and transmission coefficients. *Phys. Rev. E* **2009**, *79*, 026610. [CrossRef]

134. Basharin, A.A.; Chuguevsky, V.; Volsky, N.; Kafesaki, M.; Economou, E.N. Extremely high -factor metamaterials due to anapole excitation. *Phys. Rev. B* **2017**, *95*, 035104. [CrossRef]

135. Li, Q.; Cong, L.; Singh, R.; Xu, N.; Cao, W.; Zhang, X.; Tian, Z.; Du, L.; Han, J.; Zhang, W. Monolayer graphene sensing enabled by the strong Fano-resonant metasurface. *Nanoscale* **2016**, *8*, 17278. [CrossRef] [PubMed]

136. Srivastava, Y.K.; Manjappa, M.; Cong, L.; Cao, W.; Al-Naib, I.; Zhang, W.; Singh, R. Ultrahigh-Q Fano Resonances in Terahertz Metasurfaces: Strong Influence of Metallic Conductivity at Extremely Low Asymmetry. *Adv. Opt. Mater.* **2016**, *4*, 457. [CrossRef]

137. Cong, L.; Singh, R. Symmetry-protected dual bound states in the continuum in metamaterials. *Adv. Opt. Mater.* **2019**, *7*, 1900383. [CrossRef]

138. Wang, J.; Song, C.; Hang, J.; Hu, Z.; Zhang, F. Tunable Fano resonance based on grating-coupled and graphene-based Otto configuration. *Opt. Express* **2017**, *25*, 23880. [CrossRef]

139. Singh, R.; Al-Naib, I.A.I.; Koch, M.; Zhang, W. Sharp Fano resonances in THz metamaterials. *Opt. Express* **2011**, *19*, 6312. [CrossRef]

140. Cao, W.; Singh, R.; Al-Naib, I.A.I.; He, M.; Taylor, A.J.; Zhang, W. Low-loss ultra-high-Q dark mode plasmonic Fano metamaterials. *Opt. Lett.* **2012**, *37*, 3366. [CrossRef]

141. Offermans, P.; Schaafsma, M.C.; Rodriguez, S.R.K.; Zhang, Y.; Crego-Calama, M.; Brongersma, S.H.; Rivas, J.G. Universal Scaling of the Figure of Merit of Plasmonic Sensors. *ACS Nano* **2011**, *5*, 5151. [CrossRef]
142. Zhu, W.M.; Liu, A.Q.; Bourouina, T.; Tsai, D.P.; Teng, J.H.; Zhang, X.H.; Lo, G.Q.; Kwong, D.L.; Zheludev, N.I. Microelectromechanical Maltese-cross metamaterial with tunable terahertz anisotropy. *Nat. Commun.* **2012**, *3*, 1274. [CrossRef]
143. Huang, Y.; Yan, J.; Ma, C.; Yang, G. Active tuning of the Fano resonance from a Si nanosphere dimer by the substrate effect. *Nanoscale Horiz.* **2019**, *4*, 148. [CrossRef] [PubMed]
144. Srivastava, Y.K.; Manjappa, M.; Krishnamoorthy, H.N.S.; Singh, R. Accessing the High-Q Dark Plasmonic Fano Resonances in Superconductor Metasurfaces. *Adv. Opt. Mater.* **2016**, *4*, 1875. [CrossRef]
145. Gu, J.; Singh, R.; Liu, X.; Zhang, X.; Ma, Y.; Zhang, S.; Maier, S.A.; Tian, Z.; Azad, A.K.; Chen, H.-T.; et al. Active control of electromagnetically induced transparency analogue in terahertz metamaterials. *Nat. Commun.* **2012**, *3*, 1151. [CrossRef]
146. Xiaofei, W.; Liu, G.; Xia, S.; Meng, H.; Shang, X.; He, P.; Zhai, X. Dynamically tunable Fano resonance based on graphene metamaterials. *IEEE Photonics Technol. Lett.* **2018**, *30*, 2147–2150.
147. Li, Q.; Gupta, M.; Zhang, X.; Wang, S.; Chen, T.; Singh, R.; Han, J.; Zhang, W. Active Control of Asymmetric Fano Resonances with Graphene—Silicon—Integrated Terahertz Metamaterials. *Adv. Mater. Technol.* **2020**, *5*, 1900840. [CrossRef]
148. Lin, Z.; Xu, Z.; Liu, P.; Liang, Z.; Lin, Y.-S.; Lin, Z.; Xu, Z.; Liu, P.; Liang, Z. Polarization-sensitive terahertz resonator using asymmetrical F-shaped metamaterial. *Opt. Laser Technol.* **2020**, *121*, 105826. [CrossRef]
149. Singh, M.R. Photon transparency in metallic photonic crystals doped with an ensemble of nanoparticles. *Phys. Rev. A* **2009**, *79*, 013826. [CrossRef]
150. Singh, M.R.; Davieau, K.; Carson, J.J.L. Effect of quantum interference on absorption of light in metamaterial hybrids. *J. Phys. D. Appl. Phys.* **2016**, *49*, 445103. [CrossRef]
151. Adamo, G.; MacDonald, K.F.; Fu, Y.H.; Wang, C.M.; Tsai, D.P.; de Abajo, F.G.; Zheludev, N.I. Light well: A tunable free-electron light source on a chip. *Phys. Rev. Lett.* **2009**, *103*, 113901. [CrossRef]
152. Adamo, G.; Ou, J.Y.; So, J.K.; Jenkins, S.D.; De Angelis, F.; MacDonald, K.F.; Fabrizio, D.; Ruostekoski, E.J.; Zheludev, N.I. Electron-beam-driven collective-mode metamaterial light source. *Phys. Rev. Lett.* **2012**, *109*, 217401. [CrossRef] [PubMed]
153. Wong, L.J.; Kaminer, I.; Ilic, O.; Joannopoulos, J.D.; Soljačić, M. Towards graphene plasmon-based free-electron infrared to X-ray sources. *Nat. Photonics* **2016**, *10*, 46. [CrossRef]
154. Plettner, T.; Byer, R.L. Proposed dielectric-based microstructure laser-driven undulator. *Phys. Rev. Spec. Top. Accel. Beams* **2008**, *11*, 030704. [CrossRef]
155. Rosolen, G.; Wong, L.J.; Rivera, N.; Maes, B.; Soljačić, M.; Kaminer, I. Metasurface-based multi-harmonic free-electron light source. *Light Sci. Appl.* **2018**, *7*, 64. [CrossRef] [PubMed]
156. Lu, T.; Yang, L.; Carmon, T.; Min, B. A narrow-linewidth on-chip toroid Raman laser. *IEEE J. Quantum Electron.* **2011**, *47*, 320. [CrossRef]
157. England, R.J.; Noble, R.J.; Bane, K.; Dowell, D.H.; Ng, C.K.; Spencer, J.E.; Tantawi, S.; Wu, Z.; Byer, R.L.; Peralta, E.; et al. Demonstration of electron acceleration in a laser-driven dielectric microstructure. *Rev. Mod. Phys.* **2014**, *86*, 1337. [CrossRef]
158. Peralta, E.A.; Soong, K.; England, R.J.; Colby, E.R.; Wu, Z.; Montazeri, B.; McGuinness, C.; McNeur, J.; Leedle, K.J.; Walz, D. Demonstration of electron acceleration in a laser-driven dielectric microstructure. *Nature* **2013**, *503*, 91. [CrossRef] [PubMed]
159. Lukianova-Hleb, E.Y.; Ren, X.; Sawant, R.R.; Wu, X.; Torchilin, V.P.; Lapotko, D.O. On-demand intracellular amplification of chemoradiation with cancer-specific plasmonic nanobubbles. *Nat. Med.* **2014**, *20*, 778. [CrossRef] [PubMed]
160. Rousse, A.; Rischel, C.; Gauthier, J.C. Femtosecond x-ray crystallography. *Rev. Mod. Phys.* **2001**, *73*, 17. [CrossRef]
161. Koppens, F.H.; Chang, D.E.; García de Abajo, F.J. Graphene plasmonics: A platform for strong light–matter interactions. *Nano Lett.* **2011**, *11*, 3370. [CrossRef]
162. Jablan, M.; Buljan, H.; Soljačić, M. Plasmonics in graphene at infrared frequencies. *Phys. Rev. B* **2009**, *80*, 245435. [CrossRef]
163. Brar, V.W.; Jang, M.S.; Sherrott, M.; Lopez, J.J.; Atwater, H.A. Highly confined tunable mid-infrared plasmonics in graphene nanoresonators. *Nano Lett.* **2013**, *13*, 2541. [CrossRef] [PubMed]
164. Wenger, T.; Viola, G.; Fogelström, M.; Tassin, P.; Kinaret, J. Optical signatures of nonlocal plasmons in graphene. *Phys. Rev. B* **2016**, *94*, 205419. [CrossRef]
165. Fang, F.; Thongrattanasiri, S.; Schlather, A.; Liu, Z.; Ma, L.; Wang, Y.; Ajayan, P.M.; Nordlander, P.; Halas, N.J.; García de Abajo, F.J. Gated Tunability and Hybridization of Localized Plasmons in Nanostructured Graphene. *ACS Nano* **2013**, *7*, 2388. [CrossRef]
166. Novoselov, K.S.; Geim, A.K.; Morozov, S.V.; Jiang, D.; Zhang, Y.; Dubonos, S.V.; Grigorieva, I.V.; Firsov, A.A. Electric field effect in atomically thin carbon films. *Science* **2004**, *306*, 666–669. [CrossRef]
167. Geim, A.K.; Novoselov, K.S. The rise of graphene. *Nat. Mater.* **2007**, *6*, 183–191. [CrossRef] [PubMed]
168. Nair, R.R.; Blake, P.; Grigorenko, A.N.; Novoselov, K.S.; Booth, T.J.; Stauber, T.; Peres, N.M.R.; Geim, A.K. Fine Structure Constant Defines Visual Transparency of Graphene. Fine Structure Constant Defines Visual Transparency of Graphene. *Science* **2008**, *320*, 1308. [CrossRef]
169. Fiori, G.; Bonaccorso, F.; Iannaccone, G.; Palacios, T.; Neumaier, D.; Seabaugh, A.; Banerjee, S.K.; Colombo, L. Electronics based on two-dimensional materials. *Nat. Nanotechnol.* **2014**, *9*, 768779. [CrossRef] [PubMed]
170. Novoselov, K.S.; Geim, A.K.; Morozov, S.V.; Jiang, D.; Katsnelson, M.I.; Grigorieva, I.V.; Dubonos, S.V.; Firsov, A.A. Two-dimensional gas of massless Dirac fermions in graphene. *Nature* **2005**, *438*, 197. [CrossRef] [PubMed]

171. Bonaccorso, F.; Colombo, L.; Yu, G.; Stoller, M.; Tozzini, V.; Ferrari, A.C. Graphene, related two-dimensional crystals, and hybrid systems for energy conversion and storage. *Science* **2015**, *347*, 1246501. [CrossRef]
172. Koppens, F.; Mueller, T.; Avouris, P.; Ferrari, A.C.; Vitiello, M.S.; Polini, M. Photodetectors based on graphene, other two-dimensional materials and hybrid systems. *Nat. Nanotechnol.* **2014**, *9*, 780. [CrossRef] [PubMed]
173. Bernardi, M.; Palummo, M.; Grossman, J.C. Extraordinary sunlight absorption and one nanometer thick photovoltaics using two-dimensional monolayer material. *Nano Lett.* **2013**, *13*, 3664. [CrossRef] [PubMed]
174. Kong, X.-T.; Khan, A.A.; Kidambi, P.R.; Deng, S.; Yetisen, A.K.; Dlubak, B.; Hiralal, P.; Montelongo, Y.; Bowen, J.; Xavier, S.; et al. Graphene-Based Ultrathin Flat Lenses. *ACS Photonics* **2015**, *2*, 200–207. [CrossRef]
175. Skulason, H.S.; Gaskell, P.E.; Szkopek, T. Optical reflection and transmission properties of exfoliated graphite from a graphene monolayer to several hundred graphene layers. *Nanotechnology* **2010**, *21*, 295709. [CrossRef] [PubMed]
176. Xiang, Y.; Dai, X.; Guo, J.; Zhang, H.; Wen, S.; Tang, D. Critical coupling with graphene-based hyperbolic metamaterials. *Sci. Rep.* **2015**, *4*, 5483. [CrossRef] [PubMed]
177. Sreekanth, K.V.; ElKabbash, M.; Alapan, Y.; Rashed, A.R.; Gurkan, U.A.; Strangi, G. A multiband perfect absorber based on hyperbolic metamaterials. *Sci. Rep.* **2016**, *6*, 26272. [CrossRef]
178. Li, W.; Valentine, J. Metamaterial perfect absorber based hot electron photodetection. *Nano Lett.* **2014**, *14*, 3510–3514. [CrossRef] [PubMed]
179. Pizzi, A.; Rosolen, G.; Wong, L.J.; Ischebeck, R.; Soljačić, M.; Feurer, T.; Kaminer, I. Graphene Metamaterials for Intense, Tunable, and Compact Extreme Ultraviolet and X-Ray Sources. *Adv. Sci.* **2019**, 1901609. [CrossRef]
180. Ruoff, R. Calling all chemists. *Graphene Nat. Nanotechnol.* **2008**, *3*, 10–11. [CrossRef]
181. Dreyer, D.R.; Park, S.; Bielawski, C.W.; Ruoff, R.S. Graphite oxide. *Chem. Soc. Rev.* **2010**, *39*, 228–240. [CrossRef] [PubMed]
182. Chen, D.; Feng, H.; Li, J. Graphene oxide: Preparation, functionalization, and electrochemical applications. *Chem. Rev.* **2012**, *112*, 6027–6053. [CrossRef]
183. Dikin, D.A.; Stankovich, S.; Zimney, E.J.; Piner, R.D.; Dommett, G.H.B.; Evmenenko, G.; Nguyen, S.T.; Ruoff, R.S. Preparation and characterization of graphene oxide paper. *Nature* **2007**, *448*, 457. [CrossRef]
184. Eda, G.; Fanchini, G.; Chhowalla, M. Large-area ultrathin films of reduced graphene oxide as a transparent and flexible electronic material. *Nat. Nanotechnol.* **2008**, *3*, 270. [CrossRef]
185. Dong, L.; Yang, J.; Chhowalla, M.; Loh, K.P. Synthesis and reduction of large sized graphene oxide sheets. *Chem. Soc. Rev.* **2017**, *46*, 7306. [CrossRef]
186. Zhang, Y.-L.; Guo, L.; Xia, H.; Chen, Q.-D.; Feng, J.; Sun, H.-B. Photoreduction of Graphene Oxides: Methods, Properties, and Applications. *Adv. Opt. Mater.* **2014**, *2*, 10. [CrossRef]
187. Yang, Y.; Wu, J.; Xu, X.; Liang, Y.; Chu, S.T.; Little, B.E.; Morandotti, R.; Jia, B.; Moss, D.J. Enhanced four-wave mixing in waveguides integrated with graphene oxide. *APL Photonics* **2018**, *3*, 120803. [CrossRef]
188. Li, D.; Muller, M.B.; Gilje, S.; Kaner, R.B.; Wallace, G.G. Processable aqueous dispersions of graphene nanosheets. *Nat. Nanotechnol.* **2008**, *3*, 101–105. [CrossRef]
189. Ferrari, A.C. Raman spectroscopy of graphene and graphite: Disorder, electron–phonon coupling, doping and nonadiabatic effects. *Solid State Commun.* **2007**, *143*, 47. [CrossRef]
190. Yang, Y.; Lin, H.; Zhang, B.Y.; Zhang, Y.; Zhang, X.; Yu, A.; Hong, M.; Jia, B. Graphene-Based Multilayered Metamaterials with Phototunable Architecture for on-Chip Photonic Devices. *ACS Photonics* **2019**, *6*, 1033–1046. [CrossRef]
191. Zhang, Y.; Feng, Y.; Zhao, J. Graphene-enabled active metamaterial for dynamical manipulation of terahertz reflection/transmission/absorption. *Phys. Lett. A* **2020**, *384*, 12684. [CrossRef]
192. Banerjee, P.; Ghosh, G.; Biswas, S.K. Measurement of dielectric properties of medium loss samples at X-band frequencies. *J. Metall. Mater. Sci.* **2010**, *52*, 247–255.
193. Grenier, K.; Dubuc, D.; Poleni, P.; Kumemura, M.; Toshiyoshi, H.; Fujii, T.; Fujita, H. Integrated broadband microwave and microfluidic sensor dedicated to bioengineering. *IEEE. Trans. Microw. Theory Tech.* **2009**, *57*, 3246. [CrossRef]
194. Lee, H.J.; Lee, J.H.; Choi, S.; Jang, I.S.; Choi, J.S.; Jung, H.I. Asymmetric split-ring resonator-based biosensor for detection of label-free stress biomarkers. *Appl. Phys. Lett.* **2013**, *103*, 0537021. [CrossRef]
195. Rawat, V.; Dhobale, S.; Kale, S.N. Ultra-fast selective sensing of ethanol and petrol using microwave-range metamaterial complementary split-ring resonators. *J. Appl. Phys.* **2014**, *116*, 164106l. [CrossRef]
196. Gordon, J.A.; Holloway, C.L.; Booth, J.; Kim, S.; Wang, Y.; BakerJarvis, J.; Novotny, D.R. Fluid interactions with metafilms/metasurfaces for tuning, sensing, and microwave-assisted chemical processes. *Phys. Rev. B* **2011**, *83*, 205130. [CrossRef]
197. Awang, R.A.; Tovar-Lopez, F.J.; Baum, T.; Sriram, S.; Rowe, W.S. Meta-atom microfluidic sensor for measurement of dielectric properties of liquids. *J. Appl. Phys.* **2017**, *121*, 094506. [CrossRef]
198. Withayachumnankul, W.; Jaruwongrungsee, K.; Tuantranont, A.; Fumeaux, C.; Abbott, D. Metamaterial-based microfluidic sensor for dielectric characterization. *Sens. Actuators A Phys.* **2013**, *189*, 2331. [CrossRef]
199. Paris, V.; Grenier, K.; Mata-Contreras, J.; Dubuc, D.; Martín, F. Highly-sensitive microwave sensors based on open complementary split ring resonators (OCSRRs) for dielectric characterization and solute concentration measurement in liquids. *IEEE Access* **2018**, *6*, 48324–48338.
200. Xu, X.; Peng, B.; Li, D.; Zhang, J.; Wong, L.M.; Zhang, Q.; Wang, S.; Xiong, Q. Flexible visible–infrared metamaterials and their applications in highly sensitive chemical and biological sensing. *Nano Lett.* **2011**, *11*, 3232–3238. [CrossRef]

201. Ahn, S.H.; Guo, L.J. Large-area roll-to-roll and roll-to-plate nanoimprint lithography: A step toward high-throughput application of continuous nanoimprinting. *ACS Nano* **2009**, *3*, 2304–2310. [CrossRef] [PubMed]
202. Ho, J.S.; Li, Z. *Microwave Metamaterials for Biomedical Sensing, Reference Module in Biomedical Sciences*; Elsevier: Amsterdam, Poland, 2021; ISBN 9780128012383. [CrossRef]
203. Kayal, S.; Shaw, T.; Mitra, D. Design of metamaterial-based compact and highly sensitive microwave liquid sensor. *Appl. Phys. A* **2019**, *126*, 1–9. [CrossRef]
204. Choi, G.; Bahk, Y.-M.; Kang, T.; Lee, Y.; Son, B.H.; Ahn, Y.H.; Seo, M.; Kim, D.-S. Terahertz nanoprobing of semiconductor surface dynamics. *Nano Lett.* **2017**, *17*, 6397–6401. [CrossRef] [PubMed]
205. Lee, D.-K.; Kang, J.-H.; Kwon, J.; Lee, J.-S.; Lee, S.; Woo, D.H.; Kim, J.H.; Song, C.-S.; Park, Q.-H.; Seo, M. Nano metamaterials for ultrasensitive Terahertz biosensing. *Sci. Rep.* **2017**, *7*, 8146. [CrossRef]
206. Xu, W.; Xie, L.; Zhu, J.; Tang, L.; Singh, R.; Wang, C.; Ma, Y.; Chen, H.-T.; Ying, Y. Terahertz biosensing with a graphene-metamaterial heterostructure platform. *Carbon* **2019**, *141*, 247–252. [CrossRef]
207. Lee, S.-H.; Choe, J.-H.; Kim, C.; Bae, S.; Kim, J.-S.; Park, Q.-H.; Seo, M. Graphene assisted terahertz metamaterials for sensitive bio-sensing. *Sens. Actuators B Chem.* **2020**, *310*, 127841. [CrossRef]
208. Eleftheriades, G.V.; Iyer, A.K.; Kremer, P.C. Planar negative refractive index media using periodically LC loaded transmission lines. *IEEE Trans. Microw. Theory Technol.* **2002**, *50*, 2702–2712. [CrossRef]
209. Abdolrazzaghi, M.; Daneshmand, M.; Iyer, A.K. Strongly enhanced sensitivity in planar microwave sensors based on metamaterial coupling. *IEEE Trans. Microw. Theory Tech.* **2018**, *66*, 1843–1855. [CrossRef]
210. Ran, L.; Huangfu, J.; Chen, H.; Li, Y.; Zhang, X.; Chen, K.; Kong, J.A. Microwave solid-state left-handed material with a broad bandwidth and an ultralow loss. *Phys. Rev. B* **2004**, *70*, 07302. [CrossRef]
211. Petosa, A.; Norwood, M.A. *Artech House Antennas and Propagation Library*; Artech House Publishers: Norwood, MA, USA, 2007.
212. Kumar, J.; Gupta, N. Performance analysis of dielectric resonator antennas. *Wirel. Pers. Commun.* **2014**, *75*, 1029–1049. [CrossRef]
213. Luk, K.M.; Leung, K.W. *Both of the City University of Hong Kong*; Research Studies Press Limited: Hertforodshire, UK, 2002.
214. Kumar, J.; Gupta, N. Bandwidth and gain enhancement technique for Gammadion cross dielectric resonator antenna. *Wirel. Pers. Commun.* **2015**, *85*, 2309–2317. [CrossRef]
215. Kumar, J.; Gupta, N. Linearly polarized asymmetric dielectric resonator antenna for 5.2-GHz WLAN applications. *J. Electromagn. Waves Appl.* **2015**, *29*, 1228–1237. [CrossRef]
216. Petosa, A.; Ittipiboon, A. Dielectric resonator antennas: A historical review and the current state of the art. *IEEE Antennas Propag. Mag.* **2010**, *52*, 91–116. [CrossRef]
217. Kumar, P.; Dwari, S.; Kumar, J. Design of biodegradable quadruple-shaped DRA for WLAN/Wi-Max applications. *J. Microw. Optoelectron. Electromagn. Appl.* **2017**, *16*, 867–880. [CrossRef]
218. Kumar, A.; Kapoor, P.; Kumar, P.; Kumar, J.; Kumar, A. Metamaterial loaded aperture coupled biodegradable star—shaped dielectric resonator antenna for WLAN and broadband applications. *Microw. Opt. Technol. Lett.* **2019**, *62*, 264–277. [CrossRef]
219. Dong, Y.; Itoh, T. Metamaterial-based antennas Metamaterial-based antennas. *Proc. IEEE* **2012**, *100*, 2271–2285. [CrossRef]
220. Gong, J.Q.; Jiang, J.B.; Liang, C.H. Low-profile folded-monopole antenna with unbalanced composite right-/left-handed transmission line. *Electron. Lett.* **2012**, *48*, 813–815. [CrossRef]
221. Iizuka, H.; Hall, P.S. Left-handed dipole antennas and their implementations. *IEEE Trans. Antennas Propag.* **2007**, *55*, 1246–1253. [CrossRef]
222. Zhu, J.; George, V.E. A compact transmission-line metamaterial antenna with extended bandwidth. *IEEE Antennas Wirel. Propag. Lett.* **2008**, *8*, 295–298.
223. Herraiz-Martínez, F.J.; Hsll, P.S.; Liu, Q.; Segovia-Vargas, D. Left-handed wire antennas over ground plane with wideband tuning. *IEEE Trans. Antennas Propag.* **2011**, *59*, 1460–1471. [CrossRef]
224. Mehdipour, A.; Denidni, T.A.; Sebak, A.R. Multi-band miniaturized antenna loaded by ZOR and CSRR metamaterial structures with monopolar radiation pattern. *IEEE Trans. Antennas Propag.* **2013**, *62*, 555–562. [CrossRef]
225. Elwi, T.A. A miniaturized folded antenna array for MIMO applications. *Wirel. Pers. Commun.* **2018**, *98*, 1871–1883. [CrossRef]
226. Sanborn, A.F.; Phillips, P.K. Scaling of sound pressure level and body size in cicadas (Homoptera: Cicadidae; Tibicinidae). *Ann. Entomol. Soc. Am.* **1995**, *88*, 479–484. [CrossRef]
227. Hart, P.J.; Hall, R.; Ray, W.; Beck, A.; Zook, J. Cicadas impact bird communication in a noisy tropical rainforest. *Behav. Ecol.* **2015**, *26*, 839–842. [CrossRef]
228. Zou, H.-X.; Zhao, L.-C.; Gao, Q.-H.; Zuo, L.; Liu, F.-R.; Tan, T.; Wei, K.-X.; Zhang, W.-M. Mechanical modulations for enhancing energy harvesting: Principles, methods and applications. *Appl. Energy* **2019**, *255*, 113871. [CrossRef]
229. Tan, T.; Yan, Z.; Lei, H. Optimization and performance comparison for galloping-based piezoelectric energy harvesters with alternating-current and direct-current interface circuits. *Smart Mater. Struct.* **2017**, *26*, 075007. [CrossRef]
230. Chen, Z.S.; Guo, B.; Yang, Y.M.; Cheng, C.C. Metamaterials-based enhanced energy harvesting: A review. *Physical B* **2014**, *438*, 1–8. [CrossRef]
231. Peng, X.; Wen, Y.M.; Li, P.; Yang, A.C.; Bai, X.L. Dynamically tunable broadband mid-infrared cross polarization converter based on graphene metamaterial. *Appl. Phys. Lett.* **2013**, *103*, 4.
232. Park, C.-S.; Shin, Y.C.; Jo, S.-H.; Yoon, H.; Choi, W.; Youn, B.D.; Kim, M. Two-dimensional octagonal phononic crystals for highly dense piezoelectric energy harvesting. *Nano Energy* **2018**, *57*. [CrossRef]

233. Bin, L.; You, J.H.; Kim, Y.-J. Low frequency acoustic energy harvesting using PZT piezoelectric plates in a straight tube resonator. *Smart Mater. Struct.* **2013**, *22*, 055013. [CrossRef]

234. Liu, Z.; Zhang, X.; Mao, Y.; Zhu, Y.Y.; Yang, Z.; Chan, C.T.; Sheng, D.P. Locally resonant sonic materials. *Science* **2000**, *289*, 1734–1736. [CrossRef]

235. Song, C.; Huang, Y.; Zhou, J.; Carter, P. Recent advances in broadband rectennas for wireless power transfer and ambient RF energy harvesting. In Proceedings of the 2017 11th European Conference on Antennas and Propagation (EUCAP), Paris, France, 19–24 March 2017; pp. 341–345. [CrossRef]

236. Stuart, T.; Cai, L.; Burton, A.; Gutruf, P. Wireless and battery-free platforms for collection of biosignals. *Biosens. Bioelectron.* **2021**, *178*, 113007. [CrossRef]

237. Raza, U.; Salam, A. On-Site and External Energy Harvesting in Underground Wireless. *Electronics* **2020**, *9*, 681. [CrossRef]

238. Huang, J.; Zhou, Y.; Ning, Z.; Gharavi, H. Wireless Power Transfer and Energy Harvesting: Current Status and Future Prospects. *IEEE Wirel. Commun.* **2019**, *26*, 163–169. [CrossRef]

239. Das, R.; Basir, A.; Yoo, H. A Metamaterial-Coupled Wireless Power Transfer System Based on Cubic High-Dielectric Resonators. *IEEE Trans. Ind. Electron.* **2019**, *66*, 7397–7406. [CrossRef]

240. Yongmin, L.; Zhang, X. Metamaterials: A new frontier of science and technology. *Chem. Soc. Rev.* **2011**, *40*, 2494–2507.

241. Cummer, S.A.; Christensen, J.; Alù, A. Controlling sound with acoustic metamaterials. *Nat. Rev. Mater.* **2016**, *1*, 16001. [CrossRef]

242. Chen, H.; Chan, C.T. Acoustic cloaking in three dimensions using acoustic metamaterials. *Appl. Phys. Lett.* **2007**, *91*, 183518. [CrossRef]

243. Wang, X.; Xu, J.; Ding, J.; Zhao, C.; Huang, Z. A compact and low-frequency acoustic energy harvester using layered acoustic metamaterials. *Smart Mater. Struct.* **2019**, *28*, 025035. [CrossRef]

244. Oudich, M.; Li, Y. Tunable sub-wavelength acoustic energy harvesting with a metamaterial plate. *J. Phys. D Appl. Phys.* **2017**, *50*, 315104. [CrossRef]

245. Ma, J.; Wang, Z.-H.; Liu, H.; Fan, Y.-X.; Tao, Z.-Y. Active Switching of Extremely High-Q Fano Resonances Using Vanadium Oxide-Implanted Terahertz Metamaterials. *Appl. Sci.* **2020**, *10*, 330. [CrossRef]

246. Landy, N.I.; Sajuyigbe, S.; Mock, J.J.; Smith, D.R.; Padilla, W.J. Perfect metamaterial absorber. *Phys. Rev. Lett.* **2008**, *100*, 207402. [CrossRef]

247. Chaurasiya, D.; Ghosh, S.; Bhattacharyya, S.; Srivastava, K.V. An ultrathin quad-band polarization-insensitive wide-angle metamaterial absorber. *Microw. Opt. Technol. Lett.* **2015**, *57*, 697–702. [CrossRef]

248. Cheng, Y.Z.; Fang, C.; Zhang, Z.; Wang, B.; Chen, J.; Gong, R.Z. A compact and polarization-insensitive perfect metamaterial absorber for electromagnetic energy harvesting application. In Proceedings of the Progress in Electromagnetic Research Symposium (PIERS), Shanghai, China, 8–11 August 2016; pp. 1910–1914.

249. Yagitani, S.; Katsuda, K.; Nojima, M.; Yoshimura, Y.; Sugiura, H. Imaging radio-frequency power distributions by an EBG absorber. *IEICE Trans. Commun.* **2011**, *94*, 2306–2315. [CrossRef]

250. Li, L.; Li, B.; Liu, H.-X.; Liang, C.-H. Locally resonant cavity cell model for electromagnetic band gap structures. *IEEE Trans. Antennas Propag.* **2006**, *54*, 90–100. [CrossRef]

251. Alkurt, F.O.; Altintas, O.; Ozakturk, M.; Karaaslan, M.; Akgol, O.; Unal, E.; Sabah, C. Enhancement of image quality by using metamaterial inspired energy harvester. *Phy. Lett. A* **2020**, *384*, 126041. [CrossRef]

252. Jung, P.; Ustinov, A.V.; Anlage, S.M. Progress in superconducting metamaterials. *Supercond. Sci. Technol.* **2014**, *27*, 073001. [CrossRef]

253. Banerjee, P.; Franco, A., Jr. Role of higher valent substituent on the dielectric and optical properties of $Sr_{0.8}Bi_{2.2}Nb_2O_9$ ceramics. *Mater. Chem. Phys.* **2019**, *225*, 213–218. [CrossRef]

254. Lazarides, N.; Tsironis, G.P. Multistability and self-organization in disordered SQUID metamaterials. *Supercond. Sci. Technol.* **2013**, *26*, 084006. [CrossRef]

255. Trepanier, M.; Zhang, D.; Mukhanov, O.; Anlage, S.M. Realization and modeling of metamaterials made of rf superconducting quantum-interference devices. *Phys. Rev. X* **2013**, *3*, 041029. [CrossRef]

256. Josephson, B. Possible new effects in superconductive tunnelling. *Phys. Lett. A* **1962**, *1*, 251–255. [CrossRef]

257. Du, C.; Chen, H.; Li, S. Quantum left-handed metamaterial from superconducting quantum-interference devices. *Phys. Rev. B* **2006**, *74*, 113105. [CrossRef]

258. Lazarides, N.; Tsironis, G.P. RF superconducting quantum interference device metamaterials. *Appl. Phys. Lett.* **2007**, *90*, 163501. [CrossRef]

259. Fedotov, V.A.; Tsiatmas, A.; Shi, J.H.; Buckingham, R.; De Groot, P.; Chen, Y.; Zheludev, N.I. Temperature control of Fano resonances and transmission in superconducting metamaterials. *Opt. Express* **2010**, *18*, 9015–9019. [CrossRef] [PubMed]

260. Jung, P.; Butz, S.; Marthaler, M.; Fistul, M.V.; Leppäkangas, J.; Koshelets, V.P.; Ustinov, A.V. Multistability and switching in a superconducting metamaterial. *Nat. Commun.* **2014**, *5*, 3730. [CrossRef]

261. Zhang, D.; Trepanier, M.; Mukhanov, O.; Anlage, S.M. Tunable broadband transparency of macroscopic quantum superconducting metamaterials. *Phys. Rev. X* **2015**, *5*, 041045. [CrossRef]

262. Zhang, D.; Trepanier, M.; Antonsen, T.; Ott, E.; Anlage, S.M. Intermodulation in nonlinear SQUID metamaterials: Experiment and theor. *Phys. Rev. B* **2016**, *94*, 174507. [CrossRef]

263. Kiselev, E.I.; Averkin, A.S.; Fistul, M.V.; Koshelets, V.P.; Ustinov, A.V. Two-tone spectroscopy of a SQUID metamaterial in the nonlinear regime. *Phys. Rev. Res.* **2019**, *1*, 033096. [CrossRef]
264. Trepanier, M.; Zhang, D.; Mukhanov, O.; Koshelets, V.P.; Jung, P.; Butz, S.; Ott, E.; Antonsen, T.M.; Ustinov, A.V.; Anlage, S.M. Coherent oscillations of driven rf SQUID metamaterials. *Phys. Rev. E* **2017**, *95*, 050201.
265. Saito, S.; Zhu, X.; Amsüss, R.; Matsuzaki, Y.; Kakuyanagi, K.; Shimo-Oka, T.; Mizuochi, N.; Nemoto, K.; Munro, W.J.; Semba, K. Towards realizing a quantum memory for a superconducting qubit: Storage and retrieval of quantum states. *Phys. Rev. Lett.* **2013**, *111*, 107008. [CrossRef]
266. Shulga, K.V.; Il'chev, E.; Fistul, M.V.; Besedin, I.S.; Butz, S.; Astafiev, O.V.; Hübner, U.; Ustinov, A.V. Magnetically induced transparency of a quantum metamaterial composed of twin flux qubits. *Nat. Commun.* **2018**, *9*, 150. [CrossRef]
267. Cross, M.C.; Hohenberg, P.C. Pattern formation outside of equilibrium. *Rev. Mod. Phys.* **1993**, *65*, 851. [CrossRef]
268. Showalter, K.; Epstein, I.R. From chemical systems to systems chemistry: Patterns in space and time. *Chaos* **2015**, *25*, 097613. [CrossRef]
269. Turing, A.M. The chemical basis of morphogenesis. *Philos. Trans. R. Soc. B* **1952**, *237*, 37–72.
270. Koch, A.J.; Meinhardt, H. Biological pattern formation: From basic mechanisms to complex structures. *Rev. Mod. Phys.* **1994**, *66*, 1481. [CrossRef]
271. Lazarides, N.; Tsironis, G.P. Superconducting metamaterials. *Phys. Rep.* **2018**, *752*, 1–67. [CrossRef]
272. Hizanidis, J.; Lazarides, N.; Tsironis, G.P. Pattern formation and chimera states in 2D SQUID metamaterials. *Chaos* **2020**, *30*, 013115. [CrossRef] [PubMed]
273. Zhuravel, A.P.; Bae, S.; Lukashenko, A.V.; Averkin, A.S.; Ustinov, A.V.; Anlage, S.M. Imaging collective behavior in an rf-SQUID metamaterial tuned by DC and RF magnetic fields. *Appl. Phys. Lett.* **2019**, *114*, 082601. [CrossRef]

Article

Crystal Growth of the Quasi-2D Quarternary Compound AgCrP$_2$S$_6$ by Chemical Vapor Transport

Sebastian Selter [1], Yuliia Shemerliuk [1], Bernd Büchner [1,2] and Saicharan Aswartham [1,*]

[1] Institute for Solid State Research, Leibniz IFW Dresden, Helmholtzstr. 20, 01069 Dresden, Germany; s.selter@ifw-dresden.de (S.S.); y.shemerliuk@ifw-dresden.de (Y.S.); b.buechner@ifw-dresden.de (B.B.)

[2] Institute of Solid State and Materials Physics and Würzburg-Dresden Cluster of Excellence ct.qmat, Technische Universität Dresden, 01062 Dresden, Germany

* Correspondence: s.aswartham@ifw-dresden.de

Abstract: We report optimized crystal growth conditions for the quarternary compound AgCrP$_2$S$_6$ by chemical vapor transport. Compositional and structural characterization of the obtained crystals were carried out by means of energy-dispersive X-ray spectroscopy and powder X-ray diffraction. AgCrP$_2$S$_6$ is structurally closely related to the M_2P$_2$S$_6$ family, which contains several compounds that are under investigation as 2D magnets. As-grown crystals exhibit a plate-like, layered morphology as well as a hexagonal habitus. AgCrP$_2$S$_6$ crystallizes in monoclinic symmetry in the space group $P2/a$ (No. 13). The successful growth of large high-quality single crystals paves the way for further investigations of low dimensional magnetism and its anisotropies in the future and may further allow for the manufacturing of few-layer (or even monolayer) samples by exfoliation.

Keywords: transition metal phosphorus sulfide; van der Waals layered material; 2D material; low dimensional magnetism; magnetic chains; crystal growth; chemical vapor transport; powder X-ray diffraction; rietveld refinement

Citation: Selter, S.; Shemerliuk, Y.; Büchner, B.; Aswartham, S. Crystal Growth of the Quasi-2D Quarternary Compound AgCrP$_2$S$_6$ by Chemical Vapor Transport. *Crystals* **2021**, *11*, 500. https://doi.org/10.3390/cryst11050500

Academic Editor: Raghvendra Singh Yadav

Received: 12 April 2021
Accepted: 26 April 2021
Published: 2 May 2021

Publisher's Note: MDPI stays neutral with regard to jurisdictional claims in published maps and institutional affiliations.

1. Introduction

Among the magnetic quasi-two-dimensional materials that have recently moved in the focus of (quasi-)two-dimensional (2D) materials research [1–3], the M_2P$_2$S$_6$ class of layered materials offers a plenitude of isostructural compounds with different magnetic properties depending on M [4,5]. Thus, M_2P$_2$S$_6$ allows to investigate fundamental aspects of low dimensional magnetism and several members may be promising for future applications, e.g., complementing non-magnetic (quasi-)2D materials in heterostructures or in spintronic devices [6,7]. Furthermore, future applications in the field of catalysis are conceivable due to the structural similarity to the non-magnetic 2D materials such as graphene or the transition metal dichalcogenide compounds for which such applications are already discussed [8–10].

Regarding the crystal structure, the M_2P$_2$S$_6$ family consists of van der Waals layered compounds which share a honeycomb network of M^{2+} and, most prominently, a dominantly covalent [P$_2$S$_6$]$^{4-}$ anion located in the voids of the honeycomb [4,5]. In the bulk, such layers are stacked on top of each other only interacting via weak van der Waals forces. Consequently, these compounds can be easily exfoliated potentially down to a single layer [11,12].

Several isovalent substitution series of M^{2+} by another M'^{2+} (e.g., (Mn$_{1-x}$Fe$_x$)$_2$P$_2$S$_6$ [13], (Mn$_{1-x}$Ni$_x$)$_2$P$_2$S$_6$ [14], (Fe$_{1-x}$Ni$_x$)$_2$P$_2$S$_6$ [15,16] and (Zn$_{1-x}$Ni$_x$)$_2$P$_2$S$_6$ [4]) are reported to exhibit solid solution behavior and, thus, imply a random distribution of the substituents on the honeycomb network, as illustrated in Figure 1a. Beyond isovalent substitution, Colombet et al. [17–20] demonstrated that a substitution of M_2^{2+} by $M^{1+}M'^{3+}$ also yields several stable compounds. In contrast to the isovalent substitution series however, M^{1+} and M'^{3+} do not randomly occupy the M positions in the lattice but order either in an alternating or in a zig-zag stripe-like arrangement on the honeycomb, as illustrated in Figure 1b,c, respectively. The former arrangement is attributed to a minimization of repulsive Coloumb interactions (i.e., charge ordering). The latter is observed for compounds for which M^{1+} and M'^{3+} have notably different sizes and, thus, is dominantly driven by a minimization of lattice distortion and steric effects [4,17].

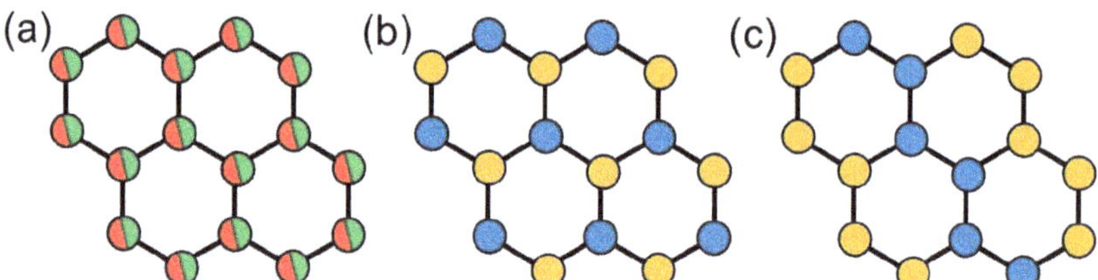

Figure 1. Schematic illustration of the different arrangements of M and M' on the honeycomb lattice of $M_2P_2S_6$. (**a**) Random distribution for $M^{2+}M'^{2+}P_2S_6$. (**b**) Alternating/triangular arrangement and (**c**) zig-zag stripe like arrangement for $M^{1+}M'^{3+}P_2S_6$.

With M'^{3+} being a magnetic ion (e.g., V^{3+} or Cr^{3+}) and M^{1+} being non-magnetic (e.g., Cu^{1+} or Ag^{1+}), the magnetic sublattices formed in $M^{1+}M'^{3+}P_2S_6$ extend the magnetic structures of the usually magnetically hexagonal $M_2P_2S_6$ compounds by an alternating/triangular and a zig-zag stripe-like magnetic arrangement [4,17,19]. The stripe-like magnetic structure is especially notable, as each stripe of magnetic ions is well isolated from the adjacent magnetic stripes by a stripe of non-magnetic ions. Although the corresponding compound still has a (quasi-)2D layered crystal structure, the magnetic structure can be expected to exhibit 1D magnetic characteristics. Indeed, several indications for such low dimensional magnetism are reported for $M^{1+}M'^{3+}P_2S_6$ with $M = $ Ag and $M' = $ Cr [19,21], making it an interesting compound for further studies.

However, until now only details on the synthesis of $AgCrP_2S_6$ via solid state synthesis are reported (although Mutka et al. [21] mention CVT grown crystals, they do not report any further details or conditions regarding the crystal growth) [19]. Although small crystals in the μm scale could be obtained by solid state synthesis, which allowed for a structural solution based on single crystal X-ray diffraction, significantly larger crystals are needed for detailed investigations of the physical properties including anisotropies.

As a crystal growth method of choice, for macroscopic $AgCrP_2S_6$ single crystals, the chemical vapor transport (CVT) technique is suitable due to the contained volatile elements such as S and P. Phosphorus and sulfur are both volatile and readily evaporate at elevated temperatures. The generation of volatile intermediate transition metal species for the vapor transport using so-called transport agents is well established [22]. CVT is the crystal growth technique of choice for virtually all ternary $M_2P_2S_6$ compounds [4,5]. For example, Taylor et al. [23] and Nitsche et al. [24] report the successful crystal growth of $M_2P_2S_6$ with $M = $ Mn, Fe, Ni, Cd and Sn by CVT using either chlorine or iodine as transport agent and we present the crystal growth of mixed transition metal phosphorus sulfides of the substitution series $(Fe_{1-x}Ni_x)_2P_2S_6$ [16] and $(Mn_{1-x}Ni_x)_2P_2S_6$ [14] by the same technique with iodine as agent. To determine a suitable temperature gradient for the CVT growth of the quarternary compound $AgCrP_2S_6$, several growth experiments with different temperature profiles were conducted. The temperature profile, which is reported hereafter, resulted in the best crystal size and quality as well as regarding impurity contributions and opens up access to macroscopic $AgCrP_2S_6$ single crystals. In addition to the crystal growth of $AgCrP_2S_6$, we also present a comprehensive compositional and structural characterization of the obtained crystals.

2. Materials and Methods

The elemental educts for the crystal growth of $AgCrP_2S_6$, as listed in Table 1, were obtained from Alfa Aesar and kept in an argon filled glove box for storage and handling.

Table 1. Elemental educts used for the CVT growth of $AgCrP_2S_6$.

Chemical	Specification	Purity
Silver	Powder, APS 4–7 micron	99.9%
Chromium	Powder, −100+325 mesh	99.99%
Red phosphorus	Lumps	99.999%
Sulfur	Pieces	99.999%
Iodine	Resublimed crystals	99.9985%

The crystals obtained from the CVT crystal growth experiments were thoroughly characterized by scanning electron microscopy (SEM) regarding their morphology and topography using a secondary electron (SE) detector and regarding chemical homogeneity via the chemical contrast obtained from a back scattered electron (BSE) detector. For this, a ZEISS EVO MA 10 scanning electron microscope was used. The chemical composition of the crystals was investigated by energy dispersive X-ray spectroscopy (EDX), which was measured in the same SEM device with an accelerating voltage of $30\,kV$ for the electron beam and using an energy dispersive X-ray analyzer.

The crystal structure of the obtained crystals was investigated by powder X-ray diffraction (pXRD), which was measured on a STOE STADI laboratory diffractometer in transmission geometry with $Cu\text{-}K_{\alpha 1}$ radiation from a curved Ge(111) single crystal monochromator and detected by a MYTHEN 1K 12.5°-linear position sensitive detector manufactured by DECTRIS. The pXRD patterns were initially analyzed by pattern matching using the HighScore Plus program suite [25]. After the crystallographic phase was identified, a structural refinement of the crystal structure model was performed based on our experimental patterns using the Rietveld method in Jana2006 [26].

3. Crystal Growth via Chemical Vapor Transport

All procedures for the preparation were performed in a glove box under argon atmosphere. The elemental educts silver, chromium, red phosphorus and sulfur were weighed out in a molar ratio of $Ag:Cr:P:S = 1:1:2:6$ and homogenized in an agate mortar. $0.5\,g$ of reaction mixture were loaded in a quartz ampule ($6\,mm$ inner diameter, $2\,mm$ wall thickness) together with approx. $50\,mg$ of the transport agent iodine. Immediately prior to use, the ampule was cleaned by washing with distilled water, rinsing with isopropanol and, subsequently, baking out at $800\,°C$ for at least $12\,h$ in an electric tube furnace. This is done to avoid contamination of the reaction volume with (adsorbed) water. The filled ampule was then transferred to a vacuum pump and evacuated to a residual pressure of $10^{-8}\,bar$. To suppress the unintended sublimation of the transport agent during evacuation, the end of the ampule containing the material was cooled with a small Dewar flask filled with liquid nitrogen. After reaching the desired internal pressure, the valve to the vacuum pump was closed, the cooling was stopped and the ampule was sealed under static pressure at a length of approximately $12\,cm$.

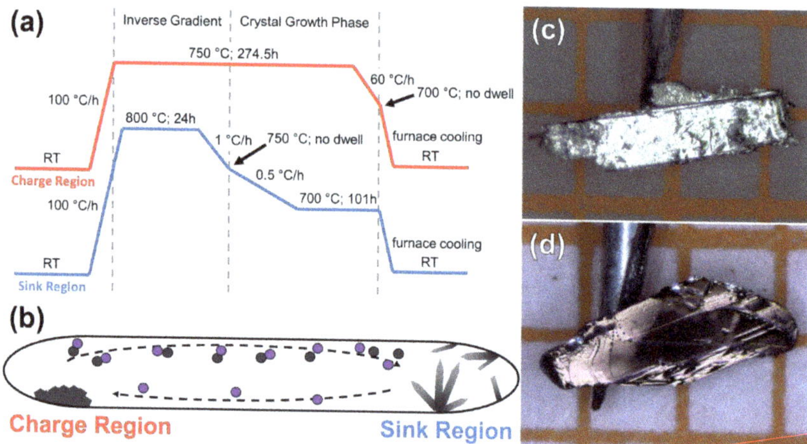

Figure 2. (**a**) Graphical illustrations of the temperature profile for the CVT growth of $AgCrP_2S_6$ and (**b**) schematic drawing of an ampule during CVT. Arrows indicate the mass flow of the volatile transport species (top) and the flow of the released transport agent back to the charge (bottom). (**c**,**d**): As-grown crystals of $AgCrP_2S_6$. A orange square in the background corresponds to $1\,\text{mm} \times 1\,\text{mm}$ for scale.

The ampule was carefully placed in a two-zone tube furnace in such a way that the reaction mixture was only at one side of the ampule which is referred to as the charge region. As illustrated in Figure 2a, the furnace was initially heated homogeneously to $750\,°C$ at $100\,°C/h$. The charge region was kept at this temperature for $274.5\,h$ while the other side of the ampule, which is the sink region (see Figure 2b), was initially heated up to $800\,°C$ at $100\,°C/h$, dwelled at this temperature for $24\,h$ and then cooled back to $750\,°C$ at $1\,°C/h$. An inverse transport gradient is formed, i.e., transport from sink to charge, to clean the sink region of particles which stuck to the walls of the quartz ampule during the previous preparation steps. This ensures improved nucleation conditions in the following step. Then the sink region was cooled to $690\,°C$ at $0.5\,°C/h$ to gradually form the thermal transport gradient resulting in a controlled nucleation. With a final gradient of $750\,°C$ (charge) to $690\,°C$ (sink), the ampule was dwelled for $100\,h$. After this period of time, the charge region was cooled to the sink temperature in $1\,h$ before both regions were furnace cooled (i.e., the heating elements were turned off) to room temperature.

Shiny plate-like crystals of $AgCrP_2S_6$ in the size of approximately $2\,\text{mm} \times 2\,\text{mm} \times 100\,\mu\text{m}$ were obtained. As example, as-grown single crystals are shown in Figure 2c,d. These crystals exhibit a layered morphology and are easily exfoliated, which is typical for bulk crystals of (quasi-)2D materials.

As shown in Figure 3a, some as-grown crystals exhibit small spherical particles (likely solidified droplets) of a secondary phase (bright) and pieces of a second secondary phase (dark). As the secondary phases are found only on the surface, exfoliating the crystals is sufficient to remove the secondary phases and results in crystals with clean surfaces, as illustrated in Figure 3b.

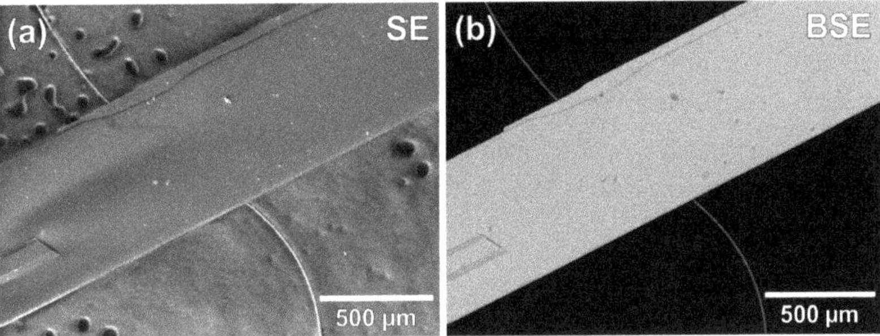

Figure 3. SEM image with chemical contrast (BSE detector) of (**a**) an as-grown crystal of $AgCrP_2S_6$ with superficial impurities and (**b**) a piece of the same crystal after exfoliation with a clean surface.

4. Crystal Morphology and Compositional Analysis

The topographical SE image of a $AgCrP_2S_6$ crystal in Figure 4a exhibits a flat crystal surface and sharp edges. The terrace close to the upper edge of the crystal is a typical feature of layered systems. Furthermore, on the upper edge of the crystal, some steps can be seen, which form $120°$ angles, indicative of a hexagonal crystal habitus. The SE image with BSE detector in in Figure 4b shows an overall homogeneous contrast over the surface of the crystal demonstrating that it is chemically homogeneous. At some small areas, a change in contrast is observed. In comparison with the SEM image in SE mode, these spots can be clearly attributed to impurity particles on top of the crystal and not to any region of intergrowth with a secondary phase.

Figure 4. SEM image of a $AgCrP_2S_6$ crystal with topographical contrast (SE mode) in (**a**) and chemical contrast (BSE mode) in (**b**).

By EDX measurements on multiple spots on several crystals, the mean elemental composition of the crystals was obtained as $Ag_{1.03(2)}Cr_{1.06(2)}P_{2.03(2)}S_{5.88(1)}$. This composition is in ideal agreement with the expected composition of $AgCrP_2S_6$ and the small standard deviations (given in parentheses) indicate a homogeneous elemental distribution and composition.

5. Structural Analysis

The pXRD pattern obtained from exfoliated $AgCrP_2S_6$ crystals, as shown in Figure 5a, was indexed in the space group $P2/a$ (No. 13), in agreement with the literature [19]. No additional reflection were observed demonstrating the intrinsic phase purity of our crystals.

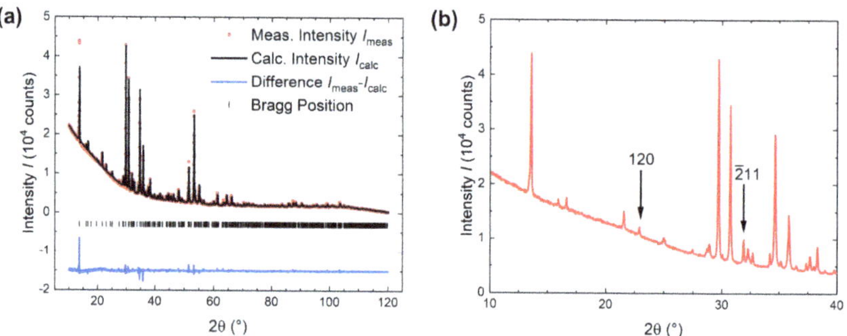

Figure 5. (**a**) pXRD pattern from powdered $AgCrP_2S_6$ crystals compared to the calculated pattern based on the refined crystal structure model. (**b**) Zoomed-in view on the low angle 2θ regime (10–40°). The marked reflections are expected to be systematically absent assuming a crystal structure in the space group of $C2/m$ instead of $P2/a$.

The $C2/m$ space group, which is typically observed for compounds of the $M_2P_2S_6$ family [23,27], including $M^{2+}M'^{2+}P_2S_6$ compounds of isovalent substitution series (e.g., $MnFeP_2S_6$ [13], $MnNiP_2S_6$ [14] and $FeNiP_2S_6$ [16]), can be ruled out. Assuming a monoclinic unit cell, several observed reflections correspond to Laue indices that are systematically absent for C centering, as they violate the reflection condition hkl: $h + k = 2n$. Examples are the reflections at $2\theta = 22.94°$ corresponding to 120 and at $2\theta = 31.92°$ corresponding to $\bar{2}11$, as shown in Figure 5b.

This implies that Ag and Cr indeed arrange as zig-zag stripes in $AgCrP_2S_6$ and do not just randomly occupy the corners of the structural honeycomb network, as it is the case for isovalent substitutions. While the former scenario breaks the mirror symmetry of the $C2/m$ space group of the $Fe_2P_2S_6$ aristotype [28], which results in a $P2/a$ space group, the latter scenario would not. Furthermore, a $C2/c$ space group, as reported, e.g., for $CuCrP_2S_6$ [17] with a triangular arrangement of the two transition element cations, can be ruled out based on the same considerations.

Starting from the crystal structure model proposed by Colombet et al. [19], a refined crystal structure model is obtained using the Rietveld method which is sufficient to describe our experimental pattern with good agreement, as shown in Figure 5a. The obtained lattice parameter and reliability factors are summarized in Table 2 (top) and the refined structural model is given in the same table on the bottom and is illustrated in Figure 6. The strongest disagreement between model and experiment is observed for the high intensity 001 reflection at $2\theta = 13.67°$. As this reflection corresponds to the stacking of layers, it is most prominently affected by any kind of disorder or defects influencing the stacking. Due to the weak structural interaction between layers, which are only based on weak van der Waals forces, the $M_2P_2S_6$ compounds are prone to stacking faults and twinning between layers. In the presence of such defects, the shape of the corresponding 001 reflection is altered, which may be a reason for the observed deviation between the experiment and the model without defects.

Table 2. Top : Summary of experimental parameters of the pXRD experiment on $AgCrP_2S_6$, extracted lattice parameters and reliability factors of the structural model obtained by the Rietveld method. Bottom: Refined crystal structure model of $AgCrP_2S_6$ and isotropic displacement parameters with standard deviations given in parentheses. All sites were treated as fully occupied.

Experiment						
Temperature (K)						293(2)
Radiation						$Cu-K_{\alpha_1}$
Wavelength (Å)						1.54059
θ_{min} (°)						10.00
θ_{step} (°)						0.03
θ_{max} (°)						120.13
Crystal Data						
Crystal System						monoclinic
Space Group						$P2/a$
a (Å)						5.8832(1)
b (Å)						10.6214(2)
c (Å)						6.7450(3)
β (°)						106.043(2)
Refinement						
Goodness-Of-Fit						2.13
R_p (%)						2.08
wR_p (%)						3.11
R_F (%)						5.45

Label	Type	Wyck	x	y	z	U_{iso} $(\times 10^{-3}\text{Å}^2)$
Ag1	Ag	$2e$	0.75	0.4364(2)	0	34(1)
Cr1	Cr	$2e$	0.25	0.9229(4)	0	20(2)
P1	P	$4g$	0.2979(6)	0.2466(4)	0.1659(6)	2(1)
S1	S	$4g$	0.9792(7)	0.2309(5)	0.2336(8)	18(2)
S2	S	$4g$	0.9880(5)	0.9233(4)	0.2165(8)	7(1)
S3	S	$4g$	0.4777(7)	0.3947(4)	0.2802(7)	34(1)

Additionally, the experimental pattern exhibits significantly altered reflection intensities compared to an initial model, which are attributed to a strongly preferred orientation of the crystallites in the investigated sample. Due to the layered structure with only weak van der Waals interactions between layers, the powder particles obtained from grinding $AgCrP_2S_6$ crystals are plate-like and tend to lie flat on the sample holder. Thus, reflections with a dominant l component (e.g., 001) exhibit higher intensities than expected for spherical crystallites in transmission geometry. To adjust for this effect in the model, the method proposed by March [29] and extended by Dollase [30] was used. However, the preferred orientation in $AgCrP_2S_6$ is strongly pronounced, such that it might be beyond the limit of what the semi-empirical March-Dollase model is capable of describing accurately. This may furthermore contribute to the deviation between model and experiment around the 001 reflection.

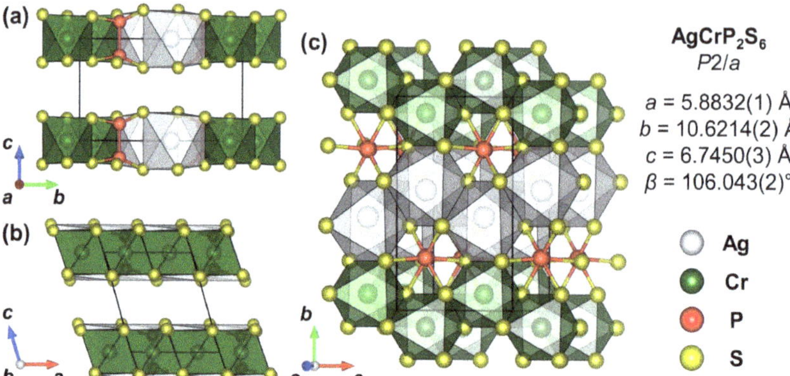

Figure 6. Refined crystal structure model of $AgCrP_2S_6$ after Rietveld refinement. View along a in (**a**), along b in (**b**) and along c^* in (**c**). The CrS_6 and AgS_6 coordination environments are shown in the color of the respective central atom.

The refined crystal structure model for $AgCrP_2S_6$ shows that the Ag–S bonds are notably longer than the Cr–S bonds, as expected based on the difference between the size of the transition element cations (e.g., ionic radii for octahedral coordination: $r(Ag^{1+}) = 1.15\,\text{Å}$ and $r(Cr^{3+}) = 0.62\,\text{Å}$ [31]). These different bond lengths result in a distortion of the structure compared to the aristotype $Fe_2P_2S_6$, which can be clearly observed, e.g., in Figure 6a. In detail, the CrS_6 coordination environment remains antiprismatic (i.e., close to octahedral with a slight trigonal elongation along c^*) with the faces above and below the shared plane of the transition elements being parallel to each other. However, the AgS_6 coordination environment as well as for the P_2S_6 environment are distorted in such a way that the faces above and below the transition element plane are not parallel to each other. In the view along the c^* direction in Figure 6c, this distortion manifests in Ag and P_2 being shifted off-center in their respective sulfur coordination environments away from the closest Cr positions. Meanwhile, Cr is located exactly in the center of the CrS_6 unit. The observation that the CrS_6 unit is closer to an ideal octahedral coordination environment than the AgS_6 unit can be understood considering the local charge density (i.e., ionic size and charge). Cr^{3+} is small and highly charged and, thus, interacts with the surrounding S atoms stronger than the comparable large and less charged Ag^{1+}. Another notable structural aspect is the strong distortion of the $[P_2S_6]^{4-}$ units that demonstrates how flexible this covalent complex anion is. This complex anion is a common and characteristic building unit in the $M_2P_2S_6$ family and its flexibility may indicate that several more compounds of the general formula $M^{1+}M'^{3+}P_2S_6$ are stable but have not been synthesized yet.

6. Summary and Conclusions

We report optimized crystal growth conditions for the quarternary compound $AgCrP_2S_6$ via Chemical Vapor Transport (CVT). A temperature profile adapted from the CVT growth of ternary $M_2P_2S_6$ compounds is sufficient to yield crystals of the target $AgCrP_2S_6$ phase in the mm-size. On some crystals, traces of a superficial impurity phase is found which could be readily removed by exfoliation.

The as-grown crystals exhibit a plate-like, layered morphology as well as a hexagonal habitus and have the expected composition of $AgCrP_2S_6$ based on EDX spectroscopy. The pXRD pattern is indexed in the space group $P2/a$ in agreement with the literature [19]. The $P2/a$ space group, on which the zig-zag type arrangement of M and M' is based on, can be well distinguished from, e.g., the $C2/m$ and $C2/c$ space groups due to reflections that are systematically absent for C centering. Starting from the model of Colombet et al. [19], a refined structural model is obtained using the Rietveld method. This model contains a notable distortion of the AgS_6 and P_2S_6 coordination environments, while the CrS_6 units remain antiprismatic with a slight trigonal distortion.

The zig-zag stripe-like arrangement in $AgCrP_2S_6$ and the alternating arrangement of M and M', which is reported, e.g., for $CuCrP_2S_6$, are promising to yield interesting magnetic and electronic structures. While only few such quarternary phosphorus sulfide compounds have been synthesized until now, many more combinations of a 1+-ion and a 3+-ion can be expected to form analogous compounds. Furthermore, the fundamental idea of replacing M^{X+} by $M_{0.5}^{(X-1)+} M'^{(X+1)+}_{0.5}$ may be adoptable to the closely related structures such as, $M_2^{3+}(Si,Ge)_2Te_6$ compounds.

The single crystals of $AgCrP_2S_6$ that were obtained using the presented growth conditions allow for studies of the low dimensional magnetic interactions including the magnetic anisotropy of this compound in the future, which may lead to a better fundamental understanding of low dimensional magnetism. Furthermore, the van der Waals layered structure makes exfoliation easily possible and, thus, our successful growth of single crystals paves the way for further manufacturing of few-layer or even monolayer samples of $AgCrP_2S_6$.

Author Contributions: Investigation, S.S., Y.S., S.A.; data curation, S.S., Y.S.; writing—original draft preparation, S.S., S.A.; writing—review and editing, S.S., S.A.; supervision, B.B., S.A.; funding acquisition, B.B., S.A. All authors have read and agreed to the published version of the manuscript.

Funding: This work is supported by the Deutsche Forschungsgemeinschaft (DFG) via Grant No. DFG A.S 523\4-1. S.S. acknowledges financial support from GRK-1621 graduate academy of the DFG. B.B. acknowledges financial support from the DFG through SFB 1143 (project-id 247310070). Y.S acknowledge the support of BMBF through UKRATOP (BMBF). S.A., B.B. and S.S. thank DFG for financial support in the frame of the joint DFG-RSF project-id 405940956.

Data Availability Statement: The refined crystal structure model and the powder X-ray diffraction dataset of $AgCrP_2S_6$ presented in this study are openly available in the Crystallography Open Database (COD), COD ID: 3000295 under https://www.crystallography.net/cod/3000295.html.

Acknowledgments: The publication of this article was funded by the Open Access Fund of the Leibniz Association.

Conflicts of Interest: The authors declare no conflict of interest.

Sample Availability: Single crystals of $AgCrP_2S_6$ are available from the corresponding author.

Abbreviations

The following abbreviations are used in this manuscript:

2D	Two-dimensional
CVT	Chemical vapor transport
SEM	Scanning electron microscopy
SE	Secondary electron
BSE	Back-scattered electron
EDX	Energy dispersive X-ray spectroscopy
pXRD	Powder X-ray diffraction

References

1. 2D magnetism gets hot. *Nat. Nanotechnol.* **2018**, *13*, 269. [CrossRef]
2. Gibertini, M.; Koperski, M.; Morpurgo, A.F.; Novoselov, K.S. Magnetic 2D materials and heterostructures. *Nat. Nanotechnol.* **2019**, *14*, 408–419. [CrossRef]
3. Samarth, N. Magnetism in flatland. *Nature* **2017**, *546*, 216–217. [CrossRef]
4. Brec, R. Review on Structural and Chemical Properties of Transition Metal Phosphorus Trisulfides MPS₃. In *Intercalation in Layered Materials*; Springer: New York, NY, USA, 1986; pp. 93–124. [CrossRef]
5. Susner, M.A.; Chyasnavichyus, M.; McGuire, M.A.; Ganesh, P.; Maksymovych, P. Metal Thio- and Selenophosphates as Multifunctional van der Waals Layered Materials. *Adv. Mater.* **2017**, *29*, 1602852. [CrossRef] [PubMed]

6. Zhong, D.; Seyler, K.L.; Linpeng, X.; Cheng, R.; Sivadas, N.; Huang, B.; Schmidgall, E.; Taniguchi, T.; Watanabe, K.; McGuire, M.A.; et al. Van der Waals engineering of ferromagnetic semiconductor heterostructures for spin and valleytronics. *Sci. Adv.* **2017**, *3*, e1603113. [CrossRef] [PubMed]

7. Song, T.; Tu, M.W.Y.; Carnahan, C.; Cai, X.; Taniguchi, T.; Watanabe, K.; McGuire, M.A.; Cobden, D.H.; Xiao, D.; Yao, W.; et al. Voltage Control of a van der Waals Spin-Filter Magnetic Tunnel Junction. *Nano Lett.* **2019**, *19*, 915–920. [CrossRef]

8. Kong, D.; Cha, J.J.; Wang, H.; Lee, H.R.; Cui, Y. First-row transition metal dichalcogenide catalysts for hydrogen evolution reaction. *Energy Environ. Sci.* **2013**, *6*, 3553. [CrossRef]

9. Ramakrishnan, S.; Karuppannan, M.; Vinothkannan, M.; Ramachandran, K.; Kwon, O.J.; Yoo, D.J. Ultrafine Pt Nanoparticles Stabilized by MoS_2/N-Doped Reduced Graphene Oxide as a Durable Electrocatalyst for Alcohol Oxidation and Oxygen Reduction Reactions. *ACS Appl. Mater. Interfaces* **2019**, *11*, 12504–12515. [CrossRef] [PubMed]

10. Yan, Y.; Shin, W.I.; Chen, H.; Lee, S.M.; Manickam, S.; Hanson, S.; Zhao, H.; Lester, E.; Wu, T.; Pang, C.H. A recent trend: Application of graphene in catalysis. *Carbon Lett.* **2020**, *31*, 177–199. [CrossRef]

11. Lee, J.U.; Lee, S.; Ryoo, J.H.; Kang, S.; Kim, T.Y.; Kim, P.; Park, C.H.; Park, J.G.; Cheong, H. Ising-Type Magnetic Ordering in Atomically Thin $FePS_3$. *Nano Lett.* **2016**, *16*, 7433–7438. [CrossRef]

12. Kim, K.; Lim, S.Y.; Lee, J.U.; Lee, S.; Kim, T.Y.; Park, K.; Jeon, G.S.; Park, C.H.; Park, J.G.; Cheong, H. Suppression of magnetic ordering in XXZ-type antiferromagnetic monolayer NiPS3. *Nat. Commun.* **2019**, *10*, 345. [CrossRef] [PubMed]

13. Masubuchi, T.; Hoya, H.; Watanabe, T.; Takahashi, Y.; Ban, S.; Ohkubo, N.; Takase, K.; Takano, Y. Phase diagram, magnetic properties and specific heat of $Mn_{1-x}Fe_xPS_3$. *J. Alloys Compd.* **2008**, *460*, 668–674. [CrossRef]

14. Shemerliuk, Y.; Wolter, A.U.B.; Cao, G.; Zhou, Y.; Yang, Z.; Büchner, B.; Aswartham, S. Tuning magnetic and transport properties in $(Mn_{1-x}Ni_x)_2P_2S_6$ Single crystals. *arXiv* **2021**, arXiv:2104.11579.

15. Rao, R.R.; Raychaudhuri, A.K. Magnetic Studies of a Mixed Antiferromagnetic System $Fe_{1-x}Ni_xPS_3$. *J. Phys. Chem. Solids* **1992**, *53*, 577–583. [CrossRef]

16. Selter, S.; Shemerliuk, Y.; Sturza, M.I.; Wolter, A.U.B.; Büchner, B.; Aswartham, S. Crystal Growth and Anisotropic Magnetic Properties of Quasi-2D $(Fe_{1-x}Ni_x)_2P_2S_6$. *arXiv* **2021**, arXiv:2104.00066.

17. Colombet, P.; Leblanc, A.; Danot, M.; Rouxel, J. Structural aspects and magnetic properties of the lamellar compound $Cu_{0.50}Cr_{0.50}PS_3$. *J. Solid State Chem.* **1982**, *41*, 174–184. [CrossRef]

18. Ouili, Z.; Leblanc, A.; Colombet, P. Crystal structure of a new lamellar compound:. *J. Solid State Chem.* **1987**, *66*, 86–94. [CrossRef]

19. Colombet, P.; Leblanc, A.; Danot, M.; Rouxel, J. Coordinance inhabituelle de l'argent dans un sufur lamellaire a sous-reseau magnetique 1D: Le compose $Ag_{0.5}Cr_{0.5}PS_3$. *Nouv. J. Chim.* **1983**, *7*, 333–338.

20. Lee, S.; Colombet, P.; Ouvrard, G.; Brec, R. A new chain compound of vanadium (III): Structure, metal ordering, and magnetic properties. *Mater. Res. Bull.* **1986**, *21*, 917–928. [CrossRef]

21. Mutka, H.; Payen, C.; Molinié, P. One-Dimensional Heisenberg Antiferromagnet with Spin S = 3/2. Experiments on $AgCrP_2S_6$. *Europhys. Lett. (EPL)* **1993**, *21*, 623–628. [CrossRef]

22. Binnewies, M.; Glaum, R.; Schmidt, M.; Schmidt, P. *Chemical Vapor Transport Reactions*; Walter de Gruyter GmbH: Berlin, Germany 2012.

23. Taylor, B.E.; Steger, J.; Wold, A. Preparation and properties of some transition metal phosphorus trisulfide compounds. *J. Solid State Chem.* **1973**, *7*, 461–467. [CrossRef]

24. Nitsche, R.; Wild, P. Crystal growth of metal-phosphorus-sulfur compounds by vapor transport. *Mater. Res. Bull.* **1970**, *5*, 419–423. [CrossRef]

25. Degen, T.; Sadki, M.; Bron, E.; König, U.; Nénert, G. The HighScore suite. *Powder Diffr.* **2014**, *29*, S13–S18. [CrossRef]

26. Petříček, V.; Dušek, M.; Palatinus, L. Crystallographic Computing System JANA2006: General features. *Z. Krist.-Cryst. Mater.* **2014**, *229*, 345–352. [CrossRef]

27. Brec, R.; Ouvrard, G.; Louisy, A.; Rouxel, J. Proprietes Structurales de Phases $MIIPX_3$ (X = S, Se). *Ann. Chim.* **1980**, *5*, 499–512.

28. Klingen, W.; Eulenberger, G.; Hahn, H. Über Hexachalkogeno-hypodiphosphate vom Typ $M_2P_2X_6$. *Die Naturwissenschaften* **1970**, *57*, 88. [CrossRef]

29. March, A. Mathematische Theorie der Regelung nach der Korngestalt bei affiner Deformation. *Z. Krist.-Cryst. Mater.* **1932**, *81*, 285. [CrossRef]

30. Dollase, W.A. Correction of intensities for preferred orientation in powder diffractometry: Application of the March model. *J. Appl. Crystallogr.* **1986**, *19*, 267–272. [CrossRef]

31. Shannon, R.D. Revised Effective Ionic Radii and Systematic Studies of Interatomic Distances in Halides and Chalcogenides. *Acta Crystallogr. Sect. A* **1976**, *32*, 751–767. [CrossRef]

Article

Entropy Generation Incorporating γ-Nanofluids under the Influence of Nonlinear Radiation with Mixed Convection

Umair Khan [1], Aurang Zaib [2], Ilyas Khan [3,*] and Kottakkaran Sooppy Nisar [4]

1   Department of Mathematics and Social Sciences, Sukkur IBA University, Sukkur 65200, Pakistan; umairkhan@iba-suk.edu.pk
2   Department of Mathematical Sciences, Federal Urdu University of Arts, Science & Technology, Gulshan-e-Iqbal Karachi 75300, Pakistan; aurangzaib@fuuast.edu.pk
3   Department of Mathematics, College of Science Al-Zulfi, Majmaah University, Al-Majmaah 11952, Saudi Arabia
4   Department of Mathematics, College of Arts and Science, Prince Sattam bin Abdulaziz University, Wadi Al-Dawaser 11991, Saudi Arabia; n.sooppy@psau.edu.sa
*   Correspondence: i.said@mu.edu.sa

Abstract: Nanofluids offer the potential to improve heat transport performance. In light of this, the current exploration gives a numerical simulation of mixed convection flow (MCF) using an effective Prandtl model and comprising water- and ethylene-based $\gamma - Al_2O_3$ particles over a stretched vertical sheet. The impacts of entropy along with non-linear radiation and viscous dissipation are analyzed. Experimentally based expressions of thermal conductivity as well as viscosity are utilized for $\gamma - Al_2O_3$ nanoparticles. The governing boundary-layer equations are stimulated numerically utilizing bvp4c (boundary-value problem of fourth order). The outcomes involving flow parameter found for the temperature, velocity, heat transfer and drag force are conferred via graphs. It is determined from the obtained results that the temperature and velocity increase the function of the nanoparticle volume fraction for $H_2O\backslash C_2H_6O_2$ based $\gamma - Al_2O_3$ nanofluids. In addition, it is noted that the larger unsteady parameter results in a significant advancement in the heat transport and friction factor. Heat transfer performance in the fluid flow is also augmented with an upsurge in radiation.

Keywords: time-dependent flow; entropy generation; non-linear radiation; γ-alumina nanoparticle; MHD; mixed convection

check for **updates**

Citation: Khan, U.; Zaib, A.; Khan, I.; Nisar, K.S. Entropy Generation Incorporating γ-Nanofluids under the Influence of Nonlinear Radiation with Mixed Convection. *Crystals* **2021**, *11*, 400. https://doi.org/10.3390/cryst11040400

Academic Editors: Rajratan Basu and Arcady Zhukov

Received: 26 February 2021
Accepted: 31 March 2021
Published: 10 April 2021

Publisher's Note: MDPI stays neutral with regard to jurisdictional claims in published maps and institutional affiliations.

1. Introduction

Several industrial processes, such as the growing of crystals, the manufacture of rubber and plastic sheets, paper and glass fiber production, and processes of polymer and metal extrusion are affected by the flow problem with heat transport provoked through stretched sheets; thus, this issues is extremely important. The cooling rate plays a significant role concerning the quality of the finished product through these procedures; where a moving sheet materializes via an incision, as a result, a boundary layer flow (BLF) emerges in the track of the surface progress. Crane [1] scrutinized the 2D steady flow of viscous fluid from a stretched sheet. After this study, the pioneering effort on the flow field through a stretched sheet achieved substantial interest; as a result, an excellent quantity of literature has been engendered on this work [2–9].

In recent times, nanotechnology has magnetized researchers' attention owing to its several distinct applications in the modern era, such as cancer therapy and diagnosis, interfaces in neuroelectronics, chemical production, and molecular and in vivo therapy applications such as kinesis and surgery, etc. In addition, there have been enhancements in the heat transfer in mechanical as well as thermal systems. Several regular fluids (ethylene glycol, oil, polymer solutions, water, etc.) have low thermal conductivity. Thus, augmenting the performance of such heat transport fluids appears imperative to achieve the expectations of scientists and researchers. Choi [10] primarily developed the concept of nanofluids for the purpose of augmenting the performance of regular fluids. Sheikholeslami et al. [11] scrutinized forced convective flow with nanofluids from a stretchable sheet with magnetic function. Mutuku and Makinde [12] examined the influences of dual stratification on time-dependent flow from a smooth sheet with nanofluid and magnetic function. The effect of entropy generation (EG) on the thin fluid flow with nanofluids via a stretched cylinder was scrutinized by Khan et al. [13]. Gireesha et al. [14] implemented a KVL (Khanafer-Vafai-Lightstone) model to explore the influence of nanofluids via dusty fluids with Hall effects. Recently, the influences of nanofluids rendering to assorted surfaces have been studied by numerous researchers [15–20].

The alumina nanofluids are another aspect that has recently attracted the attention of researchers due to their application in numerous procedures of cooling [21–26]. The alumina nanofluids are identified in accordance with their dimension, e.g., alpha and gamma aluminize, etc. The surface properties in well-described forms of gamma and eta alumina were examined in [27]. The entropy influence on the flow of ethylene- and water-based γ-alumina through stretched sheets, as determined using the effective Prandtl model, was explored by Rashidi et al. [28]. The authors claimed that the fluid temperature decelerates owing to effective Pr and accelerates without effective Pr. A comparative investigation considering $\gamma - Al_2O_3$ with distinct base fluids was scrutinized by Ganesh et al. [29]. They showed that similar nanoparticles have opposite effects on temperature. Moghaieb et al. [30] employed the $\gamma - Al_2O_3$ particles in their research as an engine coolant. Ahmed et al. [31] examined the unsteady radiative flow comprising ethylene- and water-based $\gamma - Al_2O_3$ nanomaterials through a thin slit with magnetic function. Recently, Zaib et al. [32] developed the model of effective Prandtl to examine the mixed convective flow through a wedge by nanofluids. They achieved multiple results for the opposite flow.

The second law of thermodynamics is more consistent than the first law of thermodynamics because of the restriction of the effectiveness of the first law in engineering systems of heat transport. To find the best method for thermal structures, the second law is employed through the curtailing of irreversibility [33,34]. A larger entropy generation (EG) signifies a larger scope of irreversibility. Hence, EG can be utilized to ascertain criteria for the manufacturing of devices in engineering. An assessment of EG can be used to augment the performance of a system [35–41]. In addition, entropy generation can be utilized in analysis of the brain and its diseases from both a psychiatric as well as a neurological perspective. Rashidi et al. [42] examined the stimulus of magnetic function on the fluid flow in a rotated permeable disk with nanofluid. Dalir et al. [43] surveyed the effect of entropy on the force convective flow from a stretching surface containing viscoelastic nanofluid. The Keller-box algorithm was utilized to find the numerical result. Shit et al. [44] discussed the effect of EG on convective magneto flow using nanofluid in porous medium with radiation impact. They employed FDM (finite difference method) along with Newton's technique of linearization. The influences of radiation and viscous dissipation on the flow of copper and silver nanomaterials through a rotated disk with entropy were studied by Hayat et al. [45]. Recently, Shafee et al. [46] scrutinized the stimulus of nanofluid via a tube with entropy generation by involving swirl tools of the flow.

The above-mentioned investigations were dependent on steady- state behavior. However, in certain situations, the flow depends on time, owing to unexpected changes in temperature or the heat-flux of the surface, and as a result, it becomes vital to take time-dependent (unsteady) flow conditions into consideration. In addition, the phenomena of time-dependent flow is significant in numerous areas of engineering, such as rotating parts in piston engines, the turbo machinery and aerodynamics of helicopters, etc. Thus, the intention of the current research is to explore the impact of time-dependent mixed convective flow incorporating $H_2O\backslash C_2H_6O_2$ based γ-nanofluids. The influences of nonlinear radiation and viscous dissipation with entropy are also analyzed. The Lobatto IIIA formula is used to find the numerical solutions of the transmuted ODEs (ordinary differential equations).

2. Mathematical Formulation

In the mathematical model presented herein, we incorporated the time-dependent 2D mixed convective flow of $H_2O\backslash C_2H_6O_2$ based $\gamma - Al_2O_3$ nanoparticles through a stretched vertical sheet. The viscous dissipation, non-linear radiation and non-uniform heat source/sink were taken as an extra assumption in the energy equation. It was also presumed that the flow was incompressible and that the nanoparticles and the base fluid were in thermal equilibrium. The applied magnetic field (MF) was taken to be time-dependent $B = B_0/\sqrt{1 - Ct}$ and normal to the flow of surface. In addition, there was no polarization effect, and thus the external electric field was presumed to be zero and the magnetic Reynolds number was presumed to be small (in comparison to the applied MF, the induced MF was negligible). The demarcated values of the thermo-physical properties of the aforementioned nanofluids are shown in Table 1.

Table 1. Thermo-physical properties of nanoparticle and base fluids [47].

	Water (H_2O)	Ethylene Glycol ($C_2H_6O_2$)	Alumina (Al_2O_3)
ρ' (kg/m^3)	998.3	1116.6	3970
c'_p (J/kg, K)	4182	2382	765
k' (W/m, K)	0.60	0.249	40
$\beta' \times 10^{-5}$ (K^{-1})	20.06	65	0.85
σ' (Ω, m)$^{-1}$	0.05	1.07×10^{-7}	10^{-12}
Pr	6.96	204	-

The coordinate system is assumed in Cartesian form (x, y, t), where the x-axis is run along the stretching sheet and the y-axis is orthogonal to it; t symbolizes the time. The velocity and temperature at the stretching sheet are respectively presented as $U_w = ax/(1 - Ct)$ and $T_f = T_\infty + bx^2/(1 - Ct)^2$, where a, b, are the constants and the capital letter C is used for the decelerated and accelerated sheet when $C < 0$ and $C > 0$, respectively. Under these hypotheses, the governing equations for the momentum and heat transfer of nanofluids with thermo-physical properties and unsteady boundary layer convective flow can be explained as:

$$\frac{\partial u_1}{\partial x} + \frac{\partial v_1}{\partial y} = 0 \tag{1}$$

$$\frac{\partial u_1}{\partial t} + v_1\frac{\partial u_1}{\partial y} + u_1\frac{\partial u_1}{\partial x} = \frac{\mu'_{nf}}{\rho'_{nf}}\frac{\partial^2 u_1}{\partial y^2} - \frac{\sigma'_{nf}B^2}{\rho'_{nf}}u_1 + g'\frac{(\rho'\beta')_{nf}}{\rho'_{nf}}(T_1 - T_\infty) \tag{2}$$

$$\frac{\partial T_1}{\partial t} + v_1\frac{\partial T_1}{\partial y} + u_1\frac{\partial T_1}{\partial x} = \frac{k'_{nf}}{(\rho'c'_p)_{nf}}\frac{\partial^2 T_1}{\partial y^2} - \frac{1}{(\rho'c'_p)_{nf}}\left(\frac{\partial q'_r}{\partial y}\right) + \frac{\mu'_{nf}}{(\rho'c'_p)_{nf}}\left(\frac{\partial u_1}{\partial y}\right)^2 + \frac{Q_0}{(\rho'c'_p)_{nf}} \tag{3}$$

The approximation of Rosseland for the term nonlinear radiative heat flux is given as:

$$q'_r = -\frac{4\sigma'^*}{3k'^*}\frac{\partial T_1^4}{\partial y} = -\frac{16\sigma'^*}{3k'^*}T_1^3\frac{\partial T_1}{\partial y} \tag{4}$$

Utilizing Equation (4) in Equation (3), it can defined as:

$$\frac{\partial T_1}{\partial t} + v_1\frac{\partial T_1}{\partial y} + u_1\frac{\partial T_1}{\partial x} = \frac{k'_{nf}}{(\rho'c'_p)_{nf}}\frac{\partial^2 T_1}{\partial y^2} + \frac{16\sigma'^*T_1^2}{3k'^*(\rho'c'_p)_{nf}}\left(T_1\frac{\partial}{\partial y}\left(\frac{\partial T_1}{\partial y}\right) + 3\left(\frac{\partial T_1}{\partial y}\right)^2\right) +$$
$$\frac{\mu'_{nf}}{(\rho'c'_p)_{nf}}\left(\frac{\partial u_1}{\partial y}\right)^2 + \frac{Q_0}{(\rho'c'_p)_{nf}}, \tag{5}$$

where the last term represents the erratic heat sink/source and is defined as:

$$Q_0 = \frac{k'_f\left(T_f - T_\infty\right)U_w(x,t)}{xv'_f}\left(A_0 f' + B_0\left(\frac{T_1 - T_\infty}{T_f - T_\infty}\right)\right) \tag{6}$$

The boundary conditions are:

$$-k'_{nf}\frac{\partial T_1}{\partial y} = h_f\left(T_f - T_1\right),\, u_1 = U_w(x,t),\, v_1 = 0,\, \text{at } y = 0,$$

$$T_1 \to T_\infty,\, u_1 \to 0 \text{ as } y \to \infty. \tag{7}$$

Here, T_1 is the temperature, T_∞ is the free stream or the cold temperature moving on the right side of the sheet, with a zero free stream velocity, while the left side of the sheet is heated at temperature T_f from a hot fluid owing convection, which offers a coefficient of heat transfer h_f and comprising the expression of thermo-physical properties revealed in Table 2. The interpretations of the rest of the symbols or notations and the mathematical letters in Equation (1) to Equation (7) are presented in Table 3.

Table 2. Thermo-physical properties of gamma nanofluids.

	Symbols	Expressions	Model
Effective dynamic viscosity	μ'_{nf}/μ'_f	$123\phi^2 + 7.3\phi + 1$	$\gamma Al_2O_3-H_2O$
Effective dynamic viscosity	μ'_{nf}/μ'_f	$306\phi^2 - 0.19\phi + 1$	$\gamma Al_2O_3-C_2H_6O_2$
Effective thermal conductivity	k'_{nf}/k'_f	$4.97\phi^2 + 2.72\phi + 1$	$\gamma Al_2O_3-H_2O$
Effective thermal conductivity	k'_{nf}/k'_f	$28.905\phi^2 + 2.8273\phi + 1$	$\gamma Al_2O_3-C_2H_6O_2$
Effective Prandtl number	Pr_{nf}/Pr_f	$82.1\phi^2 + 3.95\phi + 1$	$\gamma Al_2O_3-H_2O$
Effective Prandtl number	Pr_{nf}/Pr_f	$254.3\phi^2 - 3\phi + 1$	$\gamma Al_2O_3-C_2H_6O_2$
Effective dynamic density	ρ'_{nf}	$(1-\phi)\rho'_f + \phi\rho'_s$	
Heat capacitance	$(\rho'c'_p)_{nf}$	$(1-\phi)(\rho'c'_p)_f + \phi(\rho'c'_p)_s$	
Thermal expansion	$(\rho'\beta')_{nf}$	$(1-\phi)(\rho'\beta')_f + \phi(\rho'\beta')_s$	
Electrical conductivity	σ'_{nf}/σ'_f	$\left\{1 + \frac{3(\sigma'_s/\sigma'_f-1)}{(\sigma'_s/\sigma'_f+2)-\phi(\sigma'_s/\sigma'_f-1)}\right\}$	

Following the non-dimensional similarity variables are:

$$u_1 = ax(1-Ct)^{-0.5}F',\, v_1 = -\left(v'_f a(1-Ct)^{-0.5}\right)^{\frac{1}{2}}F,$$

$$\eta = y\left(\frac{a(1-Ct)^{-0.5}}{v'_f}\right)^{\frac{1}{2}},\, \theta = \frac{T_1 - T_\infty}{T_f - T_\infty}. \tag{8}$$

Using Equation (8) in Equation (2) to Equation (6), along with the boundary condition (7) we get the dimensional form of the momentum equations, as follows:

$$\left.K_1 F''' + \left[K_2\left(FF'' - F'^2 - \varepsilon\left(\tfrac{\eta}{2}F'' + F'\right)\right) - K_3 MF' + K_4\lambda\theta\right] = 0 \atop (\text{for } \gamma Al_2O_3 - H_2O)\right\} \tag{9}$$

$$\left.K_5 F''' + \left[K_2\left(FF'' - F'^2 - \varepsilon\left(\tfrac{\eta}{2}F'' + F'\right)\right) - K_3 MF' + K_4\lambda\theta\right] = 0 \atop (\text{for } \gamma Al_2O_3 - C_2H_6O_2)\right\}, \tag{10}$$

Table 3. The list of symbols used and their interpretation.

Symbols	Interpretation
$(u_1(x,y,t), v_1(x,y,t), 0)$	Velocity components
(x,y)	Cartesian Coordinates
t	Time
F	Dimensionless velocity
B	Magnetic number
$A_0 > 0, B_0 > 0$	Heat source
$A_0 < 0, B_0 < 0$	Heat sink
g'	Gravitational acceleration
Greek Symbols	**Interpretation**
μ'_{nf}	Dynamic viscosity of nanofluid
ρ'_{nf}	Density of nanofluid
σ'_{nf}	Electrical conductivity of nanofluid
$(\rho'\beta')_{nf}$	Thermal expansion of nanofluid
$(\rho'c'_p)_{nf}$	Heat capacity of nanofluid
k'_{nf}	Thermal conductivity of nanofluid
v'_f	Kinematic viscosity
σ'^*	Stefan Boltzmann constant
k'^*	Mean absorption constant
$\theta(\eta)$	Dimensionless temperature
ϕ	Nanoparticle volume fraction
Subscript	**Interpretation**
nf	Nanofluid

In which:

$$K_1 = \left(123\phi^2 + 7.3\phi + 1\right), K_2 = \left(1 - \phi + \phi\left(\frac{\rho'_s}{\rho'_f}\right)\right), K_3 = \left[\frac{3\phi\left(\frac{\sigma'_s}{\sigma'_f} - 1\right)}{\left(\frac{\sigma'_s}{\sigma'_f} + 2\right) - \left(\frac{\sigma'_s}{\sigma'_f} - 1\right)\phi} + 1\right],$$

$$K_4 = (1 - \phi) + \phi\frac{(\rho'\beta')_s}{(\rho'\beta')_f}, K_5 = \left(306\phi^2 - 0.19\phi + 1\right).$$

while the corresponding dimensional form of the energy equations for the γAl_2O_3 nanoparticle are given as:

$$\left.\begin{aligned} &\theta''\left[1 + \tfrac{4}{3}R_d K_6(1 + (\theta_w - 1)\theta)^3\right] + 4R_d K_6\left[(1 + (\theta_w - 1)\theta)^2\theta'^2(\theta_w - 1)\right] + \\ &K_7\{(F\theta' - 2F'\theta) - \varepsilon(2\theta + \tfrac{\eta}{2}\theta')\} + K_6(A_0 F' + B_0\theta) + Pr_f K_1 Ec(F'')^2 = 0 \\ &\hspace{8cm}(\text{for } \gamma Al_2O_3 - H_2O) \end{aligned}\right\} \quad (11)$$

$$\left.\begin{aligned} &\theta''\left[1 + \tfrac{4}{3}R_d K_8(1 + (\theta_w - 1)\theta)^3\right] + 4R_d K_8\left[(1 + \theta(\theta_w - 1))^2\theta'^2(\theta_w - 1)\right] + \\ &K_9\{(F\theta' - 2F'\theta) - \varepsilon(2\theta + \tfrac{\eta}{2}\theta')\} + K_8(A_0 F' + B_0\theta) + Pr_f K_5 Ec(F'')^2 = 0 \\ &\hspace{8cm}(\text{for } \gamma Al_2O_3 - C_2H_6O_2) \end{aligned}\right\} \quad (12)$$

and the appropriate boundary conditions are:

$$\left.\begin{aligned}
&\theta'(0) = -K_6\xi(1-\theta(0)), \; F'(0) = 1, \; F(0) = 0 \text{ at } \eta = 0, \\
&\theta(\eta) \to 0, \; F'(\eta) \to 0 \text{ as } \eta \to \infty. \\
&\qquad\qquad (\text{for } \gamma Al_2O_3 - H_2O)
\end{aligned}\right\}, \tag{13}$$

$$\left.\begin{aligned}
&\theta'(0) = -K_8\xi(1-\theta(0)), \; F'(0) = 1, \; F(0) = 0 \text{ at } \eta = 0, \\
&\theta(\eta) \to 0, \; F'(\eta) \to 0, \text{ as } \eta \to \infty. \\
&\qquad\qquad (\text{for } \gamma Al_2O_3 - C_2H_6O_2)
\end{aligned}\right\}. \tag{14}$$

Where:

$$K_6 = \frac{1}{4.97\phi^2+2.72\phi+1}, \; K_7 = \frac{\mathrm{Pr}_f\left(1-\phi+\phi\left(\frac{\rho'_s}{\rho'_f}\right)\right)(82.1\phi^2+3.95\phi+1)}{123\phi^2+7.3\phi+1},$$

$$K_8 = \frac{1}{28.905\phi^2+2.8273\phi+1}, \; K_9 = \frac{\mathrm{Pr}_f\left(1-\phi+\phi\left(\frac{\rho'_s}{\rho'_f}\right)\right)(254.3\phi^2-3\phi+1)}{306\phi^2-0.19\phi+1}.$$

For the above equations, the interpretations of the various dimensional parameters are given in Table 4 (for Equation (9) to Equation (14)). The remaining two parameters are the local mixed convection parameter (ratio of the Grashof number and Reynolds number) and the convective parameter and are demarcated as follows:

$$\lambda = \frac{Gr_x}{Re^2_x}, \; Re_x = \frac{xU_w}{v'_f},$$

$$Gr_x = g'\beta'_f\left(T_f - T_\infty\right)x^3/v'^2_f, \; Bi = \frac{h_f\sqrt{v'_f}(1-Ct)}{k'_f\sqrt{a}} \tag{15}$$

Table 4. The list of parameters used and their values.

Name of Parameter	Notation/Symbols	Values
Magnetic parameter	M	$\sigma'_f B_0^2/\rho'_f a$
Unsteadiness parameter	ε	C/a
Radiation parameter	R_d	$4\sigma'^* T_\infty^3/k'_f k'^*$
Temperature ratio parameter	θ_w	T_f/T_∞
Eckert number	Ec	$\mu'_f U_w^2/(c'_p)_f\left(T_f - T_\infty\right)$

In order to find the similarity solution for Equations (9)–(12), it is presumed that [48]

$$\beta'_f = m_1 x^{-1} \text{ and } h_f = m_2(1-Ct)^{-0.5} \tag{16}$$

where m_1, m_2 are the constants.

Engineering Quantities of Interest

The friction factor and the temperature gradient in mathematical structure are described as:

$$C_F = \frac{\tau'_w}{\rho'_f U_w^2}, \; Nu_x = \frac{xq'_w}{k'_f(T_f - T_\infty)}, \tag{17}$$

The wall shear stress and the heat-flux are expressed as:

$$\tau'_w = \mu'_{nf}\left(\frac{\partial u_1}{\partial y}\right)_{y=0}, q'_w = -k'_f\left(\frac{k'_{nf}}{k'_f} + \frac{16\sigma'^* T_1^3}{3k'^* k'_f}\right)\left(\frac{\partial T_1}{\partial y}\right)_{y=0}. \tag{18}$$

Utilizing Equation (18) in Equation (17), the dimensionless expressions are:

$$\left.\begin{aligned}
C_F Re_x^{0.5} &= K_1 F''(0) \\
Nu_x Re_x^{-0.5} &= -\left(\tfrac{1}{K_6} + \tfrac{4}{3}R_d(1 + (\theta_w - 1)\theta(0))^3\right)\theta'(0) \\
&\qquad \text{(for } \gamma\, Al_2O_3 - H_2O)
\end{aligned}\right\} \tag{19}$$

$$\left.\begin{aligned}
C_F Re_x^{0.5} &= K_5 F''(0) \\
Nu_x Re_x^{-0.5} &= -\left(\tfrac{1}{K_8} + \tfrac{4}{3}R_d(1 + (\theta_w - 1)\theta(0))^3\right)\theta'(0) \\
&\qquad \text{(for } \gamma\, Al_2O_3 - C_2H_6O_2)
\end{aligned}\right\} \tag{20}$$

3. Formulation of Entropy

The volumetric EG (entropy generation) for γAl_2O_3 nanoparticles is expressed as:

$$H_G = \frac{k'_f}{T_\infty^2}\left[\frac{k'_{nf}}{k'_f} + \frac{16\sigma'^* T_1^3}{3k'^* k'_f}\right]\left(\frac{\partial T_1}{\partial y}\right)^2 + \frac{\mu'_{nf}}{T_\infty}\left(\frac{\partial u_1}{\partial y}\right)^2 + \frac{\sigma'_{nf} B^2}{T_\infty} u_1^2 \tag{21}$$

The characteristic EG rate can be written as:

$$H_{g0} = \frac{k'_f (\Delta T)^2}{L^2 T_\infty^2} \tag{22}$$

By using the ratio of Equations (21) and (22), the EG number is described as:

$$H_g = \frac{H_G}{H_{g0}} \tag{23}$$

Implementing Equation (8) in Equations (21) and (22), we obtain:

$$\left.\begin{aligned}
H_g &= Re_L\left(\tfrac{1}{K_6} + \tfrac{4}{3}R_d(1 + (\theta_w - 1)\theta)^3\right)\theta'^2 + \frac{Re_L Br}{\Omega}K_1 F''^2 + K_3\frac{MBrRe_L}{\Omega}F'^2, \\
&\qquad \text{(for } \gamma Al_2O_3 - H_2O)
\end{aligned}\right\} \tag{24}$$

$$\left.\begin{aligned}
H_g &= Re_L\left(\tfrac{1}{K_8} + \tfrac{4}{3}R_d(1 + (\theta_w - 1)\theta)^3\right)\theta'^2 + \frac{Re_L Br}{\Omega}K_5 F''^2 + K_3\frac{MBrRe_L}{\Omega}F'^2, \\
&\qquad \text{(for } \gamma Al_2O_3 - C_2H_6O_2)
\end{aligned}\right\} \tag{25}$$

where the parameters $\Omega = \Delta T/T_\infty$, $Br = \mu'_f (U_w)^2/k'_f \Delta T$, $Re_L = aL^2/v'_f(1 - Ct)$ are described as the temperature difference and the Brinkman and Reynolds numbers, respectively.

The assessment of the Bejan Be number is vital in sequence to investigate the heat transfer irreversibility, and range of values is between 0 and 1. The Be number in dimensionless form is described as:

$$\left.Be = \frac{Re_L\left(\tfrac{1}{K_6} + \tfrac{4}{3}R_d(1 + (\theta_w - 1)\theta)^3\right)\theta'^2}{Re_L\left(\tfrac{1}{K_6} + \tfrac{4}{3}R_d(1 + (\theta_w - 1)\theta)^3\right)\theta'^2 + \frac{Re_L Br}{\Omega}K_1 F''^2 + K_3\frac{MBrRe_L}{\Omega}F'^2}\quad \atop {\text{(for } \gamma Al_2O_3 - H_2O)}\right\} \tag{26}$$

$$\left.Be = \frac{Re_L\left(\tfrac{1}{K_8} + \tfrac{4}{3}R_d(1 + (\theta_w - 1)\theta)^3\right)\theta'^2}{Re_L\left(\tfrac{1}{K_8} + \tfrac{4}{3}R_d(1 + (\theta_w - 1)\theta)^3\right)\theta'^2 + \frac{Re_L Br}{\Omega}K_5 F''^2 + K_3\frac{MBrRe_L}{\Omega}F'^2}\quad \atop {\text{(for } \gamma Al_2O_3 - C_2H_6O_2)}\right\} \tag{27}$$

It is concluded from the expressions mentioned above that the irreversibility of fluid friction dominates when Be differs from 0–0.5, while the heat transport irreversibility dominates when Be differs from 0.5–1. The value of Be demonstrates that the irreversibility of fluid friction and heat transfer equally contribute to EG.

4. Methodology

The momentum and energy coupled non-linear ODEs (Equations (9) and (10)) and (Equations (11) and (12)), along with the boundary conditions (BCs) in Equation (13) and Equation (14), are solved numerically via bvp4c in MATLAB, which is based on a three-stage Lobatto technique for the various comprising parameters and gamma nanofluids. The three-stage Lobatto technique is a collocation technique with fourth-order accuracy. The form of the ODEs (ordinary differential equations), along with the BCs, is altered into the group of first order IVP (intial value problem) by exercising the new variables. This process is carried forward by introducing the following variables:

$$F = Z_1, \ F' = Z_2, \ F'' = Z_3, \ \theta = Z_4, \ \theta' = Z_5, \tag{28}$$

Utilizing Equation (28) for the aforementioned ODEs and the boundary conditions, we get a system of ODEs for the model of $(\gamma Al_2O_3-H_2O)$ and $(\gamma Al_2O_3-C_2H_6O_2)$ nanofluids, respectively given as:

$$\frac{d}{d\eta}\begin{pmatrix} Z_1 \\ Z_2 \\ Z_3 \\ Z_4 \\ Z_5 \end{pmatrix} = \begin{pmatrix} Z_2 \\ Z_3 \\ \dfrac{-\left\{K_2\left(Z_1 Z_3 - Z_2 Z_2 - \varepsilon\left(\frac{\eta}{2}Z_3 + Z_2\right)\right) - MK_3 Z_2 + K_4 \lambda Z_4\right\}}{K_1} \\ Z_5 \\ \dfrac{\left\{ -4R_d K_6(1+(\theta_w-1)Z_4)^2(\theta_w-1)Z_5 Z_5 - K_7\left((Z_1 Z_5 - 2Z_2 Z_4) - \varepsilon\left(2Z_4 + \frac{\eta}{2}Z_5\right)\right) - K_6(A_0 Z_2 + B_0 Z_4) - \Pr_f K_1 Ec Z_3 Z_3 \right\}}{\left(1+\frac{4}{3}R_d K_6(1+(\theta_w-1)Z_4)^3\right)} \end{pmatrix} \tag{29}$$

with initial conditions (ICs) as follows:

$$\begin{pmatrix} Z_1(0) \\ Z_2(0) \\ Z_2(\infty) \\ Z_5(0) \\ Z_4(\infty) \end{pmatrix} = \begin{pmatrix} 0 \\ 1 \\ 0 \\ -K_6\xi(1-Z_4(0)) \\ 0 \end{pmatrix}. \tag{30}$$

Similarly,

$$\frac{d}{d\eta}\begin{pmatrix} Z_1 \\ Z_2 \\ Z_3 \\ Z_4 \\ Z_5 \end{pmatrix} = \begin{pmatrix} Z_2 \\ Z_3 \\ \dfrac{-\left\{K_2\left(Z_1 Z_3 - Z_2 Z_2 - \varepsilon\left(\frac{\eta}{2}Z_3 + Z_2\right)\right) - MK_3 Z_2 + K_4 \lambda Z_4\right\}}{K_5} \\ Z_5 \\ \dfrac{\left\{ -4R_d K_8(1+(\theta_w-1)Z_4)^2(\theta_w-1)Z_5 Z_5 - K_9\left((Z_1 Z_5 - 2Z_2 Z_4) - \varepsilon\left(2Z_4 + \frac{\eta}{2}Z_5\right)\right) - K_8(A_0 Z_2 + B_0 Z_4) - \Pr_f K_5 Ec Z_3 Z_3 \right\}}{\left(1+\frac{4}{3}R_d K_8(1+(\theta_w-1)Z_4)^3\right)} \end{pmatrix}, \tag{31}$$

with the corresponding (ICs) as follows:

$$\begin{pmatrix} Z_1(0) \\ Z_2(0) \\ Z_2(\infty) \\ Z_5(0) \\ Z_4(\infty) \end{pmatrix} = \begin{pmatrix} 0 \\ 1 \\ 0 \\ -K_8\xi(1-Z_4(0)) \\ 0 \end{pmatrix}. \tag{32}$$

The use of an efficient estimation for $F''(0)$ and $\theta'(0)$ until the boundary restriction is reached addresses these equations. The step size is fixed to $\Delta\eta = 0.01$, which is sufficient to achieve the graphical and the numerical result in tabular form. The range is taken to be $\eta_{\max} = 10$, where the finite value of the dimensional variable η for the boundary restrictions is η_{max}. The convergence criteria and the accuracy of the outcomes in all cases are up to level 10^{-10}.

5. Results and Discussion

The impacts of numerous pertinent parameters on the temperature, velocity, heat transfer and drag force are discussed and presented in tabular form and as well as graphically (see Figures 1–21). Table 5 shows the assessment of $-F''(0)$ with current outcomes through the outcomes reported by Shafie [49] and Chamkha [50].

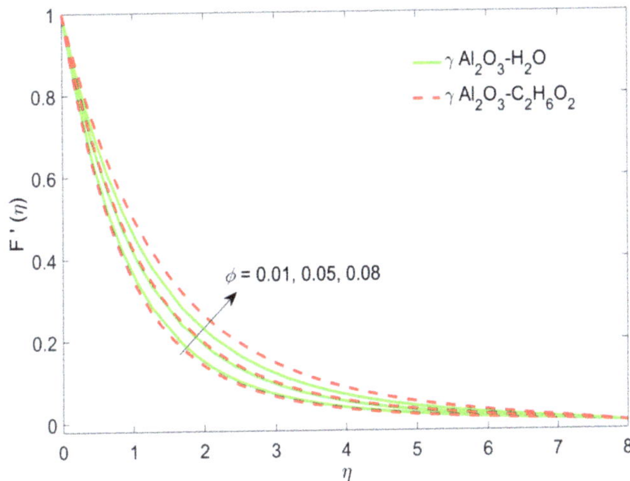

Figure 1. Impact of ϕ on $F'(\eta)$.

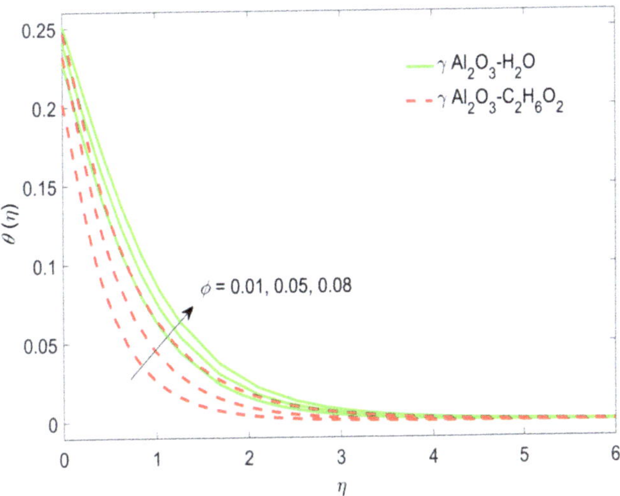

Figure 2. Impact of ϕ on $\theta(\eta)$.

The outcomes depict a superb conformity. The significant parameters for computational purposes are considered as $\phi = 0.02$, $M = 0.1$, $\varepsilon = 10$, $\xi = 0.5$, $\theta_w = 0.1$, $Ec = 0.5$, $A_0 = B_0 = 0.1$ and $R_d = 02$, with the variations shown in Figures 1–21.

Figures 1 and 2 describe the influence of volume fraction ϕ on the velocity $F'(\eta)$ and fluid temperature $\theta(\eta)$. Figures 1 and 2 confirm that the $F'(\eta)$ and $\theta(\eta)$ accelerate gradually for larger values of ϕ. Physically, the nanofluid density under consideration decreases due to the larger amount of ϕ, which consequently augments the velocity and temperature. Thus, the inter-molecular forces between the particles of nanofluids become weaker, and as a result, the fluid velocity accelerates. It is also clear from Figure 2 that the temperature is higher in the case of water and lower in case of ethylene glycol. The justification for this result is that water has a smaller Prandtl number than ethylene glycol, and as a result, the water thermal diffusivity is much superior to that of ethylene glycol. In addition, $C_2H_6O_2$ nanoliquids can be utilized for the purpose of cooling.

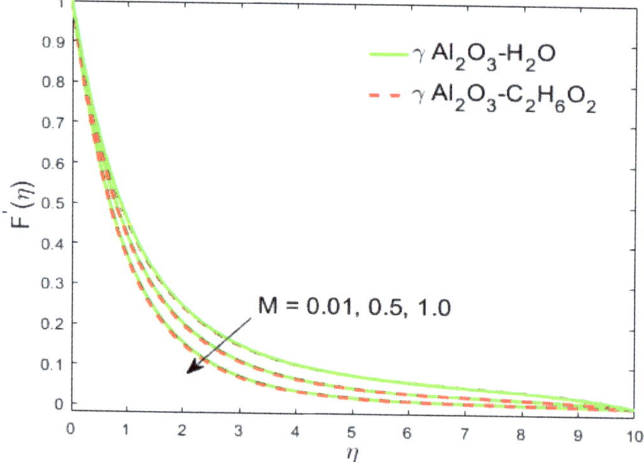

Figure 3. Impact of M on $F'(\eta)$.

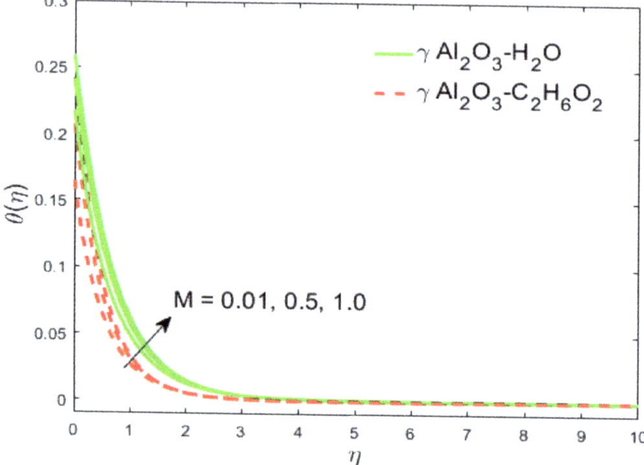

Figure 4. Impact of M on $\theta(\eta)$.

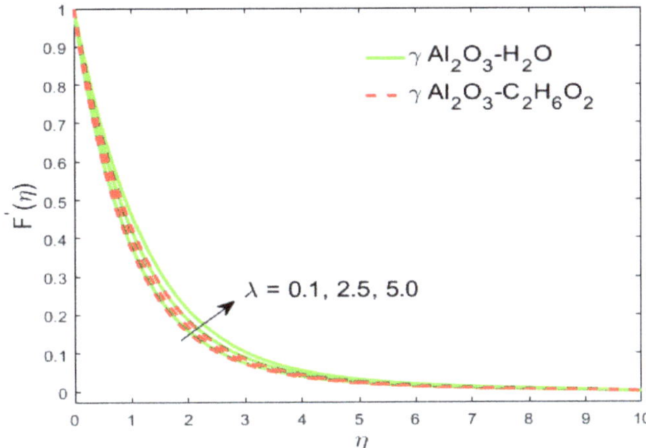

Figure 5. Impact of $\lambda > 0$ on $F'(\eta)$.

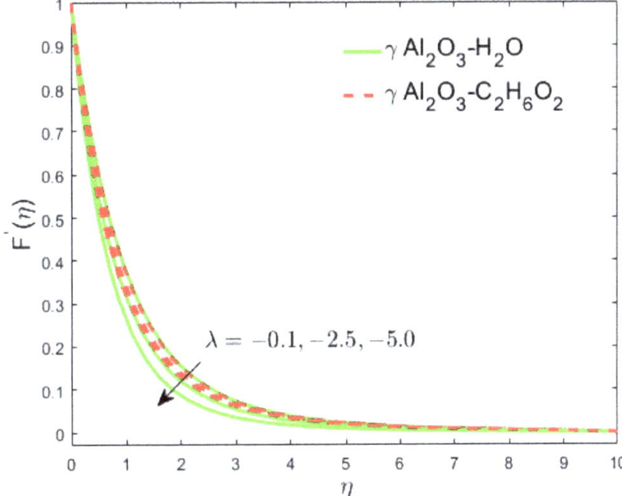

Figure 6. Impact of $\lambda < 0$ on $F'(\eta)$.

Table 5. Comparison of the values of $-F''(0)$, when $M = \phi = \lambda = 0$.

ε	Shafie et al. [49]	Chamkha et al. [50]	Current Results
0.8	1.261042	1.261512	1.2610
1.2	1.377722	1.378052	1.3777

The influence of M on $F'(\eta)$ and $\theta(\eta)$ is portrayed in Figures 3 and 4. Figure 3 suggests that the velocity declines due to M in both $H_2O\backslash C_2H_6O_2$ based nanofluids.

Physically, the existence of magnetic function engenders a type of resistive force (or Lorentz force) in the flow region, which holds the nanofluid motion. In contrast, the temperature profile (Figure 4) rises as a result of M. The physics behind this are that an enhancement in magnetic function causes an upsurge in electro-magnetic force, which controls the motion of fluid and consequently increases the temperature as well as the thickness. Figures 5–8 show the impact of λ on $F'(\eta)$ and $\theta(\eta)$ for assisting and opposing flows.

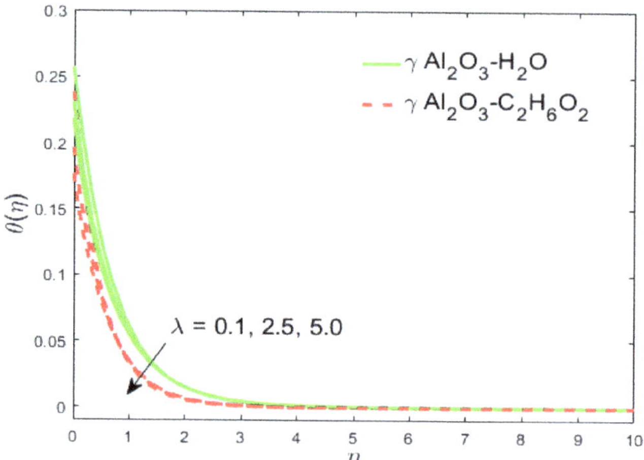

Figure 7. Impact of $\lambda > 0$ on $\theta(\eta)$.

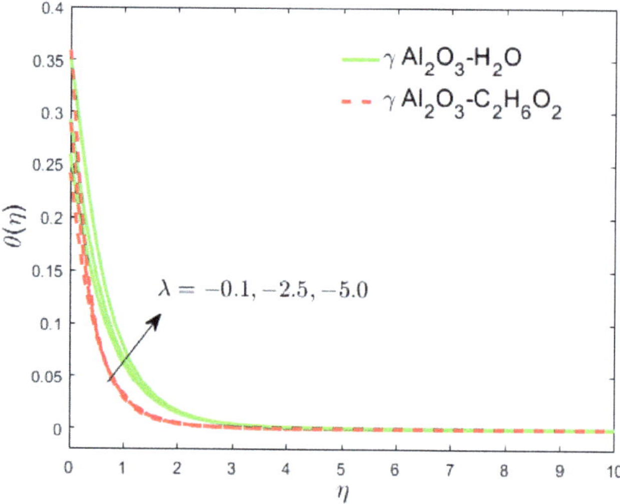

Figure 8. Impact of $\lambda < 0$ on $\theta(\eta)$.

It is clear from Figure 5 that the velocity increases with λ in the assisting flow, while the velocity as shown in Figure 6 declines in the opposing flow. Physically, a greater amount of λ generates a substantial buoyancy force that ultimately generates greater kinetic energy. The reverse is true for the opposing flow. Figure 7 shows that the temperature diminishes due to λ for assisting flow in both $\gamma Al_2O_3 - H_2O$ and $\gamma Al_2O_3 - C_2H_6O_2$ nanofluids, whereas the temperature increases in the opposing flow, as depicted in Figure 8. Physically, the fluid attains the heat from the sheet, and later on, heat energy is transmuted into different forms of energy, like kinetic energy. As expected, the temperature is lower for $\gamma Al_2O_3 - C_2H_6O_2$ than $\gamma Al_2O_3 - H_2O$ due to the greater Prandtl number. The nature of the temperature profiles is observed in Figures 9–11 for changed values of R_d, Ec and ξ.

Figure 9 confirms that temperature increases with R_d for $H_2O\backslash C_2H_6O_2$ based $\gamma - Al_2O_3$ nanofluids. The coefficient of absorption declines as radiation increases, and due to this, an enhancement occurs in the temperature distribution. Similar behavior is noticed for the Eckert number, owing to fractional heating as illustrated in Figure 10. Larger inference of Ec implies that the heat of thermal dissipation is stocked in the fluid, which ultimately increases the temperature. The convective parameter causes upsurges in the distribution of temperature (Figure 11) for $H_2O\backslash C_2H_6O_2$ based $\gamma - Al_2O_3$ nanofluids.

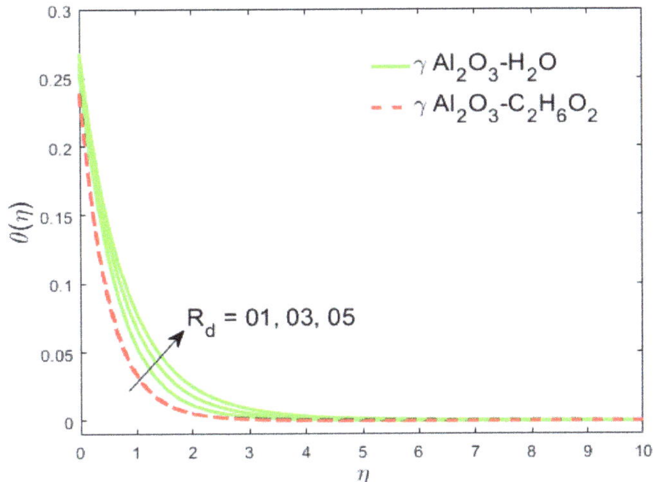

Figure 9. Impact of R_d on $\theta(\eta)$.

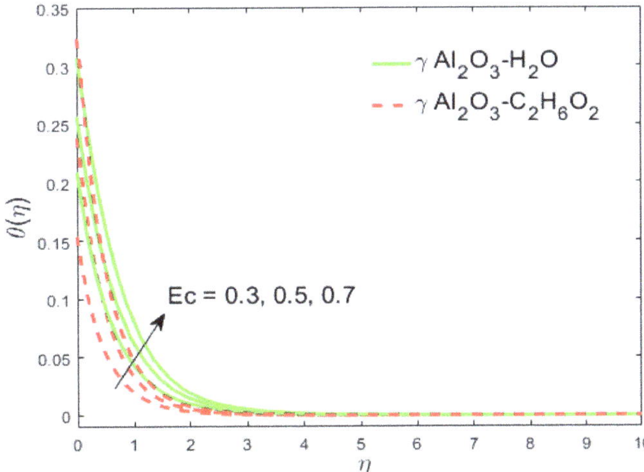

Figure 10. Impact of Ec on $\theta(\eta)$.

The sheet temperature gradient increases due to commanding convective heating. This permits the thermal influence to penetrate deeper in the sluggish fluid. Thus, the temperature increases. Figures 12 and 13 demonstrate the influence of heat sink/source on the $\theta(\eta)$ profile. It is clear from these profiles that the heat source increases the temperature, while the heat sink reduces the temperature, as expected.

Physically, the impact of the heat source $(A_0 > 0, B_0 > 0)$ adds extra energy within the boundary layer, which ultimately increases the temperature, while the heat sink $(A_0 < 0, B_0 < 0)$ absorbs the energy, which causes a reduction in the temperature.

Figures 14–16 illustrate the behavior of entropy generation for distinct parameters ϕ, Re_L and Br for $H_2O\backslash C_2H_6O_2$ based $\gamma - Al_2O_3$ nanofluids. Figure 14a,b show that the entropy increases due to ϕ in both nanofluids. It is interesting to note that ethylene-glycol-based nanofluid has greater impact on the entropy due to the huge Prandtl number and lower thermal diffusivity. Figure 15a,b suggest that the entropy enhances due to Re_L in both nanofluids owing to friction nanofluid and heat transport within the boundary layer for $\gamma Al_2O_3 - C_2H_6O_2$ and as well as $\gamma Al_2O_3 - H_2O$ nanofluids. Similarly, the impact of $\gamma Al_2O_3 - C_2H_6O_2$ on the entropy is greater than $\gamma Al_2O_3 - H_2O$. Figure 16a,b confirm that the entropy depicts the growing function of Br due to fluid friction for both nanofluids.

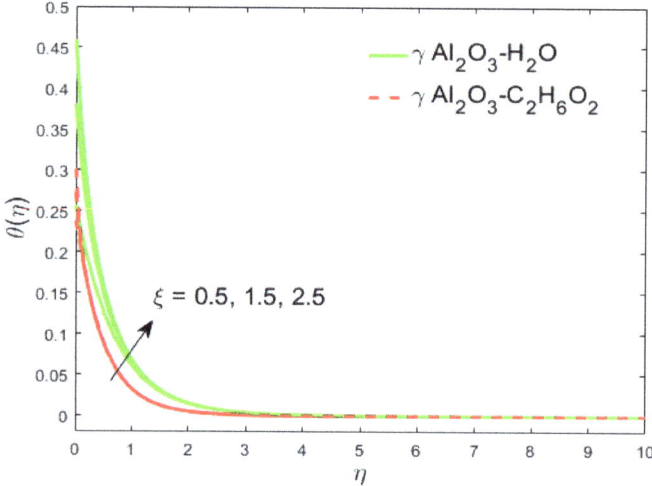

Figure 11. Impact of ξ on $\theta(\eta)$.

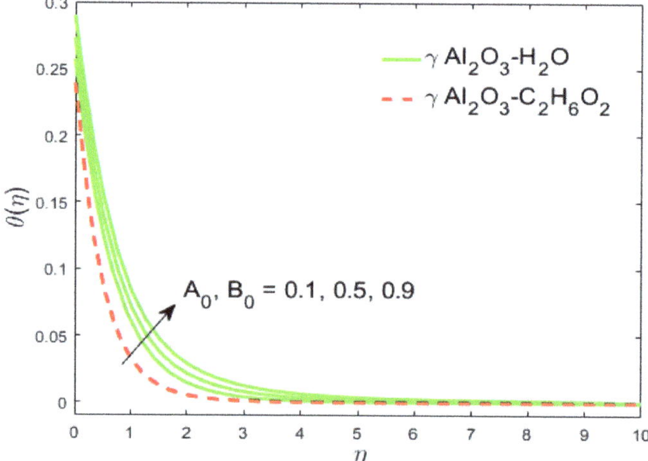

Figure 12. Impact of $A_0 > 0$, $B_0 > 0$ on $\theta(\eta)$.

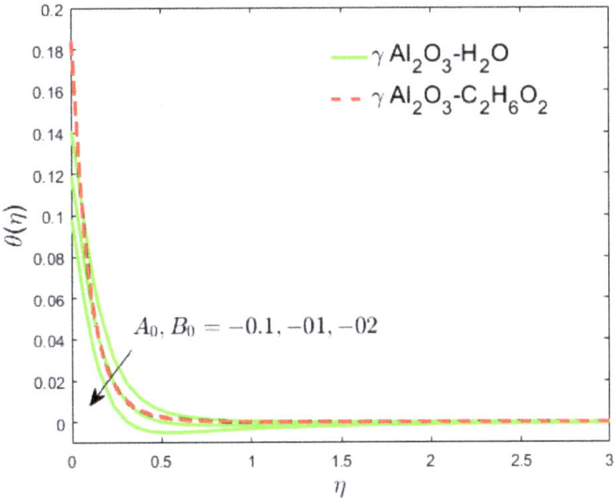

Figure 13. Impact of $A_0 < 0$, $B_0 < 0$ on $\theta(\eta)$.

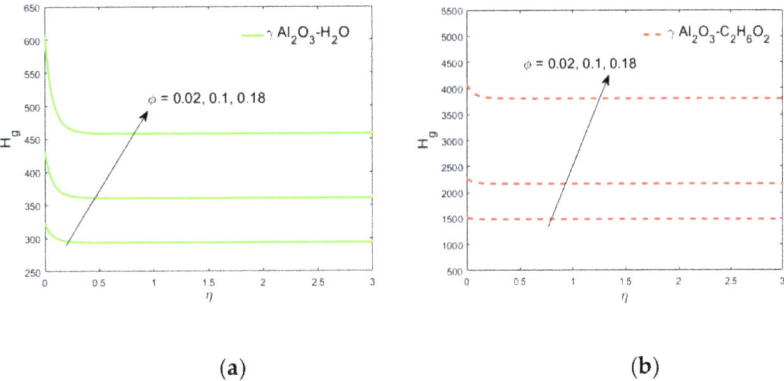

(**a**) (**b**)

Figure 14. Impact of ϕ on EG (**a**) $\gamma - Al_2O_3 - H_2O$; (**b**) $\gamma - Al_2O_3 - C_2H_6O_2$.

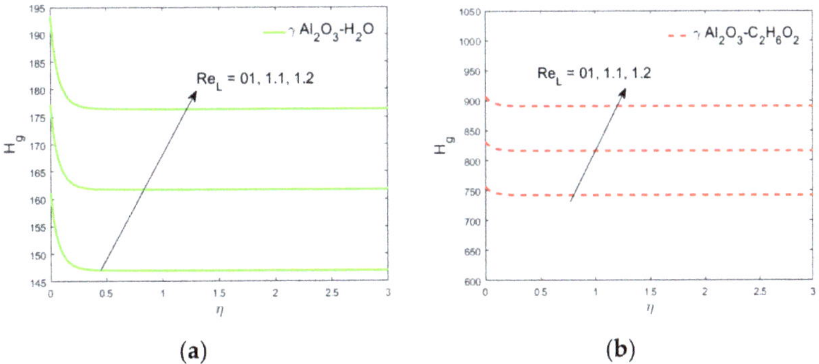

(**a**) (**b**)

Figure 15. Impact of Re_L on EG (**a**) $\gamma - Al_2O_3 - H_2O$; (**b**) $\gamma - Al_2O_3 - C_2H_6O_2$.

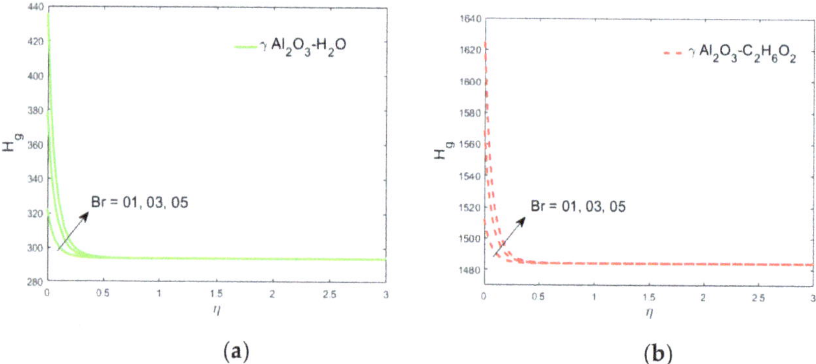

Figure 16. Impact of Br on EG. (**a**) $\gamma - Al_2O_3 - H_2O$; (**b**) $\gamma - Al_2O_3 - C_2H_6O_2$.

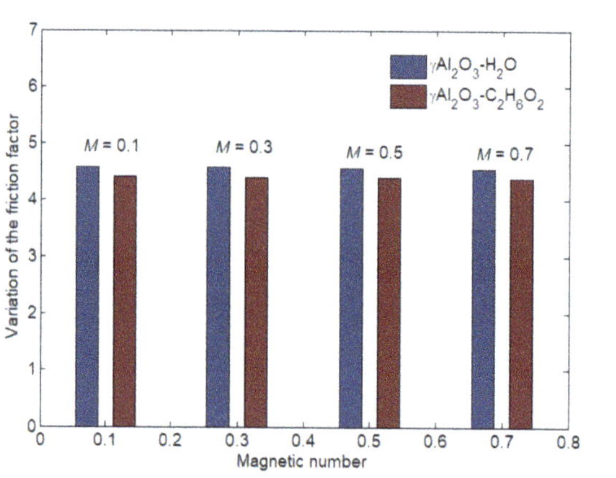

Figure 17. Impact of M on the friction factor.

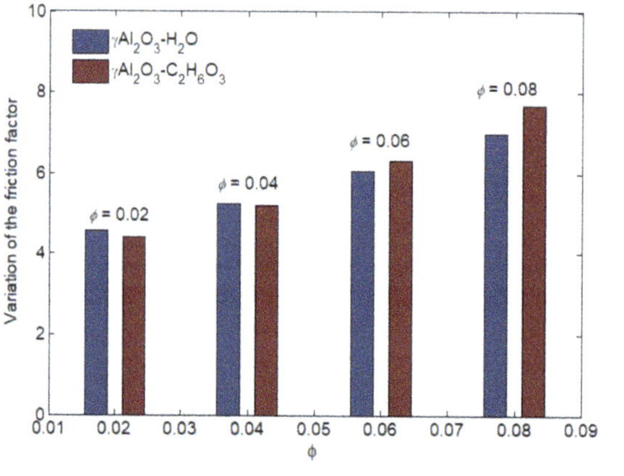

Figure 18. Impact of ϕ on the friction factor.

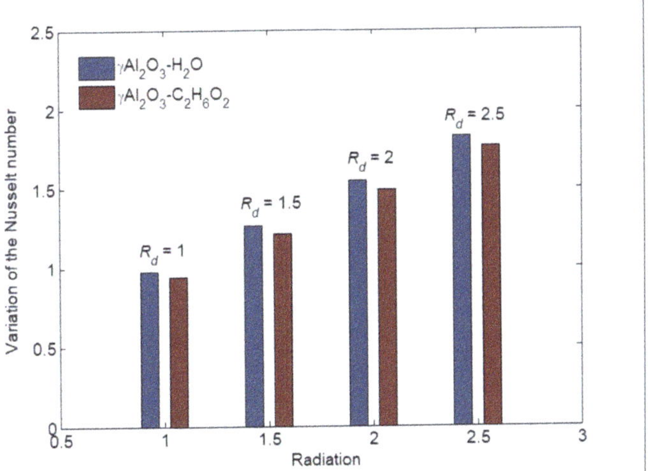

Figure 19. Impact of R_d on the Nusselt number.

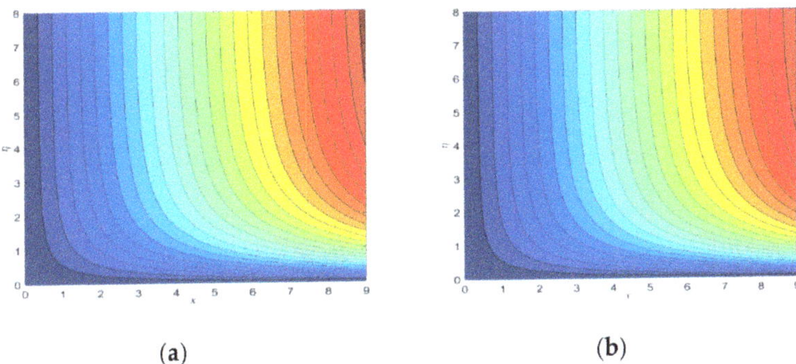

Figure 20. The streamline patterns for (**a**) $\gamma Al_2O_3 - H_2O$ and (**b**)$\gamma Al_2O_3 - C_2H_6O_2$.

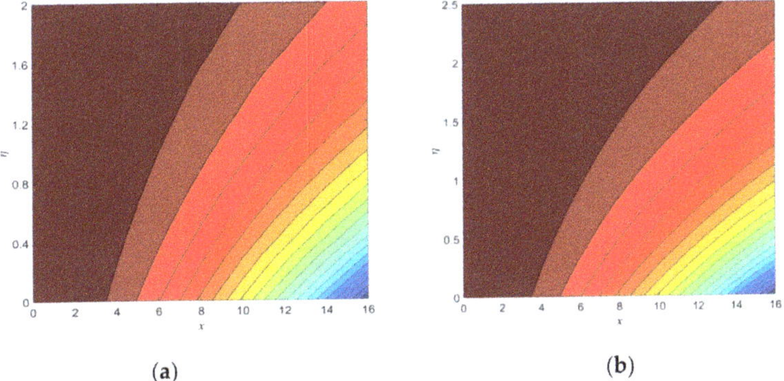

Figure 21. The isotherm patterns for (**a**) $\gamma Al_2O_3 - H_2O$ and (**b**)$\gamma Al_2O_3 - C_2H_6O_2$.

The trend of significant parameters versus $\text{Re}_x^{0.5}C_F$ and $\text{Re}_x^{-0.5}Nu_x$ for $\gamma Al_2O_3 - C_2H_6O_2$ and $\gamma Al_2O_3 - H_2O$ is seen in Tables 6 and 7.

Table 6. The numerical values of $-\mathrm{Re}_x^{0.5}C_F$ when $\lambda = 0.1$.

M	ϕ	ε	$\gamma Al_2O_3-H_2O$	$\gamma Al_2O_3-C_2H_6O_2$
0.1	0.02	20	4.5829845	4.4160221
0.3			4.5707333	4.4041249
0.5			4.5584526	4.3921989
0.7			4.5461423	4.3802439
1.0			4.5276209	4.3622570
0.1	0.02	20	4.5829845	4.4160221
	0.04		5.2573609	5.2056061
	0.06		6.0648884	6.3245067
	0.08		6.9803698	7.6700892
	0.1		7.9850461	9.1769811
0.1	0.02	10	3.3254211	3.2044599
		20	4.5829845	4.4160221
		30	5.5632504	5.3605209
		40	6.3949900	6.1619350
		50	7.1303482	6.8705020

Table 7. The numerical values of $\mathrm{Re}_x^{-0.5}Nu_x$ when $\lambda = 0.1$.

R_d	Ec	ξ	θ_w	ϕ	ε	A_0	B_0	$\gamma Al_2O_3-H_2O$	$\gamma Al_2O_3-C_2H_6O_2$
01	0.5	0.5	01	0.02	20	0.1	0.1	0.979491227	0.941840165
1.5								1.26149515	1.21399674
02								1.54320424	1.48697783
2.5								1.8244553	1.76065943
3								2.10515155	2.03495597
01	0.3	0.5	01	0.02	20	0.1	0.1	1.03436938	1.02798261
	0.5							0.979491227	0.941840165
	0.7							0.924639087	0.85574787
	1.0							0.842404491	0.726703157
	1.5							0.70546786	0.511879004
01	0.5	0.5						0.979491227	0.941840165
		0.7						1.34834268	1.31435685
-	-	0.9						1.70505352	1.68449696
		1.1						2.05020822	2.05228348
		1.3						2.38436703	2.41773846
01	0.5	0.5	01	0.02	20	0.1	0.1	0.979491227	0.941840165
-	-	-	1.5					1.12509328	1.11333832
			2.0					1.29538442	1.27500857
			2.5					1.4938953	1.46547383
			3.0					1.72502842	1.81713438

Table 7. *Cont.*

R_d	Ec	ξ	θ_w	ϕ	ε	A_0	B_0	$\gamma Al_2O_3-H_2O$	$\gamma Al_2O_3-C_2H_6O_2$
01	0.5	0.5	01	0.02	20	0.1	0.1	0.979491227	0.941840165
				0.04				0.966302546	0.923406415
				0.06				0.952597644	0.909005851
				0.08				0.939634672	0.899736613
				0.10				0.927963377	0.894148357
01	0.5	0.5	01	0.02	10	0.1	0.1	0.962640123	0.937765787
					20			0.979491227	0.941840165
					30			0.98737614	0.943605729
					40			0.992183202	0.944638809
					50			0.995503652	0.945338938
01	0.5	0.5	01	0.02	20	0.1	0.1	0.979491227	0.941840165
						0.3	0.3	0.9788224	0.941810969
						0.5	0.5	0.978152778	0.94178177
						0.7	0.7	0.97748236	0.941752571
						0.9	0.9	0.976811145	0.941723371
01	0.5	0.5	01	0.02	20	-0.1	-0.1	0.980159261	0.941869361
						-0.3	-0.3	0.980826503	0.941898555
						-0.5	-0.5	0.981492955	0.941927748
						-0.7	-0.7	0.982158618	0.94195694
						-0.9	-0.9	0.982823494	0.94198613

In addition, bar diagrams are also shown in Figures 17–19.

It is concluded from these observations that the larger values of M subdued the friction factor in both nanofluids. The major reason is that MF capitulates the flow of nanofluids through the surface of the sheet owing to the prominent magnetic impact, which subdues the friction factor. In addition, the friction factor increases owing to the ϕ in both nanofluids. In the water-based $\gamma - Al_2O_3$ nanofluid, the values of the skin factor are greater compared to the ethylene-based $\gamma - Al_2O_3$ nanofluid, due to the superior thermal diffusivity. Moreover, the Nusselt number increases with the radiation due to fact that the radiation generates superior molecular force in the flow, while the opposite trend is explored due to the Eckert number. Both the Nusselt number and the friction factor increase due to the time-dependent parameter. The streamlines and isotherms are plotted in Figure 20a,b and Figure 21a,b.

6. Conclusions

In this article we examined the time-dependent flow for an effective Prandtl model of γ nanofluids from a stretched sheet. Mixed convection, nonlinear radiation and viscous dissipation were analyzed. The significant findings are listed below:

- Both profiles of velocity and the temperature increase owing to ϕ for $\gamma Al_2O_3 - H_2O\backslash C_2H_6O_2$ nanofluids.
- The velocity increases due to the assisting flow and decline in the opposing flow for $\gamma Al_2O_3 - H_2O\backslash C_2H_6O_2$ nanofluids, while the reverse trend is seen for temperature.
- The magnetic function decreases the velocity and increases the temperature distribution.
- The temperature of nanofluids increases due to radiation, Eckert, heat source and convective parameters, while the temperature decreases due to the heat sink.

- The EG increases due to Re_L, ϕ and Br for $\gamma Al_2O_3 - C_2H_6O_2$ and $\gamma Al_2O_3 - H_2O$ nanofluids.
- The influence of ethylene-glycol-based $\gamma - Al_2O_3$ nanofluids on the temperature is lesser compared to water-based $\gamma - Al_2O_3$ nanofluids.
- The friction factor decreases due to M and increases due to ϕ in both nanofluids.
- The Nusselt number increases due to R_d and declines due to Ec in both nanofluids.
- The time-dependent parameter increases the Nusselt number as well as the friction factor.

Author Contributions: Conceptualization, A.Z. and I.K.; methodology, U.K.; software, U.K.; validation, K.S.N., I.K. and U.K.; formal analysis, A.Z.; investigation, U.K.; resources, I.K.; data curation, U.K.; writing—original draft preparation, A.Z.; writing—review and editing, I.K.; visualization, K.S.N.; supervision, K.S.N.; project administration, I.K.; funding acquisition, K.S.N. All authors have read and agreed to the published version of the manuscript.

Funding: The authors did not receive any specific funding for this work.

Data Availability Statement: Data is contained within the article.

Conflicts of Interest: The authors declare that there is no conflict of interest regarding the publication of this paper.

References

1. Crane, L.J. Flow past a stretching plate. *Zeitschrift für angewandte Mathematik und Physik* **1970**, *21*, 645–647. [CrossRef]
2. Afzal, N.; Varshney, I.S. The cooing of a low heat resistance stretching sheet moving through a fluid. *Warme Stoffubertrag* **1980**, *14*, 289–293. [CrossRef]
3. Ali, M.E. Heat transfer characteristics of a continuous stretching surface. *Heat Mass Transf.* **1994**, *29*, 227–234. [CrossRef]
4. Andersson, H.; Bech, K.; Dandapat, B. Magnetohydrodynamic flow of a power-law fluid over a stretching sheet. *Int. J. Non-linear Mech.* **1992**, *27*, 929–936. [CrossRef]
5. Magyari, E.; Keller, B. Exact solutions for self-similar boundary-layer flows induced by permeable stretching walls. *Eur. J. Mech.—B/Fluids* **2000**, *19*, 109–122. [CrossRef]
6. Sparrow, E.M.; Abraham, J.P. Universal solutions for the streamwise variation of the temperature of a moving sheet in the presence of a moving fluid. *Int. J. Heat Mass Transf.* **2005**, *48*, 3047–3056. [CrossRef]
7. Abraham, J.P.; Sparrow, E.M. Friction drag resulting from the simultaneous imposed motions of a free stream and its bounding surface. *Int. J. Heat Fluid Flow* **2005**, *26*, 289–295. [CrossRef]
8. Ishak, A.; Nazar, R.; Pop, I. Boundary layer flow and heat transfer over an unsteady stretching vertical surface. *Meccanica* **2008**, *44*, 369–375. [CrossRef]
9. Zaib, A.; Sharidan, S. Thermal diffusion and diffusion thermo effects on unsteady MHD free convection flow over a stretching surface considering Joule heating and viscous dissipation with thermal stratification, chemical reaction and Hall current. *J. Franklin Inst.* **2004**, *351*, 1268–1287. [CrossRef]
10. Choi, S.U.S. Enhancing thermal conductivity of fluids with nanoparticles. In Proceedings of the ASME International Mechanical Engineering Congress and Exposition, San Francisco, CA, USA, 12–17 November 1995; pp. 99–105.
11. Sheikholeslami, M.; Mustafa, M.T.; Ganji, D.D. Effect of Lorentz forces on forced convection nanofluid flow over a stretched surface. *Particuology* **2016**, *26*, 108–113. [CrossRef]
12. Mutuku, W.N.; Makinde, O.D. Double stratification effects on heat and mass transfer in unsteady MHD nanofluid flow over a flat surface. *Asia Pac. J. Comput. Eng.* **2017**, *4*. [CrossRef]
13. Khan, N.S.; Gul, T.; Islam, S.; Khan, I.; Alqahtani, A.M.; Alshomrani, A.S. Magneto-hydrodynamic nanoliquid thin film sprayed on a stretching cylinder with heat transfer. *Appl. Sci.* **2017**, *7*, 271. [CrossRef]
14. Gireesha, B.; Mahanthesh, B.; Thammanna, G.; Sampathkumar, P. Hall effects on dusty nanofluid two-phase transient flow past a stretching sheet using KVL model. *J. Mol. Liq.* **2018**, *256*, 139–147. [CrossRef]
15. Soomro, F.A.; Zaib, A.; Haq, R.U.; Sheikholeslami, M. Dual nature solution of water functionalized copper nanoparticles along a permeable shrinking cylinder: FDM approach. *Int. J. Heat Mass Transf.* **2019**, *129*, 1242–1249. [CrossRef]
16. Mahanthesh, B.; Lorenzini, G.; Oudina, F.M.; Animasaun, I.L. Significance of exponential space- and thermal-dependent heat source effects on nanofluid flow due to radially elongated disk with Coriolis and Lorentz forces. *J. Therm. Anal. Calorim.* **2019**, *141*, 37–44. [CrossRef]
17. Eid, M.R. Effects of NP Shapes on Non-Newtonian Bio-Nanofluid Flow in Suction/Blowing Process with Convective Condition: Sisko Model. *J. Non-Equilibrium Thermodyn.* **2020**, *45*, 97–108. [CrossRef]
18. Khan, W.A.; Ali, M.; Shahzad, M.; Sultan, F.; Irfan, M.; Asghar, Z. A note on activation energy and magnetic dipole aspects for Cross nanofluid subjected to cylindrical surface. *Appl. Nanosci.* **2019**, *10*, 3235–3244. [CrossRef]

19. Kumar, P.B.S.; Gireesha, B.J.; Mahanthesh, B.; Chamkha, A.J. Thermal analysis of nanofluid flow containing gyrotactic microorganisms in bioconvection and second-order slip with convective condition. *J. Therm. Anal. Calorim.* **2019**, *136*, 1947–1957. [CrossRef]

20. Kumar, K.G.; Rahimi-Gorji, M.; Reddy, M.G.; Chamkha, A.J.; Alarifi, I.M. Enhancement of heat transfer in a convergent/divergent channel by using carbon nanotubes in the presence of a Darcy-Forchheimer medium. *Microsyst. Technol.* **2019**, *26*, 323–332. [CrossRef]

21. Maïga, S.E.B.; Nguyen, C.T.; Galanis, N.; Roy, G. Heat transfer behaviours of nanofluids in a uniformly heated tube. *Superlattices Microstruct.* **2004**, *35*, 543–557. [CrossRef]

22. Maiga, S.E.B.; Nguyen, C.T.; Galanis, N.; Roy, G.C. Micro and nano heat transfer heat transfer enhancement in forced convection laminar tube flow by using nanofluids. In Proceedings of the CHT-04—Advances in Computational Heat Transfer III. Proceedings of the Third International Symposium, Begell House, Danbury, CT, USA, 2004; p. 24.

23. Maïga, S.E.B.; Palm, S.J.; Nguyen, C.T.; Roy, G.; Galanis, N. Heat transfer enhancement by using nanofluids in forced convection flows. *Int. J. Heat Fluid Flow* **2005**, *26*, 530–546. [CrossRef]

24. Pop, C.V.; Fohanno, S.; Polidori, G.; Nguyen, C.T. Analysis of laminar-to-turbulent threshold with water γAl2O3 and ethylene glycol-γ γAl2O3 nanofluids in free convection. In Proceedings of the 5th IASME/WSEAS Int. Conference on Heat Transfer, Thermal Engineering and Environment, Athens, Greece, 25–27 August 2007; p. 188.

25. Farajollahi, B.; Etemad, S.; Hojjat, M. Heat transfer of nanofluids in a shell and tube heat exchanger. *Int. J. Heat Mass Transf.* **2010**, *53*, 12–17. [CrossRef]

26. Sow, T.M.O.; Halelfadl, S.; Lebourlout, S.; Nguyen, C.T. Experimental study of the freezing point of γ-Al2O3water nanofluid. *Adv. Mech. Eng.* **2012**, *4*, 162961. [CrossRef]

27. Maciver, D.S.; Tobin, H.H.; Barth, R.T. Catalytic aluminas I. Surface chemistry of eta and gamma alumina. *J. Catal.* **1963**, *2*, 487–497. [CrossRef]

28. Rashidi, M.M.; Ganesh, N.V.; Hakeem, A.K.A.; Ganga, B.; Lorenzini, G. Influences of an effective Prandtl number model on nano boundary layer flow of γAl2O3–H2O andγAl2O3–C2H6O2 over a vertical stretching sheet. *Int. J. Heat Mass Transf.* **2016**, *98*, 616–623. [CrossRef]

29. Ganesh, N.V.; Hakeem, A.A.; Ganga, B. A comparative theoretical study on Al2O3 and γ-Al2O3 nanoparticles with different base fluids over a stretching sheet. *Adv. Powder Technol.* **2016**, *27*, 436–441. [CrossRef]

30. Moghaieb, H.S.; Abdel-Hamid, H.M.; Shedid, M.H.; Helali, A.B. Engine cooling using γAl2O3/water nanofluids. *Appl. Therm. Eng.* **2017**, *115*, 152–159. [CrossRef]

31. Ahmed, N.; Adnan; Khan, U.; Mohyud-Din, S.T. A theoretical investigation of unsteady thermally stratified flow of γAl2O3–H2O and γAl2O3–C2H6O2 nanofluids through a thin slit. *J. Phys. Chem. Solids* **2018**, *119*, 296–308. [CrossRef]

32. Zaib, A.; Haq, R.; Sheikholeslami, M.; Khan, U. Numerical analysis of effective Prandtl model on mixed convection flow of γAl2O3–H2O nanoliquids with micropolar liquid driven through wedge. *Phys. Scr.* **2019**, *95*, 035005. [CrossRef]

33. Bejan, A. *Entropy Generation Minimization: The Method of Thermodynamic Optimization of Finite-Size Systems and Finite-Time Processes*, 1st ed.; CRC Press; Taylor & Francis: Boca Raton, FL, USA; ISBN 9781498782920199.6.

34. Bejan, A. A Study of Entropy Generation in Fundamental Convective Heat Transfer. *J. Heat Transf.* **1979**, *101*, 718–725. [CrossRef]

35. Ko, T.; Ting, K. Optimal Reynolds number for the fully developed laminar forced convection in a helical coiled tube. *Energy* **2006**, *31*, 2142–2152. [CrossRef]

36. Hajmohammadi, M.R.; Lorenzini, G.; Shariatzadeh, O.J.; Biserni, C. Evolution in the Design of V-Shaped Highly Conductive Pathways Embedded in a Heat-Generating Piece. *J. Heat Transf.* **2015**, *137*, 061001. [CrossRef]

37. Xie, G.; Song, Y.; Asadi, M.; Lorenzini, G. Optimization of Pin-Fins for a Heat Exchanger by Entropy Generation Minimization and Constructal Law. *J. Heat Transf.* **2015**, *137*, 061901. [CrossRef]

38. Lorenzini, G.; Moretti, S. Bejan's Constructal theory and overall performance assessment: The global optimization for heat exchanging finned modules. *Therm. Sci.* **2014**, *18*, 339–348. [CrossRef]

39. Abouzar, P.; Hajmohammadi, M.R.; Sadegh, P. Investigations on the internal shape of constructal cavities intruding a heat generating body. *Therm. Sci.* **2015**, *19*, 609–618.

40. Hajmohammadi, M.R.; Campo, A.; Nourazar, S.S.; Ostad, A.M. Improvement of Forced Convection Cooling Due to the Attachment of Heat Sources to a Conducting Thick Plate. *J. Heat Transf.* **2013**, *135*, 124504. [CrossRef]

41. Govindaraju, M.; Ganesh, N.V.; Ganga, B.; Hakeem, A.A. Entropy generation analysis of magneto hydrodynamic flow of a nanofluid over a stretching sheet. *J. Egypt. Math. Soc.* **2015**, *23*, 429–434. [CrossRef]

42. Rashidi, M.; Abelman, S.; Mehr, N.F. Entropy generation in steady MHD flow due to a rotating porous disk in a nanofluid. *Int. J. Heat Mass Transf.* **2013**, *62*, 515–525. [CrossRef]

43. Dalir, N.; Dehsara, M.; Nourazar, S.S. Entropy analysis for magnetohydrodynamic flow and heat transfer of a Jeffrey nanofluid over a stretching sheet. *Energy* **2015**, *79*, 351–362. [CrossRef]

44. Shit, G.C.; Haldar, R.; Mandal, S. Entropy generation on MHD flow and convective heat transfer in a porous medium of ex-ponentially stretching surface saturated by nanofluids. *Adv. Powder Tech.* **2017**, *28*, 1519–1530. [CrossRef]

45. Hayat, T.; Khana, M.I.; Qayyuma, S.; Alsaedi, A. Entropy generation in flow with silver and copper nanoparticles. *Colloids Surfaces* **2018**, *539*, 335–346. [CrossRef]

46. Shafee, A.; Jafaryar, M.; Alsabery, A.I.; Zaib, A.; Babazadeh, H. Entropy generation of nanomaterial through a tube considering swirl flow tools. *J. Therm. Anal. Calorim.* **2020**, 1–16. [CrossRef]
47. Ganesh, N.V.; Chamkha, A.J.; Al-Mdallal, Q.M.; Kameswaran, P.K. Magneto-Marangoni nano-boundary layer flow of water and ethylene glycol based γAl2O3 nanofluids with non-linear thermal radiation effects. *Case Stud. Thermal Eng.* **2018**, *12*, 340–348.
48. Makinde, O.D.; Olanrewaju, P.O. Buoyancy Effects on Thermal Boundary Layer Over a Vertical Plate With a Convective Surface Boundary Condition. *J. Fluids Eng.* **2010**, *132*, 044502. [CrossRef]
49. Shafie, S.; Mahmood, T.; Pop, I. Similarity solutions for the unsteady boundary layer flow and heat transfer due to a stretching sheet. *Int. J. Appl. Mech. Eng.* **2006**, *11*, 647–654.
50. Chamkha, A.J.; Aly, A.M.; Mansour, M.A. Similarity solution for unsteady heat and mass transfer from a stretching surface embedded in a porous medium with suc-tion/injection and chemical reaction effects. *Chem. Eng. Commun.* **2010**, *197*, 846–858. [CrossRef]

Article

Method for Analyzing the Measurement Error with Respect to Azimuth and Incident Angle for the Rotating Polarizer Analyzer Ellipsometer

Huatian Tu [1], Yuxiang Zheng [1,*], Yao Shan [1], Yao Chen [1], Haotian Zhang [1], Rongjun Zhang [1], Songyou Wang [1], Jing Li [1], YoungPak Lee [1,2] and Liangyao Chen [1,*]

[1] Department of Optical Science and Engineering, Key Laboratory of Micro and Nano Photonic Structures, Ministry of Education, Shanghai Engineering Research Center of Ultra-Precision Optical Manufacturing, Fudan University, Shanghai 200433, China; 16110720002@fudan.edu.cn (H.T.); shanyao1754@163.com (Y.S.); 18210720008@fudan.edu.cn (Y.C.); 19210720003@fudan.edu.cn (H.Z.); rjzhang@fudan.edu.cn (R.Z.); sywang@fudan.ac.cn (S.W.); lijing@fudan.edu.cn (J.L.); yplee@hanyang.ac.kr (Y.L.)

[2] Department of Physics, Quantum Photonic Science Research Center and RINS, Hanyang University, Seoul 04763, Korea

* Correspondence: yxzheng@fudan.edu.cn (Y.Z.); lychen@fudan.ac.cn (L.C.)

Citation: Tu, H.; Zheng, Y.; Shan, Y.; Chen, Y.; Zhang, H.; Zhang, R.; Wang, S.; Li, J.; Lee, Y.; Chen, L. Method for Analyzing the Measurement Error with Respect to Azimuth and Incident Angle for the Rotating Polarizer Analyzer Ellipsometer. *Crystals* **2021**, *11*, 349. https://doi.org/10.3390/cryst11040349

Academic Editors: Valentina Domenici and Yun-Han Lee

Received: 2 March 2021
Accepted: 25 March 2021
Published: 29 March 2021

Publisher's Note: MDPI stays neutral with regard to jurisdictional claims in published maps and institutional affiliations.

Abstract: We proposed a method to study the effects of azimuth and the incident angle on the accuracy and stability of rotating polarizer analyzer ellipsometer (RPAE) with bulk Au. The dielectric function was obtained at various incident angles in a range of 55°–80° and analyzed with the spectrum of the principal angle. The initial orientations of rotating polarizing elements were deviated by a series of angles to act as the azimuthal errors in various modes. The spectroscopic measurements were performed in a wavelength range of 300–800 nm with an interval of 10 nm. The repeatedly-measured ellipsometric parameters and determined dielectric constants were recorded monochromatically at wavelengths of 350, 550, and 750 nm. The mean absolute relative error was employed to evaluate quantitatively the performance of instrument. Apart from the RPAE, the experimental error analysis implemented in this work is also applicable to other rotating element ellipsometers.

Keywords: ellipsometry; error analysis; spectroscopy; high-accuracy measurement; optical metrology; dielectric constants

1. Introduction

The rotating element ellipsometer, after continuous development in many different configurations and applications, is widely employed as a primary technique in scientific research and industry [1–9]. The typical rotating element ellipsometer system comprises various types: the rotating polarizer ellipsometer (RPE) [10,11], the rotating analyzer ellipsometer (RAE) [12–14], and the rotating polarizer analyzer ellipsometer (RPAE).

The RPAE, which allows the polarizer and the analyzer to rotate in different angular velocities simultaneously, was firstly proposed by Azzam [15]. Intensive efforts have been devoted to RPAE from different aspects in past decades [16–20]. A self-established RPAE was constructed and presented in 1987 [16], with the polarizer and the analyzer rotating synchronously at an angular velocity ratio of 1:2. The system is superior in the elimination of DC signal error and phase-shift correction. Additionally, it provided two methods for the determination of ellipsometric parameters to realize the self-consistency of the data. Subsequently, the instrument was improved in 1994 [18], which enabled a fully variable incident angle by micro-stepping techniques, and employed a fixed polarizer to eliminate the effect of residual polarization from a light source.

The system and random errors have been extensively studied as an important topic in the development of ellipsometry. The accuracy and precision of ellipsometry can be effectively improved by performing error analysis and reduction. The analyses and corrections of errors, caused by the imperfect compensator and birefringence in window, were performed by McCrackin in 1970 [21]. Aspnes systematically presented the measurement and the correction of the first-order errors [22] and the uncertainties of ellipsometric parameters [23,24]. Azzam and Bashara investigated the errors from imperfect components, cell-window birefringence, and incorrect azimuth angles [25], and performed systematic error analysis on the RAE [26]. The errors in ellipsometry have been extensively analyzed in various aspects, such as the beam deviation [27], birefringence of window [28,29], incident angle [30], azimuthal errors, and residual ellipticity [31–33]. Moreover, the systematic error analyses on different configurations have been reported, including the RAE [34], PRPSE [35], multichannel ellipsometer [36], and the Mueller matrix ellipsometer [37,38].

Although the aforementioned error analyses are conditionally applicable to the RPAE, the error investigations specifically for this type of ellipsometer are still limited. The noise effect of Fourier coefficients on the RPAE with the same configuration as in [18] was analyzed by simulation [39]. In our previous work, the systematic error reduction, induced by the analytical discrete Fourier transform, was proposed theoretically and tested experimentally [40]. Apart from the effect from the Fourier transformation, the experimental performance affected by the systematic error is worth studying further.

In this work, a method to study the error analysis on the incident angle and azimuth was presented experimentally for the self-established RPAE with bulk Au. Both spectroscopic and monochromatic repeated measurements were carried out at various incident angles. The dielectric constants were determined from the measured ellipsometric parameters to study the accuracy and stability, which were evaluated according to the differences and dispersion degrees of experimental data compared with the reference values, respectively. The initial azimuths of polarizing elements were adjusted rotationally by groups of certain angles to study the effect of azimuthal errors in three modes. The performance of RPAE was evaluated quantitatively with the mean absolute relative error (MARE). The error analysis method proposed in this work is also useful for spectroscopic ellipsometry, including temperature-dependent properties of thin polymer films and metal nanoparticles [41–44].

2. Materials and Methods

Figure 1 schematically illustrates the configuration of the RPAE system. A monochromator containing a rotatable grating for the wavelength scan was employed to disperse the light from the source. The monochromatic light from the exit slit passed through a collimator lens, a fixed polarizer, and a rotating polarizer in sequence before incidence on the sample. Subsequently, the reflected light went through a rotating analyzer and entered a detector for data acquisition. The acquired analog signal was converted to a digital one for data processing. The initial azimuths of polarizing elements were set along the direction perpendicular to the incident plane. The angular velocity of the rotating analyzer was controlled to be twice that of the rotating polarizer. The optical system was aligned and calibrated precisely by a low-power He-Ne laser to realize a continuously variable incident angle in a range of 45°–90°, with a computer-controlled resolution of 0.001° or a visual resolution of 0.005° [18]. The spectroscopic measurement was performed routinely through a wavelength scan in a spectral range of 300–800 nm with an interval of 10 nm. Au was selected as the test material for the low penetration depth in the visible range with a great optical stability in the atmospheric environment.

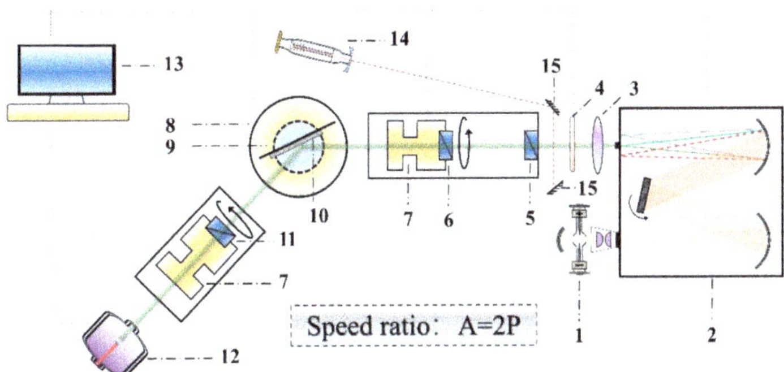

Figure 1. Schematic configuration of the RPAE optical system. (1) The continuous light source; (2) the monochromator consisting of two spherical mirrors and a rotatable plane grating; (3) light-collimating lens; (4) rotatable filters; (5) fixed polarizer; (6) rotating polarizer; (7) stepping motors; (8) rotating stage; (9) sample rotator; (10) sample; (11) rotating analyzer; (12) photomultiplier; (13) computer to control the monochromator, stepping motors, filters, rotating table, sample stage, and the photomultiplier; (14) laser used for alignment; and (15) mirrors to guide the laser beam for alignment.

The light intensity at the detector for RPAE is expressed as:

$$I(A) = I_0 + I_1 \cos A + I_2 \cos 2A + I_3 \cos 3A + I_4 \cos 4A, \tag{1}$$

where A represents the azimuth of analyzer and $I_0 - I_4$ are coefficients of one direct and four harmonic components, which are obtained by applying the discrete Fourier analysis as

$$I_k = \frac{2}{n} \sum_{i=1}^{n} I(A_i) \cos(kA_i) \quad k = 1, 2, 3, 4, \tag{2}$$

where A_i is the ith analyzer azimuth in the measurement period. Accordingly, the ellipsometric parameters are determined by [16–18]:

$$\tan \psi_1 = \left[\frac{2(I_1 + I_3 - 2I_2)}{I_1 + I_3} \right]^{1/2},$$

$$\cos \Delta_1 = \frac{I_1 - 3I_3}{[2(I_1 + I_3)(I_1 + I_3 - 2I_2)]^{1/2}}, \tag{3}$$

and

$$\tan \psi_2 = \left[\frac{9(I_1 + I_3 - 2I_2)}{2(2I_1 + I_2 + 4I_4)} \right]^{1/2},$$

$$\cos \Delta_2 = \frac{3(I_1 + I_3) - 4(I_2 + 4I_4)}{[8(I_1 + I_3)(I_1 + I_3 - 2I_2)]^{1/2}}. \tag{4}$$

The two sets of solutions are self-consistent to quantitatively verify the reliability of the results without other instruments. We prefer to use Equation (3) in the experiment, since the value of I_4 is the smallest in Equation (1). For bulk material measured at an incident angle of θ in the atmosphere, the dielectric function is determined with the well-known equation:

$$\tilde{\varepsilon} = \sin^2 \theta \left[1 + \tan^2 \theta \left(\frac{1 - \tan \psi \cdot e^{i\Delta}}{1 + \tan \psi \cdot e^{i\Delta}} \right)^2 \right]. \tag{5}$$

The accuracy and stability are evaluated by the MARE. The value of the MARE is given by:

$$\text{MARE} = \frac{1}{n} \cdot \sum_{i=1}^{n} \left| \frac{x_i^{\text{measured}} - x_i^{\text{reference}}}{x_i^{\text{reference}}} \right| \times 100\%, \tag{6}$$

where n represents the amount of data.

3. Results

3.1. Incident Angle and Principal Angle

The incident angle satisfying the condition of $\Delta = 90°$ is defined as the principal angle [45]. The error was proved theoretically to be reduced to obtain the highest precision in determining the optical constants when measured at the principal angle [17]. The spectrum of the principal angle for the Au sample was both theoretically and experimentally investigated in our previous work [46]. In this section, the ellipsometric measurements were performed by the RPAE at a series of incident angles in a range of 55° to 80°, with an interval of 5°, to evaluate the accuracy and stability.

3.1.1. Spectroscopic Measurement

The dielectric function spectra of the Au sample at various incident angles (Figure 2) were determined from the measured ellipsometric parameters with Equation (5). The spectra showed great agreement in most of the wavelength range. On the other hand, discrepancies were observed obviously in some regions, especially in the long-wavelength range. The reference dielectric function was obtained by applying the Model dielectric function [47] and Drude model [48] to the spectra of various incident angles. Accordingly, the spectrum of the principal angle was calculated with the method presented in [46], as shown in Figure 3.

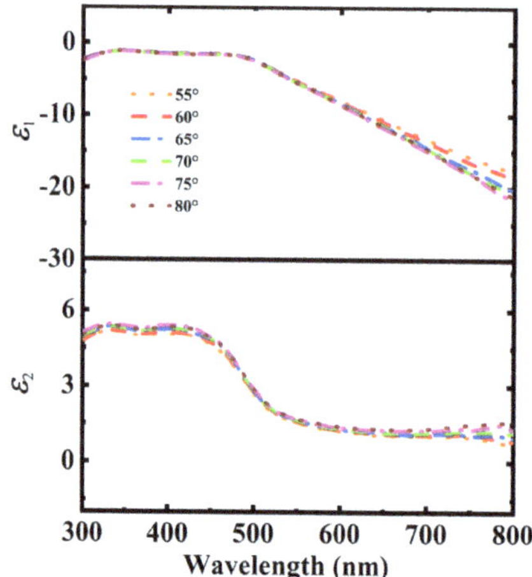

Figure 2. Dielectric functions of bulk Au determined at six incident angles.

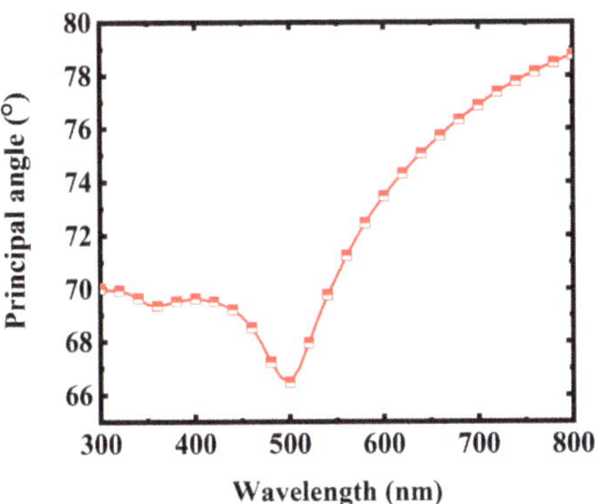

Figure 3. Spectrum of the principal angle for bulk Au.

The RPAE gave two solutions to determine the values of ψ and Δ with Equations (3) and (4). Theoretically, the results extracted by the two solutions were expected to be equal. The differences between the two results, defined as $\delta\psi = \psi_1 - \psi_2$ and $\delta\Delta = \Delta_1 - \Delta_2$, are used generally to evaluate the reliability of measurement. The values of $\delta\psi$ and $\delta\Delta$ in the spectral range are exhibited in Figure 4. For the incident angles in $65°$–$80°$, the differences between the two sets varied around 0 in the spectral range, which implied good credibility for measurement. Meanwhile, the differences of $55°$ and $60°$ were relatively large, especially in the long wavelength range. As indicated in Figure 3, the principal angle increased significantly in the long wavelength range, reaching approximately $80°$. Consequently, larger measurement errors occurred at incident angles of $55°$ and $60°$ away from the principal angle, leading to the significant discrepancy between the two solutions.

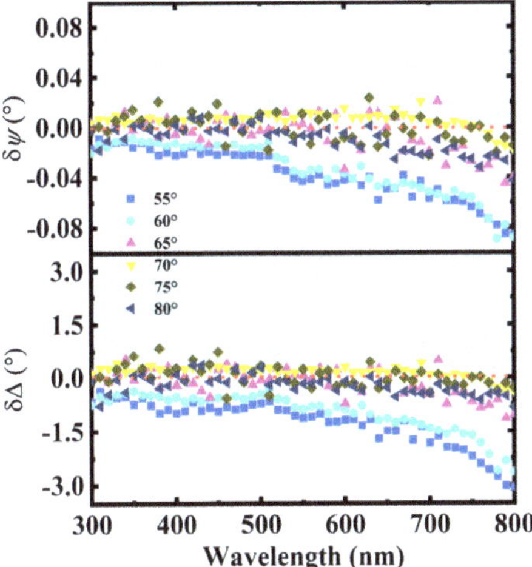

Figure 4. Discrepancies of the two sets of ellipsometric parameters at six incident angles in a spectral range.

The MARE values of the dielectric function at different incident angles are given in Table 1. The results demonstrate that the accuracy was dependent on the wavelength and corresponding principal angle. The measured results at $65°$, $70°$, and $75°$ turned out to be more accurate than those measured at $55°$, $60°$, and $80°$ in a wavelength range of 300–800 nm, which was consistent with the analysis based on the spectrum of the principal angle.

Table 1. MARE values of the dielectric functions at different incident angles.

Incident Angle (°)	55	60	65	70	75	80
MARE-ε_1 (%)	6.25	4.90	1.78	0.10	1.70	1.94
MARE-ε_2 (%)	9.12	5.45	3.8	1.39	4.19	4.19

3.1.2. Monochromatic Measurement

The monochromatic measurements at different incident angles were tested by performing 100 repeated measurements at a single wavelength of 350 nm. The real and imaginary parts of the reference dielectric constant at 350 nm were determined as $\varepsilon_1 = -1.09$ and $\varepsilon_2 = 5.2$, respectively. The principal angle of the Au sample at 350 nm was calculated to be 69.43°. Figure 5a,b display the distribution of the measured ellipsometric parameters and determined dielectric constants, respectively, at six incident angles. The data amount at each incident angle in an accurate region (relative error of ±2%, illustrated in Figure 5) is counted and listed in Table 2. The statistical values demonstrated that the results at 70°, 65°, and 75° had more accurate data compared with the others, which was consistent with the theoretical analysis of the principal angle.

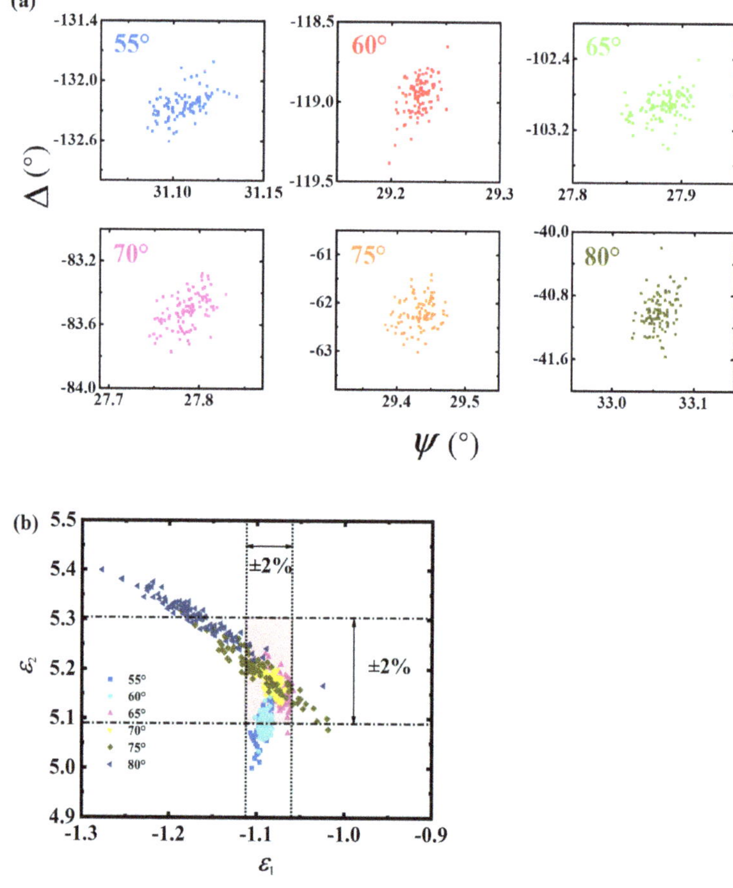

Figure 5. Distribution of the 100 (**a**) repeatedly-measured ellipsometric parameters, and (**b**) determined dielectric constants for six incident angles at a wavelength of 350 nm.

Table 2. Amounts of data points in an accurate region shown in Figure 5 in the 100 repeated measurements for different incident angles at a wavelength of 350 nm.

Incident Angle (°)	55	60	65	70	75	80
Amount of data points	30	27	96	100	68	8

The repeated measurement procedure was performed subsequently on two other wavelengths of 550 and 750 nm at different incident angles. The corresponding MAREs of the real part of the dielectric constant with varying incident angles were calculated at three wavelength points, as indicated in Figure 6. The principal angles of the Au sample at 350, 550, and 750 nm were determined to be 69.43°, 70.53°, and 78.01°, respectively. The MARE versus incident angle implies that the measurement exhibited higher accuracy and smaller error with the incident angle close to the principal angle. For example, the MARE at 750 nm decreased significantly with the increasing incident angle, which was attributed to the corresponding principal angle of 78.01°. At incident angles of 65°, 70°, and 75°, the MAREs turned out to be relatively small at all three wavelengths in Figure 6, representing the short, middle, and long wavelength parts in the spectral range. Accordingly, the results indicated that these three incident angles enabled accurate measurement for the Au sample and some other typical metals.

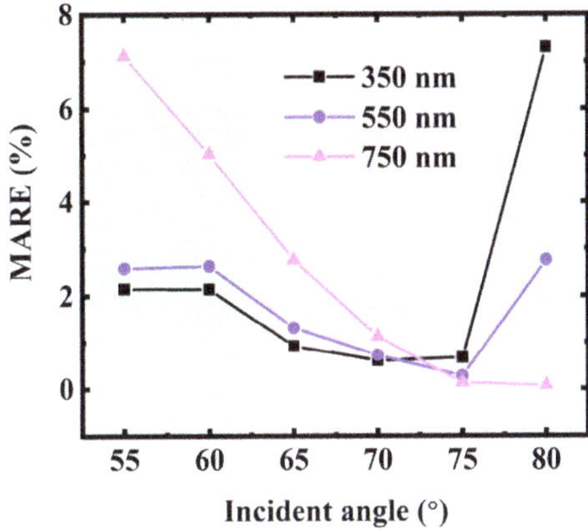

Figure 6. Values of MARE versus the incident angle at three wavelengths.

3.2. Azimuthal Error

In the rotating element ellipsometers, the measurement is fundamentally based on the detection of different polarization states, which is usually realized with the rotating elements. Consequently, the azimuthal error of the polarizing element significantly affects the performance of the instrument. In this subsection, we experimentally investigated the effect of the azimuthal error on the accuracy and stability of the results. The initial azimuths of the polarizer and analyzer were adjusted rotationally by a certain angle from the s-axis to act as the azimuthal errors, represented as $\delta\theta_P$ and $\delta\theta_A$, respectively.

3.2.1. Theoretical Analysis

For the measurement with the azimuthal error δ, assuming that the condition $A = 2P$ is still satisfied, the expression of light intensity in Equation (1) is modified as:

$$I_\delta(A) = I_0 + I_1 \cos(A + \delta) + I_2 \cos 2(A + \delta) + I_3 \cos 3(A + \delta) + I_4 \cos 4(A + \delta). \tag{7}$$

Compared with Equation (2), the four harmonic components are determined as:

$$I_{k\delta} = \frac{2}{\cos k\delta \cdot n} \sum_{i=1}^{n} I_{\delta}(A_i) \cos(kA_i) = \sec k\delta \cdot I_k \quad k = 1, 2, 3, 4. \tag{8}$$

Consequently, the ellipsometric parameters obtained in experiment are given as:

$$\tan \psi_{\delta 1} = \left[\frac{2(I_1 \cdot \sec \delta + I_3 \cdot \sec 3\delta - 2I_2 \cdot \sec 2\delta)}{I_1 \cdot \sec \delta + I_3 \cdot \sec 3\delta} \right]^{1/2},$$

$$\cos \Delta_{\delta 1} = \frac{I_1 \cdot \sec \delta - 3I_3 \cdot \sec 3\delta}{[2(I_1 \cdot \sec \delta + I_3 \cdot \sec 3\delta)(I_1 \cdot \sec \delta + I_3 \cdot \sec 3\delta - 2I_2 \cdot \sec 2\delta)]^{1/2}}, \tag{9}$$

and

$$\tan \psi_{\delta 2} = \left[\frac{9(I_1 \cdot \sec \delta + I_3 \cdot \sec 3\delta - 2I_2 \cdot \sec 2\delta)}{2(2I_1 \cdot \sec \delta + I_2 \cdot \sec 2\delta + 4I_4 \cdot \sec 4\delta)} \right]^{1/2},$$

$$\cos \Delta_{\delta 2} = \frac{3(I_1 \cdot \sec \delta + I_3 \cdot \sec 3\delta) - 4(I_2 \cdot \sec 2\delta + 4I_4 \cdot \sec 4\delta)}{[8(I_1 \cdot \sec \delta + I_3 \cdot \sec 3\delta)(I_1 \cdot \sec \delta + I_3 \cdot \sec 3\delta - 2I_2 \cdot \sec 2\delta)]^{1/2}}. \tag{10}$$

3.2.2. Experimental Results

Firstly, the ellipsometric parameters of the Au sample were measured by the RPAE at an incident angle of 70° in a wavelength range of 300–800 nm, with $\delta\theta_P$ varying from 0–20°, as shown in Figure 7a. The spectra of ψ and Δ demonstrated that the measured results deviated significantly from the standard spectra with the increasing of the azimuthal error of the polarizer. Similarly, the ellipsometric measurements were performed with $\delta\theta_A$ varying from 0–20°, as shown in Figure 7b. The spectra of ψ deviated differently compared with those of $\delta\theta_P$. Since the analyzer and polarizer rotated at an angular velocity ratio of 1:2 in measurement, the values of $\delta\theta_A$ and $\delta\theta_P$ were set at a group of angles with the same ratio. The value of $\delta\theta_A$ varied from 0–20°, compared with that of $\delta\theta_P$ in a range of 0–10°. Hence, the azimuth of the analyzer was always twice that of the polarizer during rotation. The same procedure was repeated to get the spectra of ψ and Δ (Figure 7c).

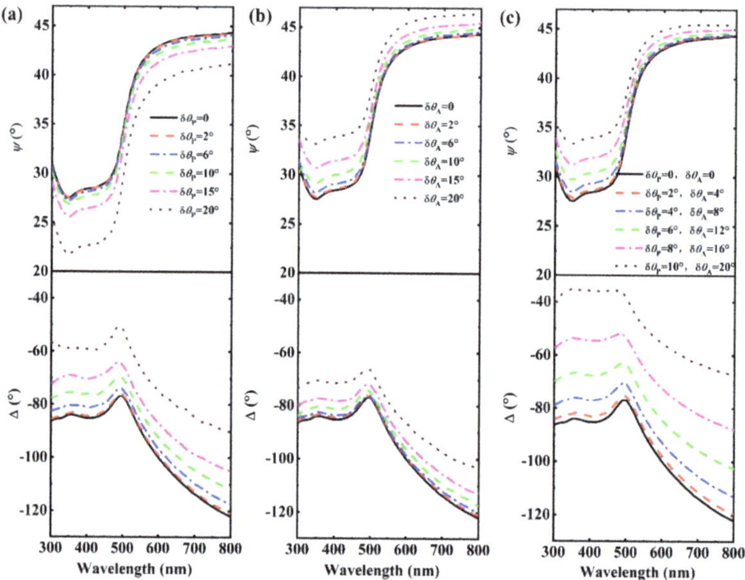

Figure 7. Measured ellipsometric parameters with various azimuthal errors of (**a**) the polarizer, (**b**) analyzer, and (**c**) both at a ratio of 1:2.

The MAREs of ψ and Δ were calculated and fitted into curves by applying the polynomial fitting method for three azimuthal error modes, as displayed in Figure 8. The comparison revealed the effect of the azimuthal error on the accuracy quantitatively. For the MARE of ψ, the three curves behaved similarly with small azimuthal errors, while the value increased significantly with a large value of $\delta\theta_P$. As for Δ, the value increased slightly for the curve of $\delta\theta_A$, while rapidly for that of $\delta\theta_P$ and $\delta\theta_P{:}\delta\theta_A = 1{:}2$. Therefore, the azimuthal error of the polarizer was indicated to more seriously affect the accuracy of the RPAE.

The polarizing elements were rotated in the opposite direction with the same absolute values of azimuthal errors to study the directional dependence on accuracy. The spectra of ψ and Δ, measured for a series of the same absolute values, showed a high consistency for both $\pm\delta\theta_P$ and $\pm\delta\theta_A$, as observed in Figure 9. Consequently, the accuracy of the RPAE was found to be less dependent on the direction of the azimuthal error in terms of the acquired spectra.

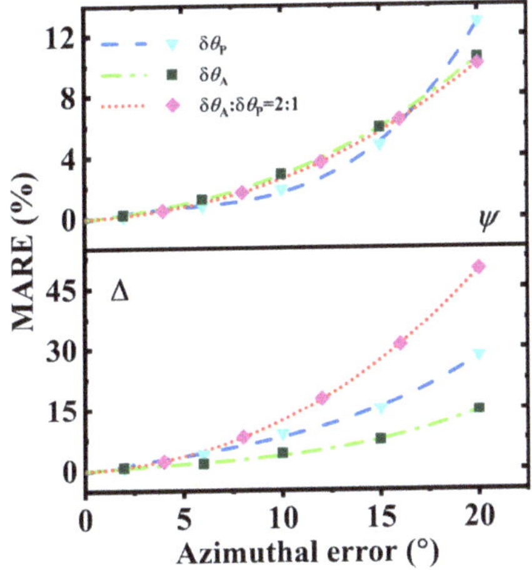

Figure 8. Experimental data (points) and fitted curves (lines) of MAREs versus azimuthal errors (for $\delta\theta_A{:}\delta\theta_P = 2{:}1$, the horizontal axis represents $\delta\theta_A$).

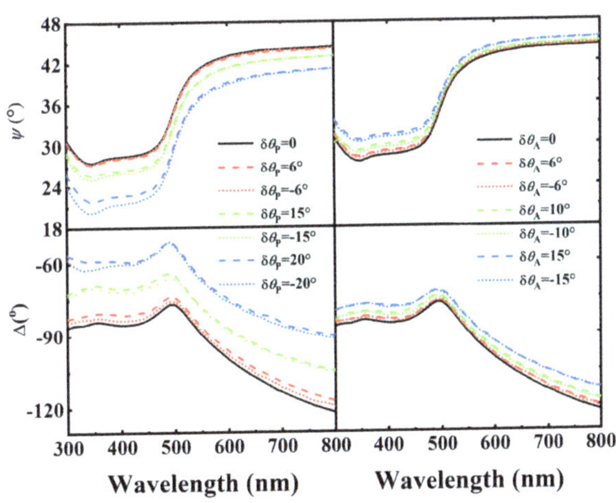

Figure 9. Ellipsometric parameters measured with azimuthal errors of the same absolute values in the opposite direction.

4. Conclusions

We presented a method to implement the experimental error analysis of azimuth and the incident angle specifically for the RPAE in this work. The dielectric functions of bulk Au were determined from the measured ψ and Δ at incident angles in a range of $55°$–$80°$, with an interval of $5°$, to study the effect on the instrument. The results acquired at an incident angle close to the principal angle were observed to exhibit higher accuracy and better stability, according to the discrepancies between the two solutions, and the values of the MARE. The azimuthal error analysis was performed experimentally with the initial orientations of polarizing elements deviating from the zero azimuth. The fitting curves of the MARE versus azimuthal error suggest that $\delta\theta_P$ more seriously affects the accuracy of the RPAE. The demonstrated error analysis reveals the relationship between the acquired data and experimental conditions, which gives access to achieving accurate and reliable measurement by using the RPAE, and is easily generalized to other rotating element ellipsometers.

Author Contributions: Conceptualization, H.T. and Y.Z.; methodology, H.T.; validation, Y.S., Y.C., and H.Z.; formal analysis, H.T. and Y.S.; investigation, Y.C. and H.Z.; data curation, Y.Z., Y.L. and L.C.; writing—original draft preparation, H.T.; writing—review and editing, Y.Z., R.Z., S.W., J.L., Y.L. and L.C.; supervision, Y.Z. and L.C.; project administration, Y.Z. and L.C.; funding acquisition, Y.Z. and L.C. All authors have read and agreed to the published version of the manuscript.

Funding: This research was funded by the National Natural Science Foundation of China, grant number 61775042 11674062 and the Fudan University CIOMP Joint Fund, grant number FC2017-003.

Institutional Review Board Statement: Not applicable.

Informed Consent Statement: Not applicable.

Data Availability Statement: Not applicable.

Acknowledgments: We would like to express our sincere appreciations to the anonymous referees for valuable suggestions and corrections.

Conflicts of Interest: The authors declare no conflict of interest.

References

1. Azzam, R.M.A.; Bashara, N.M. *Ellipsometry and Polarized Light*; North-Holland: Amsterdam, The Netherlands, 1977.
2. Hauge, P.S. Recent developments in instrumentation in ellipsometry. *Surf. Sci.* **1980**, *96*, 108–140. [CrossRef]
3. Woollam, J.A.; Snyder, P.G.; Rost, M.C. Variable angle spectroscopic ellipsometry—A non-destructive characterization technique for ultrathin and multilayer materials. *Thin Solid Film.* **1988**, *166*, 317–323. [CrossRef]
4. Collins, R.W. Automatic rotating element ellipsometers: Calibration, operation, and real-time applications. *Rev. Sci. Instrum.* **1990**, *61*, 2029–2062. [CrossRef]
5. Aspnes, D.E. Expanding horizons: New developments in ellipsometry and polarimetry. *Thin Solid Film.* **2004**, *455–456*, 3–13. [CrossRef]
6. Tompkins, H.G.; Irene, E.A. *Handbook of Ellipsometry*; William Andrew: New York, NY, USA, 2005.
7. Fujiwara, H. *Spectroscopic Ellipsometry: Principles and Applications*; John Wiley & Sons: Tokyo, Japan, 2007.
8. Aspnes, D.E. Spectroscopic ellipsometry—Past, present, and future. *Thin Solid Film.* **2014**, *571*, 334–344. [CrossRef]
9. Azzam, R.M.A. Polarization, thin-film optics, ellipsometry, and polarimetry: Retrospective. *J. Vac. Sci. Technol. B* **2019**, *37*, 060802. [CrossRef]
10. Cahan, B.D.; Spanier, R.F. A high speed precision automatic ellipsometer. *Surf. Sci.* **1969**, *16*, 166–176. [CrossRef]
11. Stobie, R.W.; Rao, B.; Dignam, M.J. Analysis of a novel ellipsometric technique with special advantages for infrared spectroscopy. *J. Opt. Soc. Am.* **1975**, *65*, 25–28. [CrossRef]
12. Aspnes, D.E. Fourier transform detection system for rotating-analyzer ellipsometers. *Opt. Commun.* **1973**, *8*, 222–225. [CrossRef]
13. Aspnes, D.E.; Studna, A.A. High Precision Scanning Ellipsometer. *Appl. Opt.* **1975**, *14*, 220. [CrossRef]
14. Schubert, M.; Rheinlander, B.; Woollam, J.A.; Johs, B.; Herzinger, C.M. Extension of rotating-analyzer ellipsometry to generalized ellipsometry: Determination of the dielectric function tensor from uniaxial TiO_2. *J. Opt. Soc. Am.* **1996**, *13*, 875. [CrossRef]
15. Azzam, R.M.A. A simple Fourier photopolarimeter with rotating polarizer and analyzer for measuring Jones and Mueller matrices. *Opt. Commun.* **1978**, *25*, 137–140. [CrossRef]
16. Chen, L.Y.; Lynch, D.W. Scanning ellipsometer by rotating polarizer and analyzer. *Appl. Opt.* **1987**, *26*, 5221–5228. [CrossRef]

17. Chen, L.Y.; Feng, X.W.; Su, Y.; Ma, H.Z.; Qian, Y.H. Improved rotating analyzer-polarizer type of scanning ellipsometer. *Thin Solid Films* **1993**, *234*, 385–389. [CrossRef]
18. Chen, L.Y.; Feng, X.W.; Su, Y.; Ma, H.Z.; Qian, Y.H. Design of a scanning ellipsometer by synchronous rotation of the polarizer and analyzer. *Appl. Opt.* **1994**, *33*, 1299–1305. [CrossRef]
19. El-Agez, T.M.; El Tayyan, A.A.; Taya, S.A. Rotating polarizer-analyzer scanning ellipsometer. *Thin Solid Film.* **2010**, *518*, 5610–5614. [CrossRef]
20. El-Agez, T.M.; Taya, S.A. Development and construction of rotating polarizer analyzer ellipsometer. *Opt. Laser. Eng.* **2011**, *49*, 507–513. [CrossRef]
21. Mccrackin, F.L. Analyses and corrections of instrumental errors in ellipsometry. *J. Opt. Soc. Am.* **1970**, *60*, 57–63. [CrossRef]
22. Aspnes, D.E. Measurement and correction of first-order errors in ellipsometry. *J. Opt. Soc. Am.* **1971**, *61*, 1077–1085. [CrossRef]
23. Aspnes, D.E. Optimizing precision of rotating-analyzer ellipsometers. *J. Opt. Soc. Am.* **1974**, *64*, 639–646. [CrossRef]
24. Aspnes, D.E. Optimizing precision of rotating-analyzer and rotating-compensator-ellipsometers. *J. Opt. Soc. Am.* **2004**, *21*, 403–410. [CrossRef] [PubMed]
25. Azzam, R.M.A.; Bashara, N.M. Unified analysis of ellipsometry errors due to imperfect components, cell-window birefringence, and incorrect azimuth angles. *J. Opt. Soc. Am.* **1971**, *61*, 600–607. [CrossRef]
26. Azzam, R.M.A.; Bashara, N.M. Analysis of systematic errors in rotating-analyzer ellipsometers. *J. Opt. Soc. Am.* **1974**, *64*, 1459–1469. [CrossRef]
27. Zeidler, J.R.; Kohles, R.B.; Bashara, N.M. Beam Deviation Errors in Ellipsometric Measurements; an Analysis. *Appl. Opt.* **1974**, *13*, 1938–1945. [CrossRef]
28. Jin, L.; Kasuga, S.; Kondoh, E. General window correction method for ellipsometry measurements. *Opt. Express* **2014**, *22*, 27811–27820. [CrossRef]
29. Nissim, N.; Eliezer, S.; Bakshi, L.; Moreno, D.; Perelmutter, L. In situ correction of windows' linear birefringence in ellipsometry measurements. *Opt. Commun.* **2009**, *282*, 3414–3420. [CrossRef]
30. Humlicek, J. Sensitivity extrema in multiple-angle ellipsometry. *J. Opt. Soc. Am. A* **1985**, *2*, 713–722.31. [CrossRef]
31. Kleim, R.; Kuntzler, L.; El Ghemmaz, A. Systematic errors in rotating-compensator ellipsometry. *J. Opt. Soc. Am. A* **1994**, *11*, 2550–2559. [CrossRef]
32. Bertucci, S.; Pawlowski, A.; Nicolas, N.; Johann, L.; El Ghemmaz, A.; Stein, N.; Kleim, R. Systematic errors in fixed polarizer, rotating polarizer, sample, fixed analyzer spectroscopic ellipsometry. *Thin Solid Films* **1998**, *313–314*, 73–78. [CrossRef]
33. Chao, Y.F.; Lee, K.Y.; Lin, Y.D. Analytical solutions of the azimuthal deviation of a polarizer and an analyzer by polarizer-sample-analyzer ellipsometry. *Appl. Opt.* **2006**, *45*, 3935–3939. [CrossRef]
34. Nijs, J.M.M.; Silfhout, A.V. Systematic and random errors in rotating-analyzer, ellipsometry. *J. Opt. Soc. Am. A* **1988**, *5*, 773–781. [CrossRef]
35. En Naciri, A.; Broch, L.; Johann, L.; Kleim, R. Fixed polarizer, rotating-polarizer and fixed analyzer spectroscopic ellipsometer: Accurate calibration method, effect of errors and testing. *Thin Solid Films* **2002**, *406*, 103–112. [CrossRef]
36. Nguyen, N.V.; Pudliner, B.S.; An, I.; Collins, R.W. Error correction for calibration and data reduction in rotating-polarizer ellipsometry applications to a novel multichannel ellipsometer. *J. Opt. Soc. Am. A* **1991**, *8*, 919–931. [CrossRef]
37. Nee, S. Error analysis for Mueller matrix measurement. *J. Opt. Soc. Am. A* **2003**, *20*, 1651–1657. [CrossRef]
38. Broch, L.; Naciri, A.E.; Johann, L. Systematic errors for a Mueller matrix dual rotating compensator ellipsometer. *Opt. Express* **2008**, *16*, 8814–8824. [CrossRef] [PubMed]
39. El-Agez, T.M.; Taya, S.A. An extensive theoretical analysis of the 1:2 ratio rotating polarizer–analyzer Fourier ellipsometer. *Phys. Scr.* **2011**, *83*, 25701. [CrossRef]
40. Mao, P.; Zheng, Y.; Cai, Q.; Zhang, D.; Zhang, R.; Zhao, H.; Chen, L. Approach to Error Analysis and Reduction for Rotating-Polarizer-Analyzer Ellipsometer. *J. Phys. Soc. Jpn.* **2012**, *81*, 124003. [CrossRef]
41. Hajduk, B.; Bednarski, H.; Trzebicka, B. Temperature-Dependent Spectroscopic Ellipsometry of Thin Polymer Films. *J. Phys. Chem. B* **2020**, *124*, 3229–3251. [CrossRef] [PubMed]
42. Nosidlak, N.; Jaglarz, J.; Danel, A. Ellipsometric studies for thin polymer layers of organic photovoltaic cells. *J. Vac. Sci. Technol. B* **2019**, *37*, 062402. [CrossRef]
43. Hajduk, B.; Bednarski, H.; Jarząbek, B.; Nitschke, P.; Janeczek, H. Phase diagram of P3HT:PC70BM thin films based on variable-temperature spectroscopic ellipsometry. *Polym. Test* **2020**, *84*, 106383. [CrossRef]
44. Jarząbek, B.; Nitschke, P.; Hajduk, B.; Domański, M.; Bednarski, H. In situ thermo-optical studies of polymer:fullerene blend films. *Polym. Test* **2020**, *88*, 106573. [CrossRef]
45. Azzam, R.M.A. Contours of constant principal angle and constant principal azimuth in the complex epsilon-plane. *J. Opt. Soc. Am. A* **1981**, *71*, 1523–1528. [CrossRef]
46. Zhu, X.; Zhao, H.; Zhang, R.; Ma, Y.; Liu, Z.; Li, J.; Shen, Z.; Wang, S.; Chen, L. Analytical study of the principal angle used in optical experiments. *Appl. Opt.* **2002**, *41*, 2592–2595. [CrossRef] [PubMed]
47. Aspnes, D.E.; Kinsbron, E.; Bacon, D.D. Optical-properties of Au: Sample effects. *Phys. Rev. B* **1980**, *21*, 3290–3299. [CrossRef]
48. Takeuchi, K.; Adachi, S. Optical properties of β-Sn films. *J. Appl. Phys.* **2009**, *105*, 073520. [CrossRef]

Article

Enhanced Properties of SAW Device Based on Beryllium Oxide Thin Films

Namrata Dewan Soni [1,*] **and Jyoti Bhola** [2]

1   Department of Physics, Hansraj College, University of Delhi, Delhi 110007, India
2   Department of Mathematics, Hansraj College, University of Delhi, Delhi 110007, India; jbhola@hrc.du.ac.in
*   Correspondence: ndsoni@hrc.du.ac.in

Abstract: The present study depicts the first-ever optimized surface acoustic wave (SAW) device based on Beryllium Oxide (BeO) thin film. The feasibility of surface acoustic wave devices based on BeO/128° YX LiNbO$_3$ layered structure has been examined theoretically. The SAW phase velocity, electromechanical coupling coefficient, and temperature coefficient of delay for BeO/128° YX LiNbO$_3$ layered structure are calculated. The layered structure is found to exhibit optimum value of phase velocity (4476 ms^{-1}) and coupling coefficient (~9.66%) at BeO over layer thickness of 0.08 λ. The BeO (0.08 λ)/128° YX LiNbO$_3$ SAW device is made temperature stable, by integrating it with negative temperature coefficient of delay (TCD) TeO$_3$ over layer of thickness 0.026λ.

Keywords: beryllium oxide; lithium Niobate; SAW devices

Citation: Soni, N.D.; Bhola, J. Enhanced Properties of SAW Device Based on Beryllium Oxide Thin Films. *Crystals* **2021**, *11*, 332. https://doi.org/10.3390/cryst11040332

Academic Editor: Raghvendra Singh Yadav

Received: 17 March 2021
Accepted: 24 March 2021
Published: 25 March 2021

Publisher's Note: MDPI stays neutral with regard to jurisdictional claims in published maps and institutional affiliations.

1. Introduction

Thin film-based surface acoustic wave (SAW) devices are enunciated to be exploited for their use in communication devices, acousto-optic devices, optoelectronics, automotive sensors and biosensors, etc., as these are efficient, compact, economical and provide the advantage of tailoring the material properties as per the need of the application relative to single crystal SAW device [1–5]. And with the advent of technology, SAW devices have become an integral part of current LTE and 5G wireless devices [6]. These devices are increasingly finding applications in the domain of life sciences and microfluidics (acoustofluidics), producing 'lab-on-a-chip' (LOC) or micro total analysis systems (μTAS) [6,7]. The important parameters that gauge the competence and use of SAW device are its phase velocity, electromechanical coupling coefficient and temperature coefficient of delay [7]. Until now, various SAW layered structures, like SiO$_2$/LiNbO$_3$, LiNbO$_3$/Sapphire, ZnO/Diamond, etc., have been investigated for their potential as acoustic wave devices [8]. It is evident from the available literature that for high frequency applications, temperature stable SAW devices with appreciable SAW phase velocity and good electromechanical coupling coefficient are required [8]. Diamond based SAW devices are reported to provide the advantage of high velocity. But these devices are expensive and one needs to compromise with coupling coefficient [8]. On the other hand, widely used LiNbO$_3$ based SAW devices have reasonable SAW velocity and good coupling coefficients [8]. Efforts are still being made to find an alternative that is suitable for high frequency applications. Beryllium Oxide (BeO) single crystal is reported to be piezoelectric material with very high acoustic velocities for bulk longitudinal and shear waves and is studied for its application in SAW device [9]. Although, there are various reports on the deposition of the crystalline BeO over layer on crystal and amorphous substrates [10,11], yet no attempt to date has been made to study the use of BeO thin films in SAW device applications. BeO is reported to have unique mechanical and thermal properties, such as hardness, high melting point, high thermal conductivity, and large elastic constants, making it suitable for large number of applications in microwave and nano devices [11,12]. All these properties make BeO a sturdy material.

So, in the present work, an attempt has been made to study the use of thin films of BeO in acoustic wave devices. The SAW propagation properties of BeO/128° YX LiNbO$_3$ layered structure have been found using the theoretical tool developed by Farnell and Alder [13,14]. The SAW software used in the present analysis has been exercised earlier by many workers, like Zhou et al., Benetti et al., etc., to find the optimum values of thickness of various layers used in the multilayered acoustic devices [15–17]. The theoretical results are in close proximity with the experimentally obtained results [15–17]. Moreover, the experimental realization of the proposed layered structure seems to be possible and supported by the report on the growth of crystalline BeO thin films irrespective of the substrate type [10]. This suggests that BeO/128°YX LiNbO$_3$ layered structure can be experimentally realized without lattice mismatch.

In the present study, the authors have considered widely used 128° YX LiNbO$_3$ SAW substrate to investigate the effect of adding BeO and subsequently TeO$_3$ thin films on it theoretically. The BeO over layer thickness is optimized and it is found that with the integration of 0.08 λ thick BeO over layer in BeO/128°YX LiNbO$_3$ layered structure, an efficient SAW device with appreciable phase velocity (~4500 ms^{-1}) along with a very high electromechanical coupling coefficient (~10%) can be realized. The bilayer BeO (0.08 λ)/128°YX LiNbO$_3$ SAW device is temperature unstable and has a high positive value of temperature coefficient of delay (TCD ~ 66 ppm °C^{-1}). The device can be made temperature stable by integrating it with negative TCD over layer. TeO$_3$ films are reported to exhibit negative temperature coefficient of delay [1,18–21] and, thus, can be used in the present layered structure to make it temperature stable. The proposed BeO (0.08 λ)/128°YX LiNbO$_3$ bilayer SAW structure is integrated with ~0.026 λ thick TeO$_3$ over layer to realize a temperature stable device and moreover, the values of SAW phase velocity and electromechanical coupling coefficient remain essentially untouched. Thus, the authors present the first-ever optimized SAW device based on BeO thin film owing to its potential use in acoustic wave device applications.

2. Materials and Methods

In the present study, the SAW propagation characteristics of proposed multi-layered structure are calculated using the SAW Analysis software (MSDOS (version 2 or later), IEEE, Montreal, QC, Canada) developed by Farnell and Adler [13,14]. The structure consists of BeO thin film integrated over 128° YX LiNbO$_3$ single crystal and TeO$_3$ over layer placed on BeO thin film. The multi-layered structure and the coordinate system are presented in Figure 1.

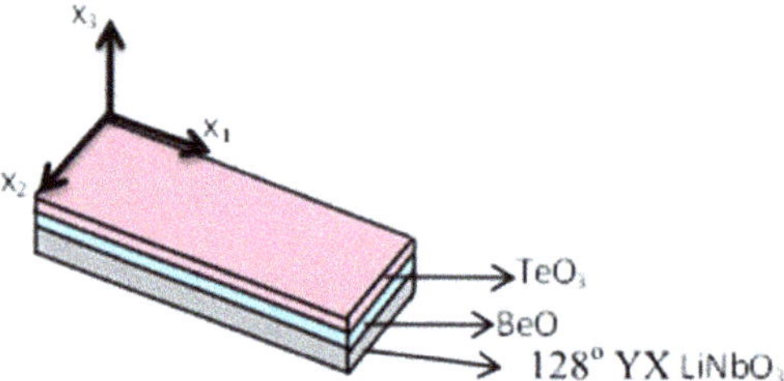

Figure 1. Illustration of TeO$_3$/Beryllium Oxide (BeO)/128° YX LiNbO$_3$ multilayer surface acoustic wave (SAW) structure and the coordinate system.

The Cartesian coordinate system is chosen in such a way that Rayleigh wave propagates along x_1—axis in which its amplitude vanishes as x_3 tends to negative infinity, and x_2—axis is parallel to the direction of particle polarization.

The electric potential ϕ and particle displacements U_k ($k = 1, 2, 3$) in a piezoelectric medium are governed by the following elastic wave equations [1,13]:

$$e_{kij} \frac{\partial^2 \phi}{\partial x_k \partial x_i} = \rho \frac{\partial^2 U_j}{\partial t^2} - C_{ijkl} \frac{\partial^2 U_k}{\partial x_i \partial x_l} , \tag{1}$$

$$e_{jkl} \frac{\partial^2 U_j}{\partial x_i \partial x_l} - \varepsilon_{jk} \frac{\partial^2 \phi}{\partial x_i \partial x_k} = 0; \qquad i, j, k, l = 1, 2, 3 \ldots\ldots , \tag{2}$$

where C_{ijkl} is the mechanical stiffness tensor, ε_{jk} is the dielectric permittivity tensor, e_{kij} is the piezoelectric tensor, and ρ is the density of the medium. The material parameters, like density, elastic constant, piezoelectric constant, and dielectric constants at a given temperature of TeO_3, Wurtzite BeO thin films, and $128°$ YX $LiNbO_3$ single crystal used in present study to estimate the SAW phase velocity of the layered structure, are taken from earlier reported data by Dewan et al., Duman et al., Cline et al., and Kovacs et al., respectively [12,18,22,23], and are presented in Table 1.

Table 1. Material constants and temperature coefficients used in simulations.

Material Constants	Materials			Temperature Coefficients ($°C^{-1}$)		
	TeO_3 [18]	BeO [12,22]	$LiNbO_3$ [23]	TeO_3 [18]	BeO [24]	$LiNbO_3$ [25]
Elastic Constants(10^{11} N/m^2)					$\times 10^{-4}$	
C_{11}	0.14	4.606	1.98	0.06	−2.63	−1.74
C_{33}		4.916	2.279		−1.34	−1.53
C_{66}		1.670	0.728		−1.98	−1.43
C_{44}	0.265	1.477	0.5965	0.0646	−0.95	−2.04
C_{12}		1.265	0.5472		−4.27	−2.52
C_{13}		0.8848	0.6513		−39.8	−1.59
C_{14}			0.0788			−2.14
Piezoelectric constant (C/m^2)						
e_{33}		0.0364	1.77			8.87
e_{31}		−0.0735	0.30			2.21
e_{15}			3.69			1.47
e_{22}			2.42			0.79
Dielectric constant (10^{-11} F/m)						
ε_{11}	23.7	3.06	45.6	1.6		3.23
ε_{33}		3.13	26.3			6.27
Density (Kg/m^3)						
ρ	4578	3010	4628			

2.1. Electromechanical Coupling Coefficient (K^2)

The effective coupling of inter digital transducer to the surface-wave is measured in terms of the electromechanical coupling coefficient, K^2 given by [1,26,27]

$$K^2(\%) = \frac{200\,(v - v')}{v} , \tag{3}$$

where v and $v\prime$ are the SAW phase velocities for electric free and short circuit conditions, respectively.

2.2. Temperature Coefficient of Delay (TCD)

The temperature dependence of SAW device is expressed in terms of TCD given by [1,26,27]

$$TCD = TCD_o - \frac{V_{35} - V_{15}}{20 \times V_{25}} , \tag{4}$$

where TCD_o is the coefficient of thermal expansion of base layer, and V_{35}, V_{25}, *and* V_{15} are the SAW velocities of the layered structure at respective temperatures. The TCD is measured in the units of ppm $°C^{-1}$.

Using the material parameters listed in Table 1, the authors obtained the SAW phase velocity for the multi-layered structure through the software. The coupling coefficient and TCD were consequently calculated using the SAW velocity so obtained in Equations (3) and (4).

3. Results and Discussion

The SAW phase velocity (V_P), electromechanical coupling coefficient (K^2) and temperature coefficient of delay (TCD) of bilayer BeO/128° YX LiNbO$_3$ SAW structure were first calculated as a function of normalized thickness (h_{BeO}/λ) of BeO over layer, where h_{BeO} is the BeO over layer thickness, and λ is the acoustic wavelength.

The change in SAW phase velocity (V_P) and electromechanical coupling coefficient (K^2) with the normalized thickness of BeO over layer is shown in Figure 2. It is found that the SAW phase velocity increases from 3800 ms^{-1} (SAW velocity of bare 128° YX LiNbO$_3$ single crystal) to 4476 ms^{-1} with increase in the BeO over layer thickness from 0 to 0.08 λ. The enhanced SAW velocity is principally due to the higher velocity (7800 ms^{-1}) of BeO film in comparison to LiNbO$_3$, and with the increasing over layer thickness of BeO, SAW energy is assembled more into BeO [9].

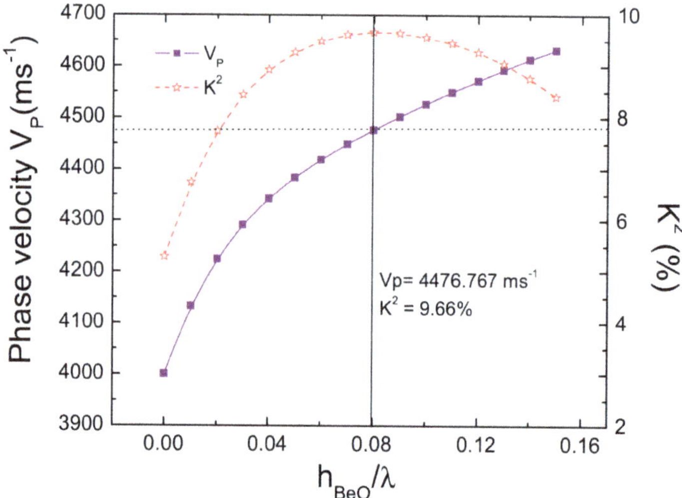

Figure 2. Variation of SAW phase velocity (V_P) and electromechanical coupling coefficient (K^2) with normalized thickness of BeO over layer in BeO/128° YX LiNBO$_3$ SAW layered structure.

Figure 2 shows the rise in the value of electromechanical coupling coefficient (K^2) for BeO/128° YX LiNbO$_3$ bilayer structure. The value of K^2 increases nearly twofold, i.e., from ~5% (K^2 of bare 128° YX LiNbO$_3$ single crystal) to ~9.66% with the change in BeO over layer thickness from 0 to 0.08 λ. And with further increase in the BeO over layer thickness (beyond 0.08 λ), its value decreases. The significant rise in the value of coupling coefficient with the integration of a BeO over layer (in the range 0 to 0.08 λ) on the top of LiNbO$_3$ single crystal is accounted to the stiffening effect produced by the over layer [28]. The stiffened layer escalates the stress and raises the potential at the interface [29]. Thus, in addition to the piezoelectric coefficients, stress also makes a noteworthy contribution in raising the electric potential and thereby augments the coupling coefficient. With the increase in the over layer thickness from 0.08 λ to 0.15 λ, the value of K^2 reduces to ~8% from 9.7% because, at greater thickness, the impact of mass loading influences the propagation [28].

The temperature coefficient of delay for the BeO/128° YX LiNbO$_3$ bilayer structure is calculated using Equation (4), and its dispersion with normalized BeO over layer thickness is shown in the Figure 3. It is found to reduce a little from 76 to ~66 ppm °C^{-1} with an increase in the BeO over thickness from 0 λ to 0.08 λ. The small reduction in the value of TCD is credited to the fact that BeO film has comparatively lower but positive TCD value than for LiNbO$_3$ crystal [24]. And with the addition of greater BeO over layer thickness, the SAW energy is more accumulated in BeO; hence, it exhibits the reduced value of TCD for BeO/128° YX LiNbO$_3$ bilayer structure. It can be inferred from Figure 3 that the BeO/128° YX LiNbO$_3$ bilayer structure is thermally unstable as both LiNbO$_3$ and BeO are positive TCD materials. The positive TCD BeO/128° YX LiNbO$_3$ bilayer structure can be made temperature stable by integrating it with an over layer (i.e., SiO$_2$ and TeO$_3$) possessing negative TCD [21,29–31]. Previously reported results show that in comparison to SiO$_2$, TeO$_3$ thin films possess high value of negative TCD [18,20,21,30,31]. So, with the integration of relatively less thick TeO$_3$ over layer, a positive TCD device can be made temperature stable [18,19].

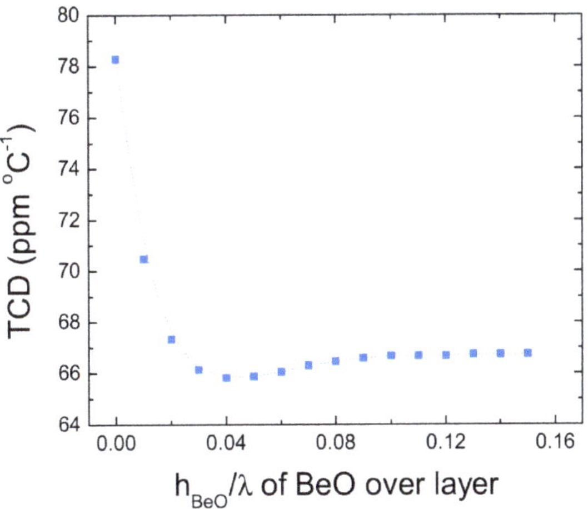

Figure 3. Variation of temperature coefficient of delay (TCD) with normalized thickness of BeO over layer in BeO/128° YX LiNBO$_3$ SAW layered structure.

Therefore, in the present study, the result of adding TeO$_3$ over layer over BeO/128° YX LiNbO$_3$ bilayer structure on its SAW propagation characteristics have been investigated further. In TeO$_3$/BeO/128° YX LiNbO$_3$ multi- layered SAW structure, the thickness of BeO layer is fixed at 0.08 λ because it is observed in Figure 2 that, at this BeO over layer thickness, the BeO/128° YX LiNbO$_3$ bilayer structure has maximum value of K^2(~9.66%) and appreciable phase velocity (~4467 ms^{-1}).

Figure 4 presents the TCD variation of TeO$_3$/BeO (0.08 λ)/128° YX LiNbO$_3$ multi-layered SAW structure as a function of normalized thickness (h_{TeO_3}/λ) of TeO$_3$ over layer, where h_{TeO_3} is the TeO$_3$ over layer thickness. The TCD of TeO$_3$/BeO(0.08 λ)/128° YX LiNbO$_3$ layered structure reduces to 0 from 66 ppm °C^{-1} with an increase in the (negative TCD) TeO$_3$ over layer thickness from 0 to 0.026 λ. Thus, a temperature stable TeO$_3$/BeO(0.08 λ)/128° YX LiNbO$_3$ multi-layered SAW device can be achieved with the integration of 0.026 λ thick TeO$_3$ over layer.

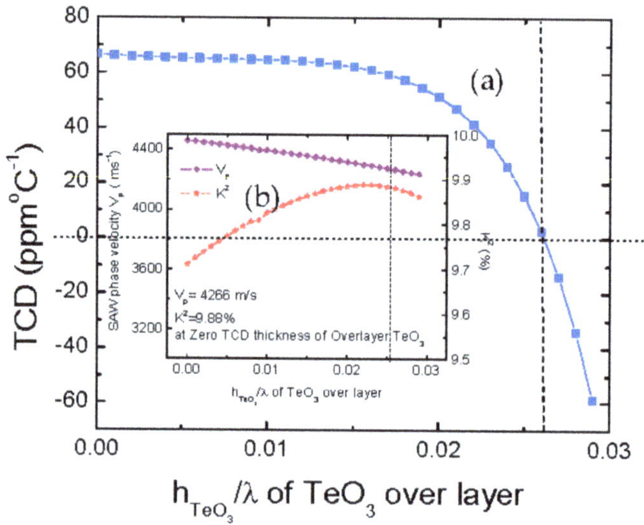

Figure 4. (**a**): Variation of TCD of $TeO_3/BeO(0.08\,\lambda)/128°$ YX $LiNbO_3$ SAW structure with normalized thickness of TeO_3 over layer. (**b**) Variation of phase velocity and K^2 for $TeO_3/BeO(0.08\,\lambda)/128°$ YX $LiNbO_3$ SAW structure with normalized thickness of TeO_3 over layer.

The effect of integrating TeO_3 over layer is examined on SAW phase velocity and K^2, as well. The inset in Figure 4b shows the change of K^2 and SAW phase velocity for TeO_3/BeO $(0.08\,\lambda)/128°$ YX $LiNbO_3$ with the normalized thickness of TeO_3 over layer. It may be seen that the value of SAW phase velocity declines faintly from 4467 to 4266 ms^{-1} with an increase in the TeO_3 over layer thickness from 0 to 0.026 λ owing to the point that TeO_3 has lower SAW phase velocity in comparison to $BeO(0.08\,\lambda)/128°$ YX $LiNbO_3$ bilayer structure. Moreover, with the rise in the TeO_3 over layer thickness from 0 to 0.026 λ, the value of K^2 increases further from 9.66% to ~9.85%. This is because the thickness of TeO_3 over layer is much less than the acoustic wavelength leading to the stiffening and further increase in the potential and hence electromechanical coupling coefficient [21,28]. It may be noted that the introduction of 0.026 λ thick TeO_3 over layer in BeO $(0.08\,\lambda)/128°$YX $LiNbO_3$ bilayer structure not only marginally increases the value of K^2 (from 9.70% to 9.85%) but also makes the device temperature stable. With the further increase in the TeO_3 over layer thickness, the TCD of TeO_3/BeO $(0.08\,\lambda)/128°$ YX $LiNbO_3$ multi-layered SAW structure becomes negative, making the device temperature unstable again. Hence the optimum thickness of TeO_3 over layer is taken to be 0.026 λ. Therefore, a temperature stable TeO_3 $(0.026\,\lambda)/BeO(0.08\,\lambda)/128°$ YX $LiNbO_3$ multi-layered SAW structure with high value of K^2 (~9.85%) and the phase velocity (~4266 ms^{-1}) is proposed, which is suitable for its applications in narrow band filters in GHz range.

4. Conclusions

In the present work the SAW propagation properties of $TeO_3/BeO/128°$ YX $LiNbO_3$ multi-layered SAW structure have been investigated theoretically. The optimized thickness values of TeO_3 and BeO thin films are obtained. It is established that with the integration of (0.026 λ) TeO_3 and (0.08 λ) BeO layer on 128° YX $LiNbO_3$ single crystal, an efficient, temperature stable, high velocity (~4266 ms^{-1}) multi-layered SAW structure can be made. Because the proposed device is based upon sturdy wurtzite BeO layer, it can work efficiently in the severe environment conditions. The proposed temperature stable SAW structure can be conceived effortlessly for possible applications in both high, as well as low, frequency regions.

Author Contributions: Conceptualization, N.D.S.; Formal analysis, N.D.S.; Investigation, N.D.S.; Methodology, N.D.S.; Software, J.B.; Writing—original draft, N.D.S.; Writing—review & editing, J.B. All authors have read and agreed to the published version of the manuscript.

Funding: This research received no external funding.

Institutional Review Board Statement: Not applicable.

Informed Consent Statement: Not applicable.

Conflicts of Interest: The authors declare no conflict of interest.

References

1. Luo, J.T.; Quan, A.J.; Liang, G.X.; Zheng, Z.H.; Ramadan, S.; Fu, C.; Li, H.L.; Fu, Y.Q. Love-mode surface acoustic wave devices based on multilayers of TeO_2/ ZnO(1 1 2 0)/Si(1 0 0) with high sensitivity and temperature stability. *Ultrasonics* **2017**, *75*, 63–70. [CrossRef] [PubMed]
2. Fan, L.; Zhang, S.; Ge, H.; Zhang, H. Theoretical investigation of acoustic wave devices based on different piezoelectric films deposited on silicon carbide. *J. Appl. Phys.* **2013**, *114*, 024504. [CrossRef]
3. Naumenko, N.; Nicolay, P. AlN/Pt/LN structure for SAW sensors capable of operating at high temperature. *Appl. Phys. Lett.* **2017**, *111*, 073507. [CrossRef]
4. Liang, J.; Yang, X.; Zheng, S.; Sun, C.; Zhang, M.; Zhang, H.; Zhang, D.; Pang, W. Modulation of acousto-electric current using a hybrid on-chip AlN SAW/GFET device. *Appl. Phys. Lett.* **2017**, *110*, 243504. [CrossRef]
5. Gillinger, M.; Shaposhnikov, K.; Knobloch, T.; Schneider, M.; Kaltenbacher, M.; Schmid, U. Impact of layer and substrate properties on the surface acoustic wave velocity in scandium doped aluminum nitride based SAW devices on sapphire. *Appl. Phys. Lett.* **2016**, *108*, 231601. [CrossRef]
6. Go, D.B.; Atashbar, M.Z.; Ramshani, Z.; Chang, H.-C. Surface acoustic wave devices for chemical sensing and microfluidics: A review and perspective. *Anal. Methods* **2017**, *9*, 4112–4134. [CrossRef]
7. Delsing, P.; Cleland, A.N.; Schuetz, M.J.; Knörzer, J.; Giedke, G.; Cirac, J.I.; Srinivasan, K.; Wu, M.; Balram, K.C.; Bäuerle, C.; et al. The 2019 surface acoustic waves roadmap. *J. Phys. D Appl. Phys.* **2019**, *52*, 353001. [CrossRef]
8. Campbell, C. *Surface Acoustic Wave Devices for Mobile and Wireless Communications, Four-Volume Set*; Academic Press: San Diego, CA, USA; Toronto, ON, Canada, 1998.
9. Collins, H.; Hagon, P.J.; Pulliam, G.R. Evaluation of new single crystal piezoelectric materials for surface acoustic –wave applications. *Ultrasonics* **1970**, *8*, 218–226. [CrossRef]
10. Lee, S.M.; Jang, Y.; Yum, J.H.; Larsen, E.S.; Lee, W.C.; Kim, S.K.; Bielawski, C.W.; Oh, J. Crystal Properties of atomic- layer deposited beryllium oxide on crystal and amorphous substrates. *Semicond. Sci. Technol.* **2019**, *34*, 115021. [CrossRef]
11. Koh, D.; Yum, J.H.; Banerjee, S.K.; Hudnall, T.W.; Bielawski, C.; Lanford, W.A.; French, B.L.; French, M.; Henry, P.; Li, H.; et al. Investigation of atomic layer deposited beryllium oxide material properties for high K dielectric applications. *J. Vac. Sci. Technol. B* **2014**, *32*, 03D117. [CrossRef]
12. Duman, S.; Sutlu, A.; Bagci, S.; Tutuncu, H.M.; Srivastava, G.P. Structural, elastic, electronic and phonon properties of zinc blende and wurtzite BeO. *J. Appl. Phys.* **2009**, *105*, 033719. [CrossRef]
13. Farnell, G.W.; Adler, E.L. Elastic Wave Propagation in Thin Layers. In *Physical Acoustics Principles and Methods*; Mason, W.P., Thurston, R.N., Eds.; Academic Press: New York, NY, USA, 1972; Volume 9, pp. 35–127.
14. Farnell, G.W.; Adler, E.L. Multilayer acoustic surface wave program. *Proc. Inst. Electr. Eng.* **1995**, *122*, 470–471.
15. Nakahata, H.; Fujii, S.; Higaki, K.; Hachigo, A.; Kitabayashi, H.; Shikata, S.; Fujimori, N. Diamond-based surface acoustic wave devices. *Semicond. Sci. Technol.* **2003**, *18*, 96–104. [CrossRef]
16. Zhou, C.; Yang, Y.; Zhan, J.; Ren, T.; Wang, X.; Tian, S. Surface acoustic wave characteristics based on c-axis (006) $LiNbO_3$/diamond/silicon layered structure. *Appl. Phys. Lett.* **2011**, *99*, 022109. [CrossRef]
17. Benetti, M.; Cannat'a, D.; Pietrantonio, F.D.; Verona, E. Growth of AlN piezoelectric film on diamond for high-frequency surface acoustic wave devices. *IEEE Trans. Ultrason. Ferroelectr. Freq. Control* **2005**, *52*, 1806. [CrossRef] [PubMed]
18. Dewan, N.; Sreenivas, K.; Gupta, V. Temperature-Compensated Devices Using Thin TeO_2 Layer with Negative TCD. *IEEE Electron Device Letters* **2006**, *27*, 752–754. [CrossRef]
19. Cacho VD, D.; Siarkowski, A.L.; Morimoto, N.I.; Borges, B.V.; Kassab, L.R. Fabrication and Characterization of TeO2-ZnO Rib Waveguides. *ECS Trans.* **2010**, *31*, 225–229. [CrossRef]
20. Dewan, N.; Sreenivas, K.; Gupta, V.; Katiyar, R.S. Growth of amorphous TeOx ($2 \leq x \leq 3$) thin film by radio frequency sputtering. *J. Appl. Phys.* **2007**, *101*, 084910. [CrossRef]
21. Dewan, N.; Tomar, M.; Gupta, V.; Sreenivas, K. Temperature stable LiNbO3 surface acoustic wave device with diode sputtered amorphous TeO_2 over-layer. *Appl. Phys. Lett.* **2005**, *86*, 223508. [CrossRef]
22. Cline, C.F.; Dunegan, H.L.; Henderson, G.W. Elastic Constants of Hexagonal BeO, ZnS and CdSe. *J. Appl. Phys.* **1967**, *38*, 1944–1948. [CrossRef]
23. Kovacs's, G.; Anhorn, M.; Engan, H.E.; Visintini, G.; Ruppel, C.C.W. Improved material constants for $LiNbO_3$ and $LiTaO_3$. *IEEE Symp. Ultrason.* **1990**, *1*, 435–438.

24. Sirota, N.N.; Kuzmina, A.M.; Orlova, N.S. Debye- Waller factors and Elastic constants for Beryllium oxide at temperature between 10 and 720 K. *Cryst. Res. Technol.* **1992**, *27*, 711–716. [CrossRef]

25. Smith, R.T.; Welsh, F.S. Temperature Dependence of the Elastic, Piezoelectric, and Dielectric Constants of Lithium Tantalate and Lithium Niobate. *J. Appl. Phys.* **1971**, *42*, 2219–2230. [CrossRef]

26. Soni, N.D. SAW propagation characteristics of TeO3/3C-SiC/LiNbO3 layered structure. *Mater. Res. Express* **2018**, *5*, 046309.

27. Rana, L.; Gupta, V.; Soni, N.D.; Tomar, M. SAW field and acousto-optical interaction in ZnO/AlN/sapphire structure. In Proceedings of the 2016 Joint IEEE International Symposium on the Applications of Ferroelectrics, European Conference on Application of Polar Dielectrics, and Piezoelectric Force Microscopy Workshop (ISAF/ECAPD/PFM), Darmstadt, Germany, 21–25 August 2016.

28. Shih, W.C.; Wang, T.L.; Hsu, L.L. Surface acoustic wave properties of aluminum oxide films on lithium Niobate. *Thin Solid Film.* **2010**, *518*, 7143–7146.

29. Tomar, M.; Gupta, V.; Sreenivas, K.; Mansingh, A. Temperature stability of ZnO thin film SAW device on fused quartz. *IEEE Trans. Device Mater. Reliab.* **2005**, *5*, 494–500. [CrossRef]

30. Gupta, V.; Tomar, M.; Sreenivas, K. Improved temperature stability of $LiNbO_3$ surface acoustic wave devices with sputtered SiO_2 over-layers. *Ferroelectrics* **2005**, *329*, 57–60. [CrossRef]

31. Zhou, F.M.; Li, Z.; Fan, L.; Zhang, S.Y.; Shui, X.J.; Wasa, K. Effects of TeO_x films on temperature coefficients of delay of Love type wave devices based on TeO_x/ 36° YX—$LiTaO_3$ structures. *Vaccuum* **2010**, *84*, 986–991. [CrossRef]

Article

Studies on the Characteristics of Nanostructures Produced by Sparking Discharge Process in the Ambient Atmosphere for Air Filtration Application

Tewasin Kumpika [1,2], Stefan Ručman [1,2,*], Siwat Polin [3], Ekkapong Kantarak [2], Wattikon Sroila [2], Wiradej Thongsuwan [2], Arisara Panthawan [2], Panupong Sanmuangmoon [4], Niwat Jhuntama [5] and Pisith Singjai [1,2,*]

[1] Materials Science Research Center, Faculty of Science, Chiang Mai University, Chiang Mai 50200, Thailand; tewasin@gmail.com
[2] Department of Physics and Materials Science, Faculty of Science, Chiang Mai University, Chiang Mai 50200, Thailand; ekkapong_k@hotmail.com (E.K.); ple_wasan@hotmail.com (W.S.); wiradej.t@cmu.ac.th (W.T.); vampire_601@hotmail.com (A.P.)
[3] Faculty of Science and Technology, Rajamangala University of Technology Suvarnabhumi, Phranakhon Si Ayutthaya 12160, Thailand; siwat.p@rmutsb.ac.th
[4] Faculty of Science, Chiang Mai University, Chiang Mai 50200, Thailand; panupong.s@cmu.ac.th
[5] Faculty of Science and Agricultural Technology, Rajamangala University of Technology Lanna, Chiang Mai 50200, Thailand; niwat_jhuntama@rmutl.ac.th
* Correspondence: stefan_rucman@cmu.ac.th (S.R.); pisith.s@cmu.ac.th (P.S.)

check for **updates**

Citation: Kumpika, T.; Ručman, S.; Polin, S.; Kantarak, E.; Sroila, W.; Thongsuwan, W.; Panthawan, A.; Sanmuangmoon, P.; Jhuntama, N.; Singjai, P. Studies on the Characteristics of Nanostructures Produced by Sparking Discharge Process in the Ambient Atmosphere for Air Filtration Application. *Crystals* **2021**, *11*, 140. https://doi.org/10.3390/cryst11020140

Academic Editor: Raghvendra Singh Yadav

Received: 20 January 2021
Accepted: 26 January 2021
Published: 29 January 2021

Abstract: Among the various methods for the preparation of nanoparticles, a sparking process at atmospheric pressure is of interest because it is a simple method for producing nanoparticles ranging from a few nanometer-sized particles to agglomerated film structures. In this research, we studied the effects of metal electrode properties on nanoparticle sizes. The experiments were carried out by applying a high voltage to different metal sparkling tips. The transfer of energies from positive ions and electron bombardments induced the melting and vaporization of electrode metals. Based on this research, we have developed a model to describe the formation of a nanoparticle film on the substrate, placed under the sparking gap, and the nanostructure produced by metal vapor on the sparking electrodes. The model provides a realistic tool that can be used for the design of a large-scale coating and the application of nanoparticles developed by this process for the filtration of PM2.5 mask fabric by air.

Keywords: sparking process; surface energy; nanoparticle nucleation; vapor deposition

1. Introduction

Nanoscale particles can have structural, thermal, electromagnetic, optical and mechanical properties that are significantly different from those of larger particles [1]. Particle properties are highly size-dependent and can be exploited in a variety of applications. Therefore, the control of nanoparticle size is very important and desirable, but it also represents a challenge. Nanoparticles can be prepared using various methods, such as laser ablation [2], ultrafine bubbles and pulsed ultrasound [3], corona discharge [4] and sparking processes [5–11]. The methods and conditions used to prepare nanoparticles strongly influence their size and shape [4,12]. The sparking process is of interest because it is performed in atmospheric air, is inexpensive and does not require a vacuum system. Moreover, this method is flexible regarding the material used, can be up-scaled, and is environmentally friendly as it does not produce any waste and does not require a chemical precursor. Although there have been several publications describing the production of nanoparticles using the sparking process in atmospheric pressure, the effects of the wire electrode properties on the nanoparticle size and formation pattern of the deposited films have not been reported previously. During the sparking process, the applied voltage induces high-temperature arcing plasma in the air gap via the field ionization process. Electrons and ions in the plasma bombarded the two tip surfaces, resulting in the vaporization and liquefaction of the metal electrodes. In our previous work [5], the formation of nanoparticles was modeled using the Young–Laplace relation by considering the relative surface energies and different pressures inside and outside the molten layers on the electrode surfaces. For the low-pressure atmosphere, Tabrizi et al. [13] produced gold nanoparticles using the spark discharge at the pressure of 1–2.5 bar. They explained that the electrode materials were evaporated, and the nanoparticles were nucleated and agglomerated from the vapor. In this report, the effects of the electrode properties on the generated nanoparticles from both vaporized and molten metals were described and applied in forms of aerosol for disposable face mask testing of filtration efficiency.

2. Experimental

To study the size of nanoparticles, various metal wires were used as sparking tips. The metals used were zinc (d = 0.38 mm), aluminum, silver, gold, nickel, cobalt, titanium, vanadium and molybdenum (purity ≥ 99.5%, d = 0.25 mm, Advent Research Materials Ltd., Eynsham, Oxford, England). The sparking tips were placed horizontally, 5 mm above the center of the glass substrate, with a gap of 2 mm between them. Sparks were produced by the discharge of a 25 nF capacitor at 10 kV. Before the nanoparticles were formed on a glass slide substrate ($10 \times 10 \times 1$ mm^3), the substrate was sonically cleaned in acetone, ethanol and distilled water and dried under nitrogen gas. To eliminate the effect of initial conditions, the tips were sparked 100 times before collected the nanoparticles. The spark was created once in an ambient air at atmospheric pressure to study the deposited particle sizes.

Scheme of sparking discharge apparatus is represented in Figure 1. The Supplementary Information depicts all of the sparking apparatus used in our research; from the first prototype that was used for the fabrication of nanoparticles for carbon nanotube growing, to this study. The principle of the work was described previously [14], in brief, our research technique consisted of a power supply that was connected to a 220 V alternative current source. The power supply unit consisted of an AC/DC converter and a controlled source of voltage and current, which was connected to a 7 KV DC boost step-up power module high-voltage generator, which was in turn connected to a capacitor that directs electrical power to a circuit breaker with changeable heads that hold metal wires. These wires were placed in a sealed container, through which gasses flowed, and in which a substrate for nanoparticle collection was placed.

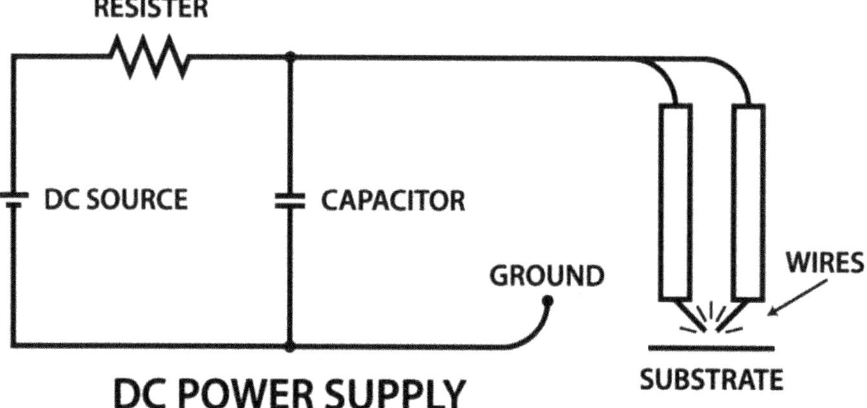

Figure 1. Conceptual representation of sparking machine apparatus. For pictures of the machine please see the Supplementary Information.

To study the particle sputtering patterns resulting from the sparking process (and therefore to study the directions in which the particles were expelled from the electrode wire's melting surface), gold and cobalt wires were used as the sparking electrodes, since they produce sparked nanoparticles of easily observable color. A voltage of 10 kV was applied across a 1 mm gap, causing the spark discharge. White paper substrate was placed 5 mm below the sparking gap and examined after sparking 1000 times.

The sparked electrodes and the deposited nanoparticles were characterized using scanning electron microscopy (SEM, JEOL JSM 6335F, Akishima, Japan). The Raman spectra were obtained with a 514.5 nm argon ion laser at a room temperature (Jobin Yvon Horiba T64000, Chiyoda-ku, Japan) to determine crystallinity of the films.

Surface characterization was done by using an atomic force microscope (AFM) in the tapping mode (Digital Instruments, Inc., Santa Barbara, CA) equipped with a standard Si tip and operated at a scan size of 1 x 1 μm^2 in air at room temperature. Section analysis was carried out to estimate sizes of the NDs (nanodots) using the Nanoscope IIIa 5.12 r3 (Veeco, New York, NY, USA) software. The primary particles size can be measured by sparking a very low surface coverage of nanoparticles on a substrate for AFM. After approximately one second of sparking onto the substrate, we noticed films with primary particles as an isolated island, consisting of primary particles that were deposited very far away from the other particles.

3. Results and Discussions

Nanoparticles produced using various sparking metals were imaged using AFM, as shown in Figure 2a. Because the lateral (horizontal) sizes of the nanoparticles estimated by AFM show a broadening effect due to a tip shape convolution [15], the nanoparticle sizes were determined by measuring their vertical heights. Due to the random nature of the nanoparticle agglomeration, some particles attached to others and created a secondary particle deposition on the substrate. To determine the primary size of the particles, only primary particles that appeared in the images were used. It can be clearly seen, that a low energy level is needed to melt the metals (Zn, Ag, Au), resulting in a higher density of the particles in the area under observation. The reason for this is that these metals require a lower energy to melt the electrode, and in turn, to melt a nano-droplet. The energy applied to the tips (E_{app}) is separated to the energy lost to the environment (E_{loss}) and the energy used to melt the metal tips, as described by Equation (1)

$$E_{app} = E_{loss} + m\left[c_p(T_m - T_r) + L_f\right] \tag{1}$$

where c_p is the specific heat capacity, m is the effective mass of the metal tip, T_m is the melting point, T_r is the room temperature and L_f is the heat of fusion. As shown in Table 1, the melting point of aluminum (933 K) is lower than that of silver (1235 K) and gold (1337 K), however, there a reduced amount of sparked aluminum observed. This is due to aluminum having a remarkably high heat fusion of 0.90 Jg^{-1}K^{-1}, whereas the high heat fusions of zinc, silver and gold are in the range of 0.13–0.39 Jg^{-1}K^{-1}. Moreover, aluminum is always covered with a microscopic thin layer of Al$_2$O$_3$ that is extremely hard and inert. Therefore, this oxide also reduces the creation of nanoparticles during the discharge. Figure 2b shows the height measurement of zinc nanoparticle using section analysis of Nanoscope III (Veeco, New York, NY, USA) software.

The relationship between the nanoparticle heights and the surface energy at melting point of the electrode metals is shown in Figure 3. It can be clearly seen, that the particle heights decreased as surface energy increased. Nevertheless, the heights of all primary particles were less than 4.5 nm, which corresponds with the transmission electron microscopy (TEM) images of sparked nanoparticles published elsewhere [10]. The relationship between the height of the particles and the surface energy is given by a linear formula H = 1.6915 γ + 4.4783, where H is the average particle height (nm) and γ is the surface energy (J·m^{-2}). This result strongly confirms that nanoparticles on the substrates were nucleated by the molten metal electrodes.

(a)

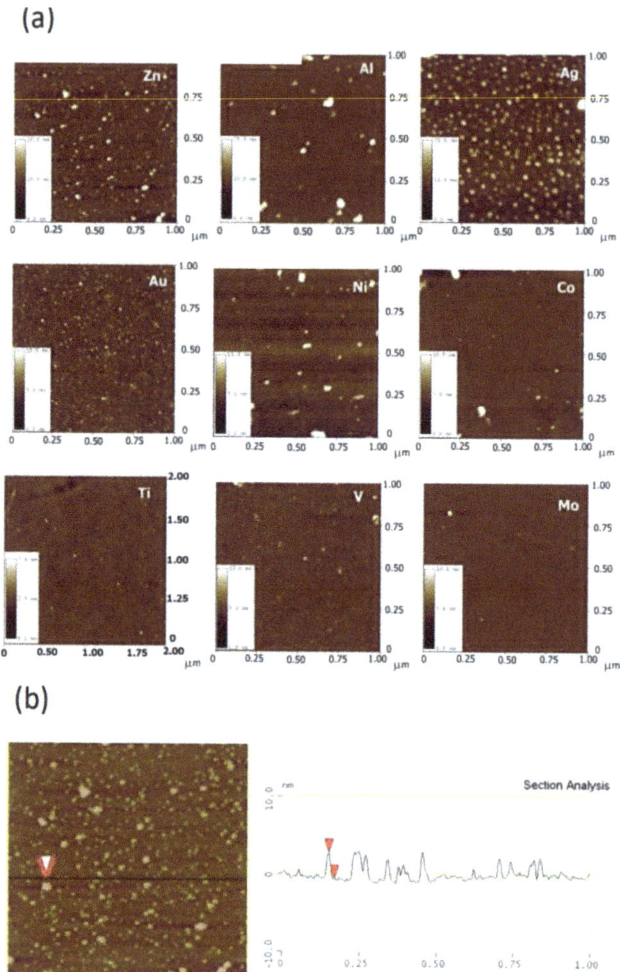

(b)

Figure 2. (a) Atomic force microscope (AFM) images of various metal nanoparticles prepared by single spark on glass substrates **(b)** Section analysis to measured height of zinc nanoparticles.

Table 1. Metal properties and experimental data [16].

Metals.	T_m (K)	σ_{LV} (Jm^{-1})	c_p (Jg^{-1}K^{-1})	k (Wcm^{-1}K^{-1})	L_f (Jg^{-1})	L_v (Jg^{-1})	Height (nm)
Zn	693	0.789	0.39	1.16	244	3,843	3.30
Al	933	0.871	0.9	2.37	830	22,569	3.15
Ag	1235	0.925	0.24	4.29	240	5,319	2.71
Au	1337	1.145	0.13	3.17	158	4,233	2.63
Ni	1718	1.796	0.44	0.91	623	13,228	2.02
Co	1811	1.881	0.42	1.00	599	13,944	1.50
Ti	1941	1.525	0.52	0.22	702	19,136	1.23
V	2183	1.855	0.49	0.31	908	19,652	1.08
Mo	2896	2.080	0.25	1.38	761	14,238	0.92

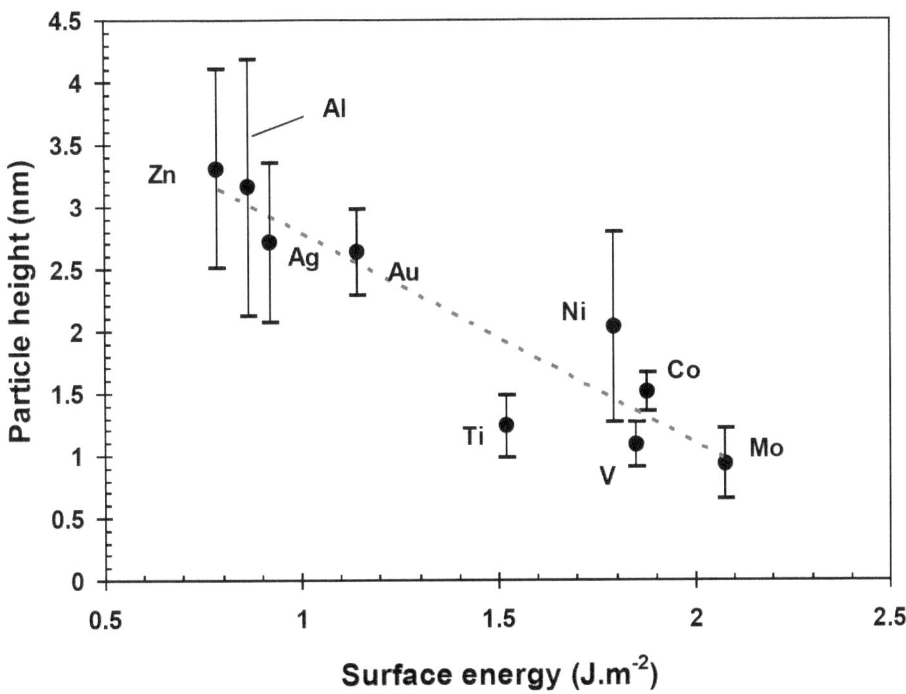

Figure 3. Plot of the height of sparked nanoparticles as a function of surface energy at melting temperature.

Figure 4a,d show sparked particle films of gold and cobalt, respectively, on white paper substrates. The film formed in two semicircle shapes with a gap of 1 mm between them. The semicircle on the right-hand side was formed under the cathode and is larger than the semicircle on the left that formed under the anode. This is caused by the process when, during the sparking, atmospheric gas breaks down and creates the positive ions and free electrons. The bombardment of the anode by the electrons and the cathode by the positive ions melts the electrode surfaces. The thermal energy expands, and a shock wave forms. The resulting high air pressure pushes the melted metal layer away from the electrode gap, and the molten metal therefore splashes away from the gap in a cone-shaped pattern, while the nanoparticles are nucleated. The cone shaped splashing produces the two semicircles on the substrate. The black color of the sparked cobalt film in Figure 4a indicates the presence of cobalt oxide. Furthermore, the sparked gold film is pink, which is a feature of gold nanoparticles [17].

Figure 4b,c show the cobalt anode and cathode electrode tips, respectively, after the sparking process. The erosion of the tip results in the round shape at both tip ends. The results are not the same for the gold anode and cathode tips, as shown in Figure 4e,f, respectively. The sparked gold tips are cone shaped. The cone angle of the anode tip is less than that of the cathode tip. The reduced cone angle of the anode tip corresponds with the smaller area of the particle film produced under the anode (Figure 4d). The shape of the anode is caused by the expansion of the plasma diameter at the anode surface [18]. In addition, the sides of the cobalt tips were covered by nanoporous films. It is believed that some of the metal evaporated as a gas phase and then deposited on the tip surfaces. The sides of the gold electrodes are different from those of the cobalt tips and no deposited nanoporous film was observed because the gold vapor could not oxidize in this setup.

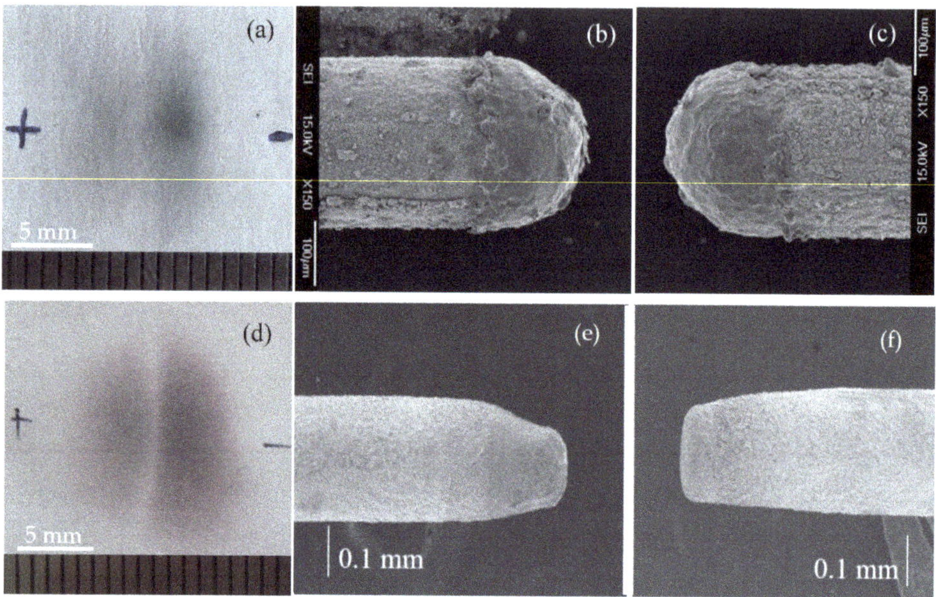

Figure 4. (**a**,**d**) Nanoparticle films on papers which were placed under the sparking gap of cobalt and gold, respectively. The electrode after 1000 cycles of spark: (**b**) Co-anode (**c**) Co-cathode (**e**) Au-anode and (**f**) Au-cathode.

More studies are needed in order to better understand the formation of particle coating on the electrode wires. In this study, titanium wires were used as sparked electrodes. The anode and cathode titanium wires were sparked 1000 times. The sparked electrode wires are shown in Figure 5a,b for the anode and cathode, respectively. In the pictures, the nanostructure films are seen to almost fully cover the anode surface. On the other hand, most of the surface area of the cathode is covered by the freeze splashing of molten metal droplets. As previously explained, positive ions have a larger mass and transfer more energy to the cathode compared to the energy transferred by electrons to the anode. Therefore, the cathode shows more surface damage than the anode. The molten electrode splashed and then froze very quickly due to the rapid cooling of the room atmosphere. Furthermore, the nanoporous film coverage observed on the tip is rather patchy. In addition, during the sparking, the glass substrate was placed 2 mm under the sparking gap to collect the nanoparticles from the metal droplets (the result shown in Figure 7).

Figure 5 shows high magnification SEM images displaying the tip surface located 200 µm, 400 µm and 600 µm from the tip end (the positions of the tips are shown in Figure 4). Nanostructures grown on the anode tip (Figure 5a,c,e for the position of 200 µm, 400 µm and 600 µm, respectively) are sponge-like structures composed of numerous irregular-shaped and randomly-oriented grains. It is believed that, the nanostructures are the deposition of titanium vapor. The difference in the size and shape of the nanostructures is explained by the influence of different substrate temperature [19,20]. On the cathode tip (Figure 5b,d,f for the position of 200 µm, 400 µm and 600 µm, respectively) no nanostructures can be observed, since the splashing of the melting titanium prevents the deposition of titanium vapor on the cathode tip surface. The results are the opposite for the cobalt, where the cathode tip is fully covered in cobalt oxide nanostructures, as shown in Figure 4c. Therefore, based on the results for gold, cobalt and titanium, it can be concluded that the nanoparticles were nucleated by the droplets of molten metals and deposited on the substrates under the sparking gap for all sparking metal electrodes. However, the nanostructure nucleation caused by the deposition of metal vapor on the sparking tip depends on the properties of each metal.

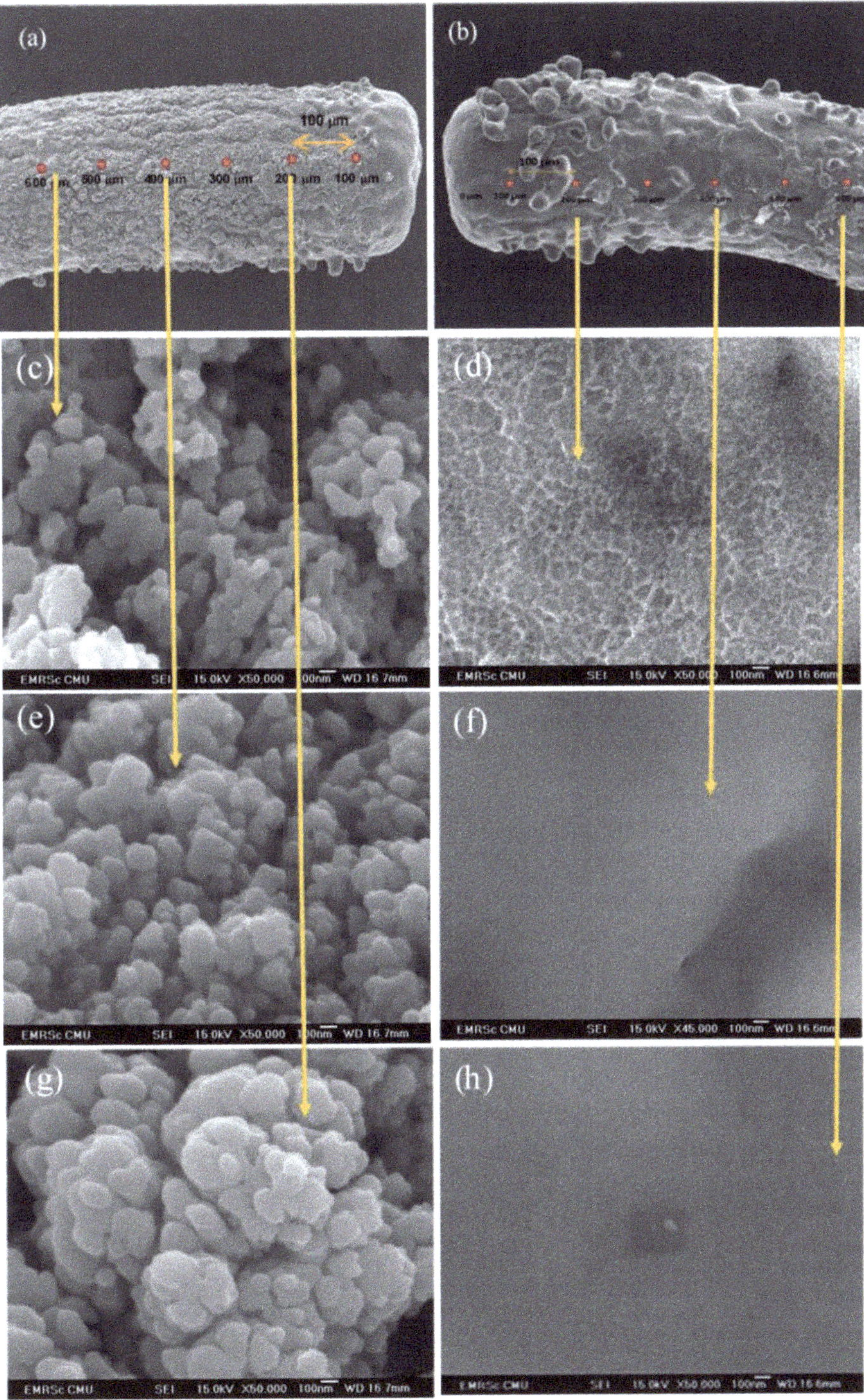

Figure 5. (**a**) Anode-Ti (**b**) Cathode-Ti after 1000 cycles of spark and surface structures on the sparked tips at the position of 200 μm (**c,d**), 400 μm (**e,f**) and 600 μm (**g,h**) from the tip ends for anode and cathode titanium tips, respectively.

Figure 6a,b illustrate the SEM images and EDS line scan result of the sparked anode and cathode titanium tips, respectively. At the tip of the anode electrode, a large titanium ratio and a small amount of oxide cover were found. Given that there was no oxygen found at the aperture created by the focus ion beam (FIB), we can conclude that the oxygen was only on the surface. The oxygen content increased in the nanoporous film area due to the formation of titanium oxide in that area. As was the case with the anode, the oxygen content also increased in the cluster of nanoporous material on the cathode.

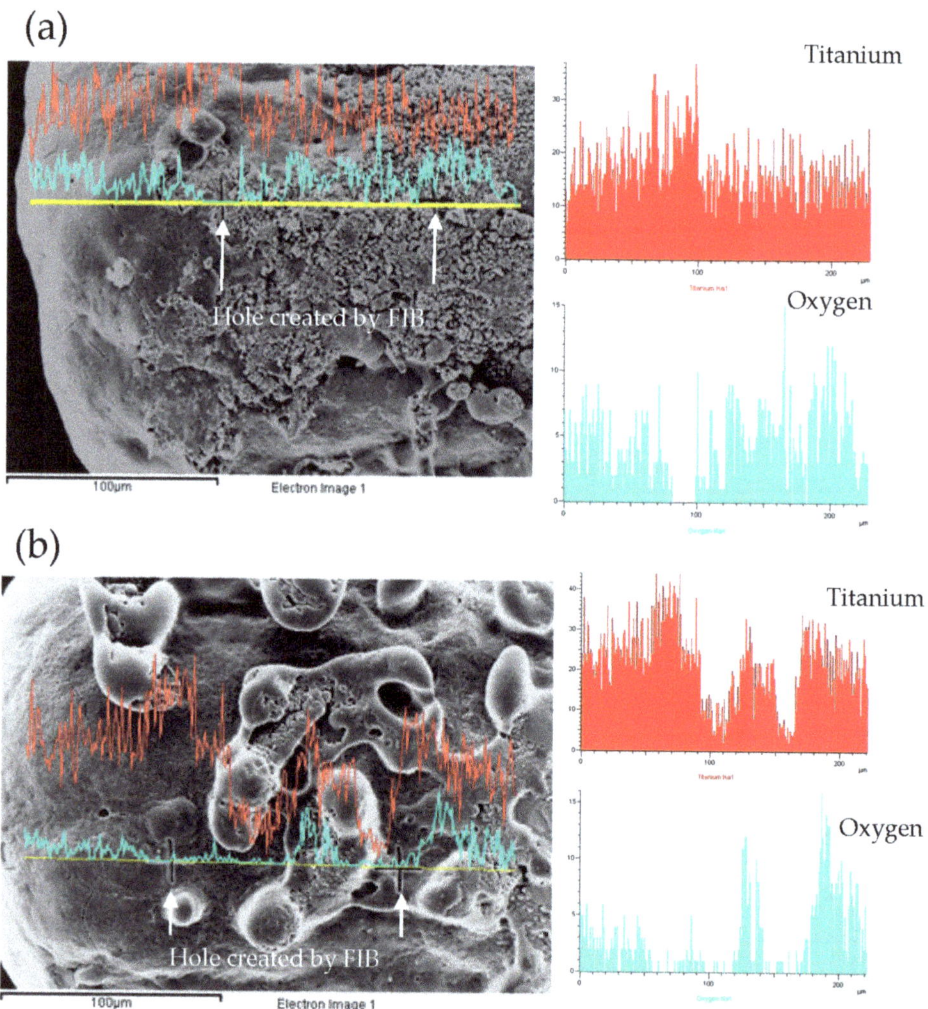

Figure 6. SEM–EDX line scans of sparked titanium tip (**a**) anode (**b**) cathode.

Figure 7 shows the SEM image of the sparked titanium particles that were deposited on the glass substrate, 5 mm under the sparking gap of titanium tips presented in Figure 5. The films were an agglomeration of nanoparticles with an approximate size of 20 nm. The size of the nanoparticles was almost ten times smaller than the size of the grain that occurred due to the vapor deposition on the tips electrodes. The crystallinity of the sparked films was initially amorphous and then gradually transitioned into the anatase phase of TiO_2 during the annealing at 400 °C for 1–5 h, as shown in Raman spectra in Figure 8.

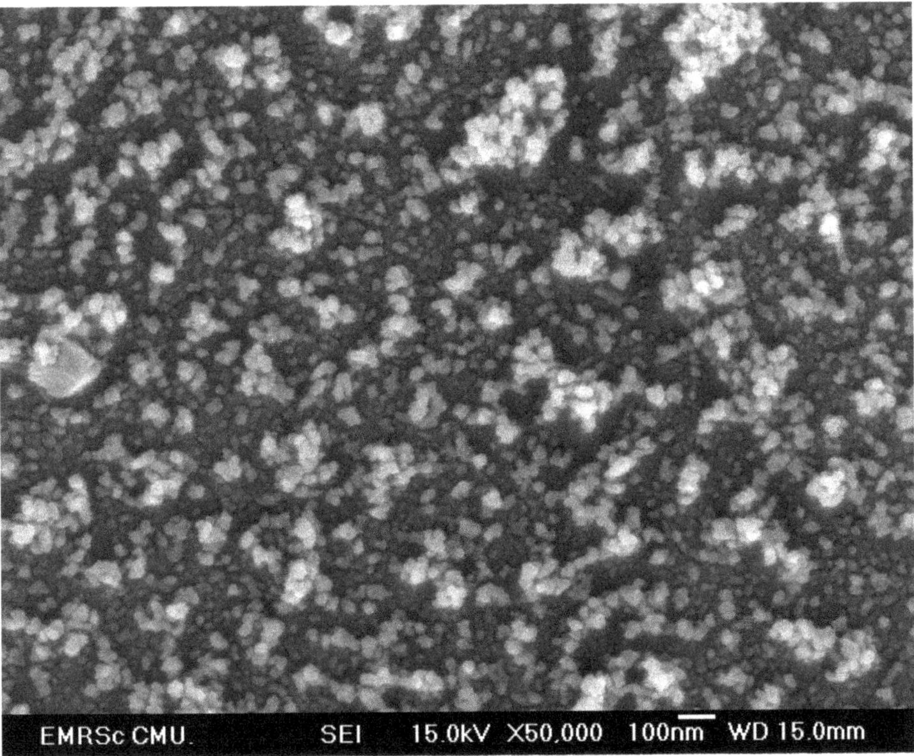

Figure 7. Scanning electron microscopy (SEM) of sparked titanium nanoparticles.

Unless wires are sparked inside of a magnetic field [21], in ambient conditions at atmospheric pressure, nanoparticles collected on the substrate during sparking discharge process will be amorphous. High energy will also break down gas molecules in the air between the sparking gap (even nitrogen) and will create different oxides, nitrides and carbonates [22]. Because of obtaining thermodynamically unstable products, annealing in controlled atmosphere will affect the crystallinity and phase of the sparked nanoparticles [23]. This is the reason why we see different phases in the Raman results of the sparked titanium.

To further explain the action of the sparking process in the air pressure atmosphere, the schematic illustrations of the nucleation of nanostructures were developed and are shown in Figure 9. When the voltage applied to the gaps is higher than the break down voltage, gas molecules in the sparking gap were ionized, and the electrons and ions produced from the neutral molecules migrated towards the anode and cathode, respectively. The bombardment of high energy electrons and ions melted and evaporated the metal tips. The expansion of the electron stream at the anode increased the bombardment area and therefore the sparked anode tips exhibited a low cone angle. For the cathode, positive ions have a larger mass and therefore transfer more energy to the cathode, resulting in a larger melting volume than the anodes. The nucleated molten nano-droplets were blown in the cone shape around the tip and deposited on the substrate which were placed under, next to, and over the gap. The reduced density and hemisphere area of the films under the tips of the anode are consequences of the reduced cone angle and lower molten volume. The growth of the nanoporous structures from the metal vapor on the sparking tips depends on the material and thermal properties and surface reactivity of the metal tips.

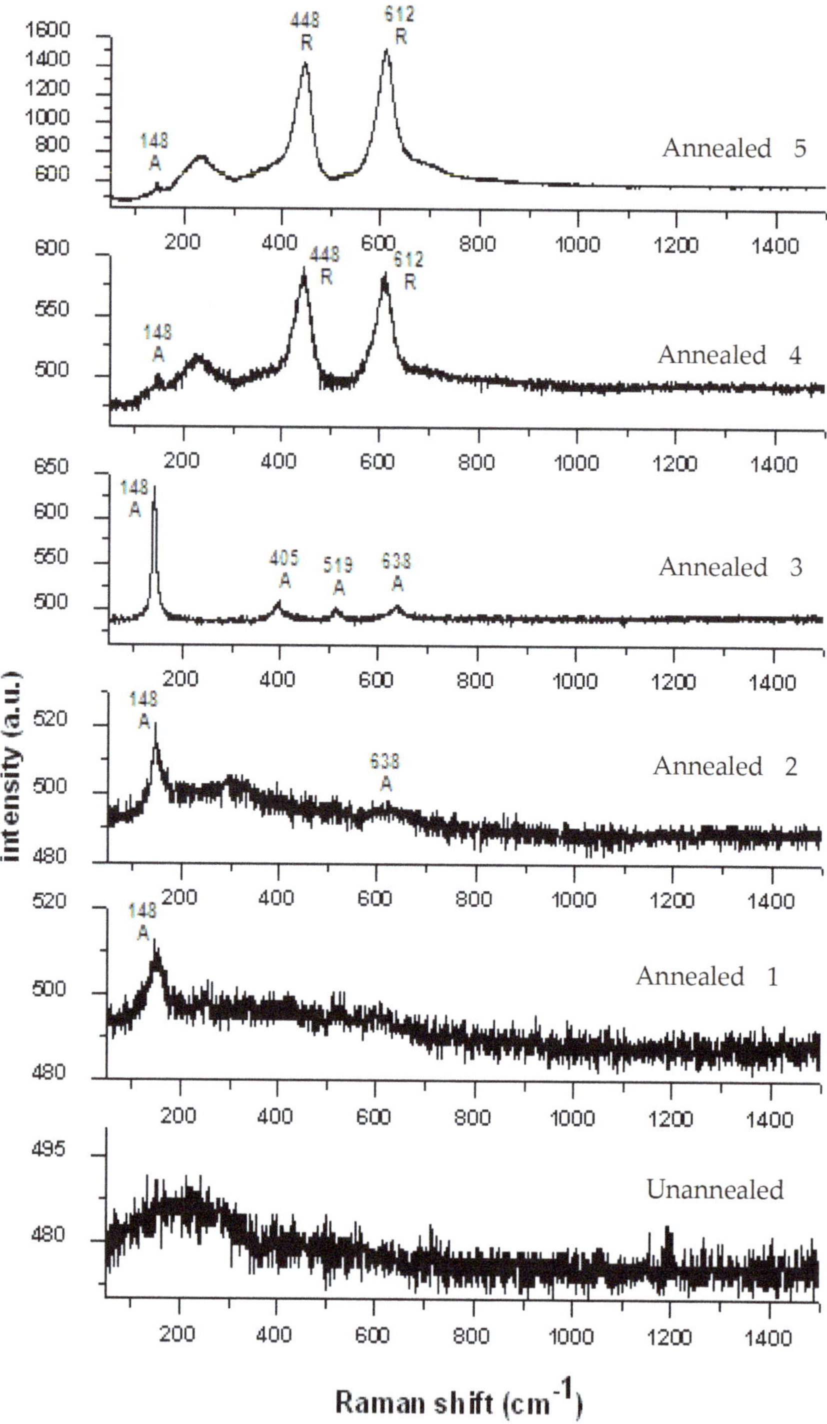

Figure 8. Raman spectra of the sparked titanium nanoparticles films annealed at different temperatures.

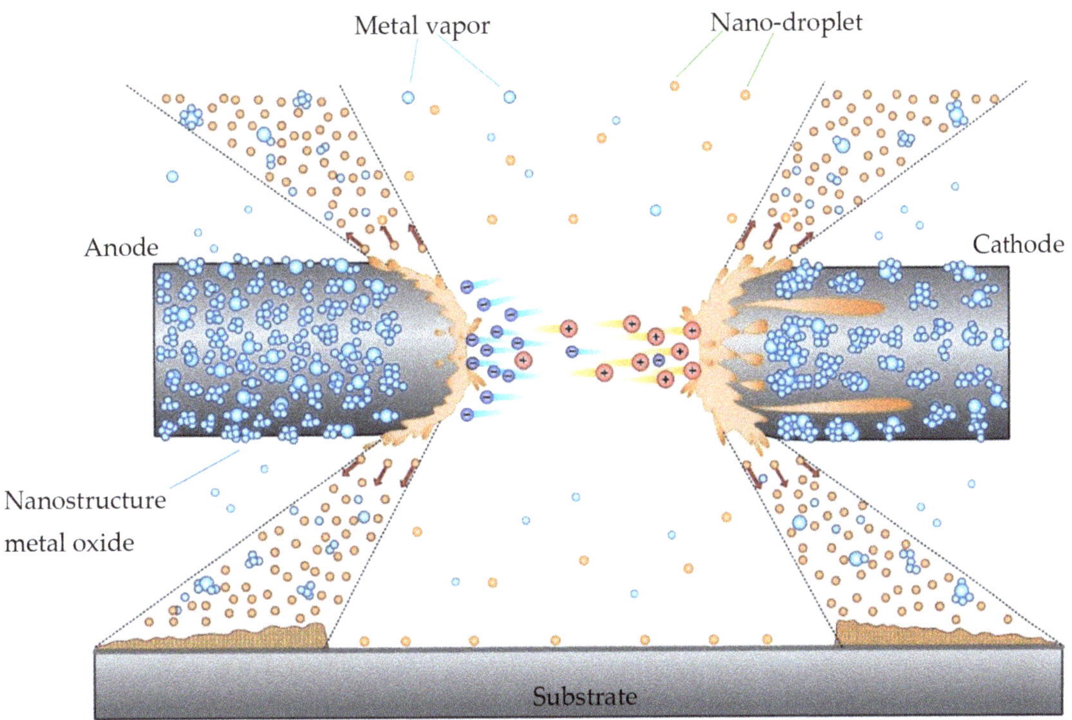

Figure 9. Schematic diagram of the nucleation mechanism of the nanostructure deposited by the sparking method.

There are a substantial number of applications involving the sparking discharge process in the field of aerosol science [24]. Even though we provide evidence in this research that most of the nanoparticles are made by the metal melt approach, the sparking discharge process can still be used for aerosol production and fabric filtration calibration applications, which are now more important than ever in light of the current COVID-19 crisis [25].

In Figure 10, the results with the filtration barrier and without the filtration barrier are both presented. The experiment was conducted as previously described, [26], using aluminum wires, with the modification of one output, in which a commercially available surgical mask as a filter is attached onto the aerosol outlet, leading to a differential mobility analyzer (DMA). The sparking discharge process suitability for the aerosol calibration was compared with the dispersing particle size standards^{TM} (from Polysciences, Inc. Nanobead NIST Traceable Particle Size Standards). The results indicate that filtration capacity of the sparking discharge is 99.6%; obtained using the following equation.

$$E_f = 100 - \frac{number\ of\ particles\ measured\ with\ DMA\ with\ filter}{number\ of\ particles\ measured\ with\ DMA\ without\ filter} \tag{2}$$

This was compared by using a 50 nm particle size standard and the results obtained are not significantly different. A 50 nm particle size was used because this was the size that was generated from the sparking aluminum wires with the machine described in our experiment.

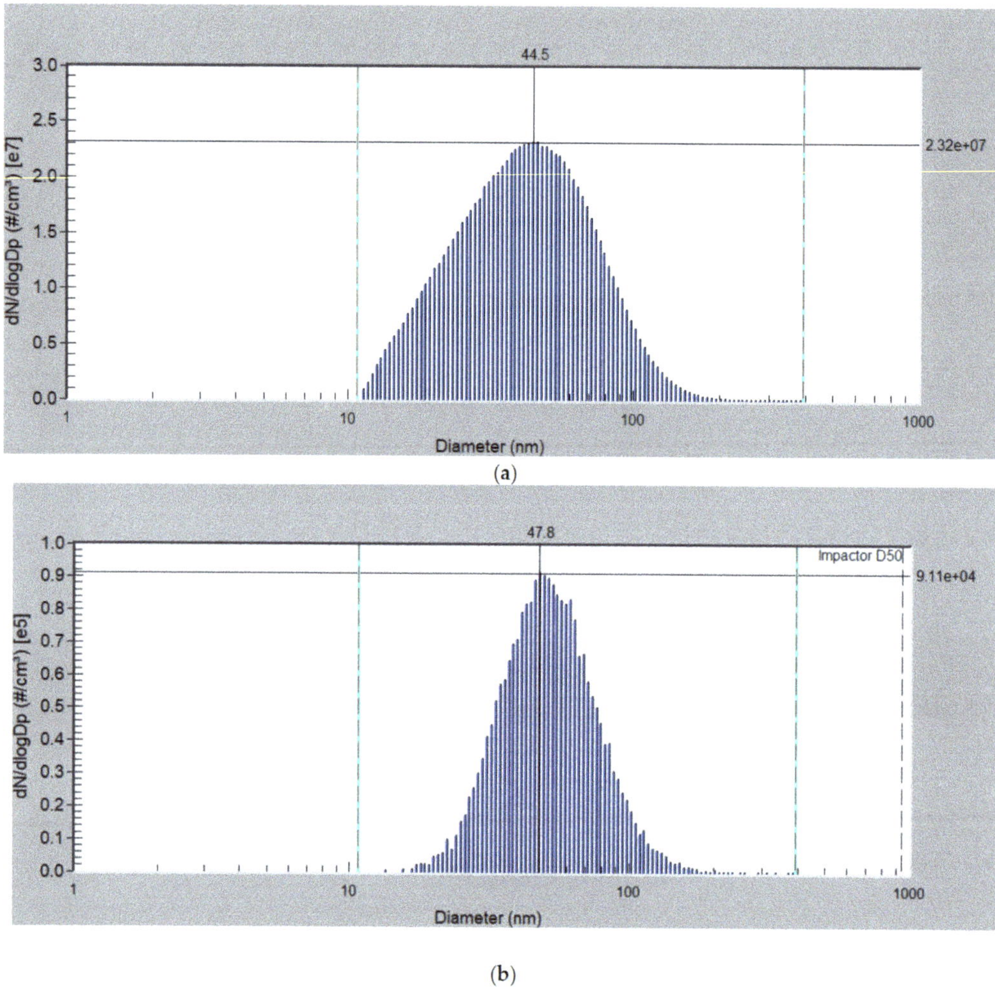

Figure 10. Filtration experiment measured by the differential mobility analyzer. (**a**) The aluminum sparked wire without a surgical mask as a filter. The concentration of nanoparticles produced by sparking was 2.32×10^7 particles at 1 cm^3 at size of 44.5 nm (**b**) Results of differential mobility analyzer (DMA) from aluminum sparked wire with a filter from a commercially available surgical mask. Concentration of nanoparticle produced by sparking was 9.11×10^4 particles at 1 cm^3 at size of 47.8 nm.

4. Conclusions

There are two approaches that are utilized in the generation of nanoparticles from a sparking discharge process. The first is an aerosol-based process [27], and the second is a metal melt process, that we used to modify the surface and create nanomaterials [28]. The sparking discharge studies by Kohut et al. [29] describe craters, undulated areas, and dendritic areas of nickel and copper electrodes (rode) in low numbers (one to three) as opposed to our study (100). Their results also support our approach to sparking discharge for nanoparticle generation trough metal melt, with only a small portion of the electrode content being aerosolized for nanoparticle synthesis. This confirms the production of nanoparticles by the sparking discharge methodology and spark discharge apparatus as mainly coming from the metal melt process.

The findings by Domaschke et al. [30] are significant because they support our exclusive use of wires as the electrode material. In their work regarding the sparking discharge process, a high capacitance of the apparatus is also expected in order to improve the energy efficiency of the sparking discharge process. They also measured the effect of the gas flow during sparking discharge on particle size and the concentration of the produced particles, which we did not deal with because we did not use the gas flow during the synthesis of nanoparticles. Our sparking discharge process was done in an ambient atmosphere and under normal pressure conditions, while metal wires were discharged with no air flow.

In summary, we have demonstrated the preparation of nanoparticles using the sparking process of various metal tips at an atmospheric air pressure. It was observed that particle sizes are related to the surface tension of molten metals. The film coated under the sparking gap was produced by molten metal droplets, whereas the nanostructure covered on the sparking tips was formed by the deposition of metal vapors. The model was created to explain the coating pattern, and this model provides a practical tool that can be used to design a sparking machine for large scale application (see Supplementary Information for examples). This sparking discharge machine can be used for the calibration of fabric filtration barriers as a replacement for dispersing particle size standards.

Supplementary Materials: The following are available online at https://www.mdpi.com/2073-4352/11/2/140/s1.

Author Contributions: T.K. wrote, analyzed, and collected the data, S.R., wrote, collected and analyzed the data, and edited the manuscript E.K., collected the data, N.J., collected and analyzed the data, W.S. collected the data, W.T., collected the data. S.P. collected the data, A.P. collected the data, P.S. (Panupong Sanmuangmoon) collected the data, P.S. (Pisith Singjai) supervised the edit and acquire funding. All authors have read and agreed to the published version of the manuscript.

Funding: Postdoctoral Fellowship of Office of Research Administration, Chiang Mai University.

Acknowledgments: This research was supported by Chiang Mai. Stefan Ručman would like to thank CMU Presidential Scholarship for Post-Doctoral Fellowship. Special thanks to Assoc. Panich Intra, Research Unit of Applied Electric Field in Engineering (RUEE), https://www.facebook.com/ruee.rmutl/ Rajamangala University of Technology Lanna for helping us and allowing us to use the differential mobility analyzer (DMA).

Conflicts of Interest: The authors declare no conflict of interest.

References

1. Hosokawa, M.; Nogi, K.; Naito, M.; Yokoyama, T. Basic properties and measuring methods of nanoparticles. In *Na-Noparticle Technology Handbook*; Masuo, H., Kiyoshi, N., Mario, N., Toyokaz, U.Y., Eds.; Elsevier: Amsterdam, The Netherlands, 2008; pp. 33–48.
2. Hamad, A.; Lin, L.; Zhu, L.; Xinag, Z.; Hong, L.; Tao, W. Generation of silver titania nanoparticles from an Ag–Ti alloy via picosecond laser ablation and their antibacterial activities. *RSC Adv.* **2015**, *5*, 72981–72994. [CrossRef]
3. Yasuda, K.; Sato, T.; Asakura, Y. Size-controlled synthesis of gold nanoparticles by ultrafine bubbles and pulsed ul-trasound. *Chem. Eng. Sci.* **2020**, *217*, 15527. [CrossRef]
4. Jung, J.H.; Kim, S.B.; Kim, S.S. Nanoparticle generation using corona discharge ions from a supersonic flow in low pressure. *Powder Technol.* **2008**, *185*, 58–66.
5. Kumpika, T.; Thongsuwan, W.; Singjai, P. Atomic force microscopy imaging of ZnO nanodots deposited on quartz by sparking off different tip shapes. *Surf. Interface Anal.* **2006**, *39*, 58–63. [CrossRef]
6. Kumpika, T.; Thongsuwan, W.; Singjai, P. Optical and electrical properties of ZnO nanoparticle thin films deposited on quartz by sparking process. *Thin Solid Films* **2008**, *516*, 5640–5644. [CrossRef]
7. Kumpika, T.; Kantarak, E.; Sroila, W.; Panthawan, A.; Jhuntama, N.; Sanmuangmoon, P.; Thongsuwan, W.; Singjai, P. Superhydrophilic/superhydrophobic surfaces fabricated by spark-coating. *Surf. Interface Anal.* **2018**, *50*, 827–834. [CrossRef]
8. Kumpika, T.; Kantarak, E.; Sroila, W.; Panthawan, A.; Sanmuangmoon, P.; Thongsuwan, W.; Singjai, P. Fabrication and composition control of porous ZnO-TiO$_2$ binary oxide thin films via a sparking method. *Optik* **2017**, *133*, 114–121. [CrossRef]
9. Tippo, P.; Thongsuwan, W.; Wiranwetchayan, O.; Kumpika, T.; Tuantranont, A.; Singjai, P. Investigation of NiO film by sparking method under a magnetic field and NiO/ZnO heterojunction. *Mater. Res. Express* **2020**, *7*, 056403. [CrossRef]
10. Thongsuwan, W.; Kumpika, T.; Singjai, P. Effect of high roughness on a long aging time of superhydrophilic TiO$_2$ nanoparticle thin films. *Curr. Appl. Phys.* **2011**, *11*, 1237–1242. [CrossRef]

11. Thongsuwan, W.; Kumpika, T.; Singjai, P. Photocatalytic property of colloidal TiO_2 nanoparticles prepared by sparking process. *Curr. Appl. Phys.* **2008**, *8*, 563–568. [CrossRef]
12. Wang, H.; Xie, C.; Zeng, D. Controlled growth of ZnO by adding H_2O. *J. Cryst. Growth* **2005**, *277*, 372–377. [CrossRef]
13. Tabrizi, N.S.; Ullmann, M.; Vons, V.A.; Lafont, U.; Schmidt-Ott, A. Generation of nanoparticles by spark discharge. *J. Nanoparticle Res.* **2009**, *11*, 315–332. [CrossRef]
14. Stefan, R.; Jakmunee, J.; Punyodom, W.; Singjai, P. A Novel Strategy for Longevity Prolongation of Iron-Based Nano-particle Thin Films by Applied Magnetic Force. *New J. Chem.* **2018**, *42*, 4807–4810.
15. Castle, J.E.; Zhdan, P.A.; Singjai, P. Enhanced morphological reconstruction of SPM images. *J. Phys. D Appl. Phys.* **1998**, *31*, 3437–3445. [CrossRef]
16. Vinet, B.; Magnusson, L.; Fredriksson, H.; Desré, P.J. Correlations between Surface and Interface Energies with Respect to Crystal Nucleation. *J. Colloid Interface Sci.* **2002**, *255*, 363–374. [CrossRef]
17. Ung, T.; Liz-Marza, L.M.; Mulvaney, P. Gold nanoparticle thin films. *Colloids Surf.* **2002**, *202*, 119–126. [CrossRef]
18. Barrufet, M.; Patel, M.; Eubank, P. Novel computations of a moving boundary heat conduction problem applied to EDM technology. *Comput. Chem. Eng.* **1991**, *15*, 609–618. [CrossRef]
19. Talib, M.; Tabassum, R.; Islam, S.S.; Mishra, P. Influence of growth temperature on titanium sulphide nanostructures: From trisulphide nanosheets and nanoribbons to disulphide nanodiscs. *RSC Adv.* **2019**, *9*, 645–657. [CrossRef]
20. Arachchige, H.M.M.M.; Zappa, D.; Poli, N.; Gunawardhana, N.; Attanayake, N.H.; Comini, E. Seed-Assisted Growth of TiO_2 Nanowires by Thermal Oxidation for Chemical Gas Sensing. *Nanomaterials* **2020**, *10*, 935. [CrossRef]
21. Ručman, S.S.; Punyodom, W.; Jakmunee, J.; Singjai, P. Inducing Crystallinity of Metal Thin Films with Weak Mag-netic Fields without Thermal Annealing. *Crystals* **2018**, *8*, 362. [CrossRef]
22. Ručman, S.; Boonruang, C.; Singjai, P. The Effect of a Weak Magnetic Field (0 T to 0.4 T) on the Valence Band and Intramolecular Hydrogen of Inorganic Aerosol Metal–Nitrogen Gas Chemical Reactions in a Sparking Discharge Process. *Crystals* **2020**, *10*, 1141. [CrossRef]
23. Hallberg, R.T.; Ludvigsson, L.; Preger, C.; Meuller, B.O.; Dick, K.A.; Messing, M.E. Hydrogen-Assisted Spark Dis-charge Generated Metal Nanoparticles to Prevent Oxide Formation. *Aerosol Sci. Technol.* **2017**, *52*, 347–358. [CrossRef]
24. Meuller, B.O.; Messing, M.E.; Engberg, D.L.J.; Jansson, A.M.; Johansson, L.I.M.; Norlén, S.M.; Tureson, N.; Deppert, K. Review of Spark Discharge Generators for Production of Nanoparticle Aerosols. *Aerosol Sci. Technol.* **2012**, *46*, 1256–1270. [CrossRef]
25. Zangmeister, C.D.; Radney, J.G.; Vicenzi, E.P.; Weaver, J.L. Filtration Efficiencies of Nanoscale Aerosol by Cloth Mask Materials Used to Slow the Spread of SARS-CoV2. *ACS Nano.* **2020**, *14*, 9188–9200. [CrossRef] [PubMed]
26. Ručman, S.; Intra, P.; Kantarak, E.; Sroila, W.; Kumpika, T.; Jakmunee, J.; Punyodom, W.; Arsić, B.; Singjai, P. Influence of the Magnetic Field on Bandgap and Chemical Composition of Zinc Thin Films Prepared by Sparking Discharge Process. *Sci. Rep.* **2020**, *10*, 1–11.
27. *Spark Ablation*; CRC Press: Boca Raton, FL, USA, 2019.
28. Pooseekheaw, P.; Thongpan, W.; Panthawan, A.; Kantarak, E.; Sroila, W.; Singjai, P. Porous V_2O_5/TiO_2 Nanohetero-structure Films with Enhanced Visible-Light Photocatalytic Performance Prepared by the Sparking Method. *Molecules* **2020**, *25*, 3327. [CrossRef] [PubMed]
29. Kohut, A.; Wagner, M.; Seipenbusch, M.; Geretovszky, Z.; Galbács, G. Surface features and energy considerations related to the erosion processes of Cu and Ni electrodes in a spark discharge nanoparticle generator. *J. Aerosol Sci.* **2018**, *119*, 51–61. [CrossRef]
30. Domaschke, M.; Schmidt, M.; Peukert, W. A model for the particle mass yield in the aerosol synthesis of ultrafine monometallic nanoparticles by spark ablation. *J. Aerosol Sci.* **2018**, *126*, 133–142. [CrossRef]

Article

Computational Approach to Dynamic Systems through Similarity Measure and Homotopy Analysis Method for Renewable Energy

Noor Saeed Khan [1,2,3,*], **Poom Kumam** [1,2,4,*] **and Phatiphat Thounthong** [5]

[1] KMUTTFixed Point Research Laboratory, Room SCL 802 Fixed Point Laboratory, Science Laboratory Building, Department of Mathematics, Faculty of Science, King Mongkut's University of Technology Thonburi (KMUTT), Bangkok 10140, Thailand

[2] Center of Excellence in Theoretical and Computational Science (TaCS-CoE), Science Laboratory Building, Faculty of Science, King Mongkut's University of Technology Thonburi (KMUTT), 126 Pracha-Uthit Road, Bang Mod, Thrung Khru, Bangkok 10140, Thailand

[3] Department of Mathematics, Division of Science and Technology, University of Education, Lahore 54000, Pakistan

[4] Department of Medical Research, China Medical University Hospital, China Medical University, Taichung 40402, Taiwan

[5] Renewable Energy Research Centre, Department of Teacher Training in Electrical Engineering, Faculty of Technical Education, King Mongkut's University of Technology North Bangkok, 1518 Pracharat 1 Road, Wongsawang, Bangsue, Bangkok 10800, Thailand; phatiphat.t@ieee.org

* Correspondence: noor.saeed@ue.edu.pk (N.S.K.); poom.kum@kmutt.ac.th (P.K.)

Received: 6 October 2020; Accepted: 10 November 2020; Published: 27 November 2020

Abstract: To achieve considerably high thermal conductivity, hybrid nanofluids are some of the best alternatives that can be considered as renewable energy resources and as replacements for the traditional ways of heat transfer through fluids. The subject of the present work is to probe the heat and mass transfer flow of an ethylene glycol based hybrid nanofluid (Au-ZnO/$C_2H_6O_2$) in three dimensions with homogeneous-heterogeneous chemical reactions and the nanoparticle shape factor. The applications of appropriate similarity transformations are done to make the corresponding non-dimensional equations, which are used in the analytic computation through the homotopy analysis method (HAM). Graphical representations are shown for the behaviors of the parameters and profiles. The hybrid nanofluid (Au-ZnO/$C_2H_6O_2$) has a great influence on the flow, temperature, and cubic autocatalysis chemical reactions. The axial velocity and the heat transfer increase and the concentration of the cubic autocatalytic chemical reactions decreases with increasing stretching parameters. The tangential velocity and the concentration of cubic autocatalytic chemical reactions decrease and the heat transfer increases with increasing Reynolds number. A close agreement of the present work with the published study is achieved.

Keywords: Au-ZnO/$C_2H_6O_2$; heat transfer; rotating systems; analytical solution

---

1. Introduction

Energy has a crucial role in the prosperity and development of any country. The daily consumed energy resources like natural gas, oil, and coal are certain to vanish with the passage of time because these are huge sources of energy and are being depleted due to their limited availability. To cope with such a situation, the replenishment of the world's energy is of utmost concern, making it is a basic requirement to search for some reliable and affordable energy alternatives. Such problems apply to renewable energy systems. Nanoparticles have been shown to solve such constraints because of their remarkable heat transfer capabilities. The application of nanoparticles in the industrial, biomedical, and energy sectors is due to their thermophysical properties. Nanoparticles have seen applications in energy conversion (e.g., fuel cells, solar cells, and thermoelectric devices), energy storage (e.g., rechargeable batteries and super capacitors), and energy saving (e.g., insulation such as aerogels and smart glazes, efficient lightning like light emitting diodes and organic light-emitting diodes). To combat climate change, clean and sustainable energy sources need to be rapidly developed. Solar energy technology converts solar energy directly into electricity, for which high performance cooling, heating, and electricity generation are among the inevitable requirements. In solar collectors, the absorbed incident solar radiation is converted to heat. The working fluid conveys the generated heat for different uses [1]. Ettefaghi et al. [2] worked on a bio-nanoemulsion fuel based on biodegradable nanoparticles to improve diesel engines' performance and reduce exhaust emissions. Gunjo et al. [3] investigated the melting enhancement of a latent heat storage with dispersed Cu, CuO, and Al_2O_3 nanoparticles for a solar thermal application. Khanafer and Vafai [4] presented a review on the applications of nanofluids in the solar energy field.

Nanofluids reduce the process time, enhance the heating rates, and improve the lifespan of machinery [5]. Nanofluids have seen applications in power saving, manufacturing, transportation, healthcare, microfluidics, nano-technology, microelectronics, etc. Recently, nano-technology has attracted great attraction from scientists [6]. Nanoparticles are the most interesting technology to introduce novel, environmentally friendly chemical and mechanical polishing slurries to fabricate effective materials [7]. Thermal conductivity is of great importance and is enhanced by the incorporation of nanoparticles in the base fluid [8]. Hamilton and Crosser [9] studied the thermal conductivity of a heterogeneous two component system. Nanofluids were obtained by the addition of nanoparticles to the base fluids, and they have gained popularity since the work of Choi and Eastman [10]. Vallejo et al. [11] analyzed the internal aspects of the fluid for six carbon-based nanomaterials in a rotating rheometer with a double conic shape containing a typical sheet. Alihosseini and Jafari [12] investigated a three-dimensional computational fluid dynamics model for an aluminum foam and nanoparticles with heat transfer using a number of cylinders having different configurations through a permeable medium. Sheikholeslami et al. [13], working with a ethylene glycol nanofluid, discussed the electric field, thermal radiation, and nanoparticle shape factors of a ferrofluid by showing that the platelet shape led to enhanced convective flow. Al-Kouz et al. [14] applied computational fluid dynamics to analyze entropy generation in a rarefied time dependent, laminar two-dimensional flow of an air-aluminum oxide nanofluid in a cavity with a square shape having more than one solid fin at the heated wall where the optimization procedure was adopted to show the conditions by which the overall entropy generation was reduced. Atta et al. [15] modified the asphaltenes isolated from crude oil to work as capping agents for the synthesis of hydrophobic silica to investigate the surface charge of hydrophobic silica nanoparticles, the chemical structure, the particle size, and the surface morphology. Rout et al. [16] presented the three and higher order nonlinear thin film study and optics fabricated with gold nanoparticles. They obtained the solution via spin-coating techniques to achieve the highest values of nonlinear absorption coefficient, nonlinear refractive index and saturation intensity. Alvarez-Regueiro et al. [17] experimentally determined the heat transfer coefficients and pressure drops of four functionalized graphene nanoplatelet nanofluids for heat transfer enhancement to discuss the nanoadditive loading, temperature and Reynolds number. Alsagri et al. [18] elaborated the heat and mass transfer flow of single walled and multi walled carbon nanotubes past a stretchable cylinder by investigating that the heat transfer enhances with the high values of nanoparticles concentration of single walled carbon nanotubes compared to that of multi walled carbon nanotubes. Working on transverse vibration, Mishra et al. [19] comparatively investigated a computational fluid dynamic model for water based nanofluid through a pipe subject to superimposed vibration, applied to the wall to increase the heat transfer in axial direction while vibration effect is decreased for pure liquid and is increased for nanofluid. Abbas et al. [20] achieved the results that in the heat and mass transfer flow of

Cross nanofluid, the Bejan number was intensified for the high values of thermal radiation parameter. Some discussion on nanofluids and other relevant studies can be found in the references [21–55].

Mono-nanofluids represent enhanced thermal conductivity and good rheological characteristics, but still they have some weak characteristics necessary for a particular purpose. By the hybridization process, different nanoparticles are added in a base fluid to make the hybrid nanofluid which has enhanced thermophysical properties and thermal conductivity as well as rheological properties. Ahmad et al. [56] investigated the hybrid nanofluid with activation energy and binary chemical reaction through a moving wedge taken into account the Darcy law of porous medium, heat generation, thermal slip, radiation, and variable viscosity. Dinarvand and Rostami [57] presented the ZnO-Au hybrid nanofluid when 15 gm of nanoparticles are added into the 100 gm base fluid, the heat transfer enhances more than 40% compared to that of the regular fluid.

Homogeneous-heterogeneous chemical reactions have important applications in chemical industries. Ahmad and Xu [58] worked on homogeneous-heterogeneous chemical reactions in which the reactive species were of regular size reacting with other species in a nanofluid to show more realistic mathematical model physically. Hayat et al. [59] elaborated the Xue nanofluid model to study the carbon nanotubes nanofluids in rotating systems incorporating Darcy–Forchheimer law, homogeneous-heterogeneous chemical reactions and optimal series solutions. Suleman et al. [60] addressed the homogeneous-heterogeneous chemical reactions in Ag-H_2O nanofluid flow past a stretching sheet with Newtonian heating to prove that concentration field was decreased for the increasing strength of homogeneous-heterogeneous chemical reactions.

In the literature, interesting studies exists like [5] which investigates the electrical conductivity, structural and optical properties of ZnO. In study [6], the theoretical and experimental results of electric current and thermal conductivity of H_2O-ethylene glycol based TiO_2 have been obtained. The study [7] relates to the oxide-ethylene glycol nanofluid with different sizes of nanoparticles. Due to the applications of the above studies, it is desire to investigate the ethylene glycol based Au-ZnO hybrid nanofluid flow with heat transfer and homogeneous-heterogeneous chemical reactions in rotating system. The present study has the applications in renewable energy technology, thermal power generating system, spin coating, turbo machinery etc. The solution of the problem is obtained through an effective technique known as homotopy analysis method [61]. Investigations are shown through graphs and discussed in detail.

2. Methods

A rotating flow of hydromagnetic, time independent and incompressible hybrid nanofluid between two parallel disks in three dimensions is analyzed. Homogeneous-heterogeneous chemical reactions are also considered. The lower disk is supposed to locate at $z = 0$ while the upper disk is at a constant distance H apart. The velocities and stretching on these disks are (Ω_1, Ω_2) and (a_1, a_2), respectively while the temperatures on these disks are T_1 and T_2, respectively. A magnetic field of strength B_0 is applied in the direction of z-axis (please see Figure 1). Ethylene glycol is chosen for the base fluid in which zinc oxide and gold nanoparticles are added.

For cubic auto-catalysis, the homogeneous reaction is

$$2C + B \rightarrow 3C, \quad rate = c^2 bk_c. \tag{1}$$

The first order isothermal reaction on the surface of catalyst is

$$B \rightarrow C, \quad rate = bk_s, \tag{2}$$

where B and C denote the chemical species with concentrations b and c, respectively. k_c and k_s are the rate constants.

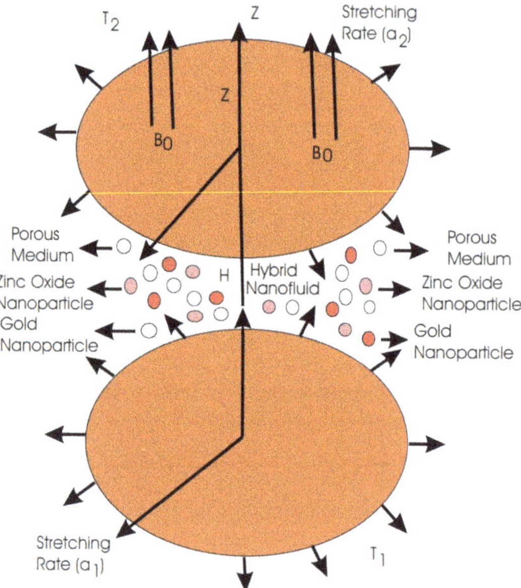

Figure 1. Geometry of the problem.

Cylindrical coordinates (r, ϑ, z), are applied to provide the thermodynamics of hybrid nanofluid as [57–59]

$$\frac{\partial w}{\partial z} + \frac{\partial u}{\partial r} + \frac{u}{r} = 0, \tag{3}$$

$$\rho_{hnf}\left(-\frac{v^2}{r} + \frac{\partial u}{\partial r}u + \frac{\partial u}{\partial z}w\right) = \mu_{hnf}\left(\frac{\partial^2 u}{\partial z^2} + \frac{\partial^2 u}{\partial r^2} - \frac{u}{r^2} + \frac{\partial u}{\partial r}\frac{1}{r}\right) - \sigma_{hnf}B_0^2 u - \frac{\mu_{hnf}}{S}u^2 - S_1 u^2 - \frac{\partial P}{\partial r}, \tag{4}$$

$$\rho_{hnf}\left(\frac{uv}{r} + w\frac{\partial v}{\partial z} + u\frac{\partial v}{\partial r}\right) = \mu_{hnf}\left(\frac{\partial^2 v}{\partial z^2} + \frac{\partial^2 v}{\partial r^2} - \frac{v}{r^2} + \frac{1}{r}\frac{\partial v}{\partial r}\right) - \sigma_{hnf}B_0^2 v - \frac{\mu_{hnf}}{S}v^2 - S_1 v^2, \tag{5}$$

$$\rho_{hnf}\left(w\frac{\partial w}{\partial z} + u\frac{\partial w}{\partial r}\right) = -\frac{\partial P}{\partial z} + \mu_{hnf}\left(\frac{\partial^2 w}{\partial z^2} + \frac{\partial^2 w}{\partial r^2} + \frac{1}{r}\frac{\partial w}{\partial r}\right) - \frac{\mu_{hnf}}{S}w^2 - S_1 w^2, \tag{6}$$

$$(\rho c_p)_{hnf}\left(w\frac{\partial T}{\partial z} + u\frac{\partial T}{\partial r}\right) = \left(k_{hnf} + \frac{16T_1^3\sigma_1}{3k_0}\right)\left(\frac{\partial^2 T}{\partial z^2} + \frac{\partial^2 T}{\partial r^2} + \frac{1}{r}\frac{\partial T}{\partial r}\right) + \sigma_{hnf}B_0^2(v^2 + u^2), \tag{7}$$

$$w\frac{\partial b}{\partial z} + u\frac{\partial b}{\partial r} = -c^2 b k_c + D_B\left(\frac{\partial^2 b}{\partial z^2} + \frac{\partial^2 b}{\partial r^2} + \frac{1}{r}\frac{\partial b}{\partial r}\right), \tag{8}$$

$$w\frac{\partial c}{\partial z} + u\frac{\partial c}{\partial r} = c^2 b k_c + D_C\left(\frac{\partial^2 c}{\partial z^2} + \frac{\partial^2 c}{\partial r^2} + \frac{1}{r}\frac{\partial c}{\partial r}\right). \tag{9}$$

The boundary conditions are

$$at \quad z = 0, \quad D_C\frac{\partial c}{\partial z} = k_s c, \quad D_B\frac{\partial b}{\partial z} = k_s b, \quad T = T_1, \quad w = 0, \quad v = r\Omega_1, \quad u = ra_1, \tag{10}$$

$$at \quad z = H, \quad c \to 0, \quad b \to b_0, \quad T = T_2, \quad P \to \infty, \quad w = 0, \quad v = r\Omega_2, \quad u = ra_2, \tag{11}$$

where $u(r, \vartheta, z)$, $v(r, \vartheta, z)$ and $w(r, \vartheta, z)$ are the components of velocity, P is the pressure. S is the permeability of porous medium, $S_1 = \frac{C_b}{rS^{\frac{1}{2}}}$ is the non-uniform inertia coefficient of porous medium with C_b as the drag coefficient. Temperature of hybrid nanofluid is T and $B = (0, 0, B_0)$ is the magnetic field. σ_1 is the Stefan Boltzmann constant and k_0 is the absorption coefficient. For the hybrid nanofluid, the important quantities are ρ_{hnf} (density), μ_{hnf} (dynamic viscosity), σ_{hnf} (electrical conductivity), $(c_p)_{hnf}$ (heat capacity) and k_{hnf} (thermal conductivity). The subscript "hnf" shows the hybrid nanofluid. For the thermal conductivity, the mathematical formulation is obtained via Hamilton–Crosser model [9] as

$$\frac{k_{nf}}{k_f} = \frac{k_1 + (n_1 - 1)k_f - (n_1 - 1)(k_f - k_1)\phi_1}{k_1 + (n_1 - 1)k_f + (k_f - k_1)\phi_1}, \tag{12}$$

where n is the empirical shape factor for the nanoparticle whose value is given in Table 1.

Table 1. Values of shape factor of different shapes of nanoparticles.

Shapes of Nanoparticle	n	Aspect Ratio
Spherical	3	-
Brick	3.7	1:1:1
Cylinder	4.8	1:8
Platelet	5.7	1:1/8

The subscript "f" denotes the base fluid namely ethylene glycol and the subscript "nf" is used for nanofluid. ρ_s and $(c_P)_s$ are the density and heat capacity at specified pressure of nanoparticles, respectively. ϕ_1 is the first nanoparticle volume fraction while ϕ_2 is the second nanoparticle volume fraction which can be formulated as [57].

$$\rho_s = \frac{(\rho_1 \times m_1) + (\rho_2 \times m_2)}{m_1 + m_2}, \tag{13}$$

$$(c_P)_s = \frac{((c_P)_1 \times m_1) + ((c_P)_2 \times m_2)}{m_1 + m_2}, \tag{14}$$

$$\phi_1 = \frac{\frac{m_1}{\rho_1}}{\frac{m_1}{\rho_1} + \frac{m_2}{\rho_2} + \frac{m_f}{\rho_f}}, \tag{15}$$

$$\phi_2 = \frac{\frac{m_2}{\rho_2}}{\frac{m_1}{\rho_1} + \frac{m_2}{\rho_2} + \frac{m_f}{\rho_f}}, \tag{16}$$

$$\phi = \phi_1 + \phi_2, \tag{17}$$

where m_1, m_2 and m_f are, respectively the mass of first nanoparticle, mass of the second nanoparticle and mass of the base fluid. ϕ is the total volume fraction of zinc oxide and gold nanoparticles.

The thermophysical properties of $C_2H_6O_2$ as well as nanoparticles are given in Table 2.

Table 2. Thermophysical properties of ethylene glycol and nanoparticles.

Properties	Ethylene Glycol ($C_2H_6O_2$)	Zinc Oxide (ZnO)	Gold (Au)
ρ (kg/m^3)	$\rho_f = 116.6$	$\rho_{s_1} = 5600$	$\rho_{s_2} = 19{,}282$
c_P (J/kg K)	$(c_P)_f = 2382$	$(c_P)_{s_1} = 495.2$	$(c_P)_{s_2} = 192$
k (W/m K)	$k_f = 0.249$	$k_{s_1} = 13$	$k_{s_2} = 310$
σ (Um)$^{-1}$	$\sigma_f = 3.14$	$\sigma_{s_1} = 7.261 \times 10^{-9}$	$\sigma_{s_2} = 4.11 \times 10^7$
Nanoparticle measurement/nm	-	29 and 77	3–40

The mathematical formulations for ρ_{hnf} (density), μ_{hnf} (dynamic viscosity), σ_{hnf} (electrical conductivity), $(c_P)_{hnf}$ (heat capacity) are given in Table 3 where ϕ_s shows the particle concentration.

Table 3. Mathematical forms of thermophysical properties.

Properties	$ZnO/C_2H_6O_2$
Density (ρ)	$\rho_{nf} = (1 - \phi_1)\rho_f + \phi_1\rho_s$
Heat capacity (ρc_P)	$(\rho c_P)_{nf} = (1 - \phi_1)(\rho c_P)_f + \phi_1(\rho c_P)_s$
Dynamic viscosity (μ)	$\dfrac{\mu_{nf}}{\mu_f} = \dfrac{1}{(1-\phi_1)^{2.5}}$
Thermal conductivity (k)	$\dfrac{k_{nf}}{k_f} = \dfrac{k_1+(n_1-1)k_f-(n_1-1)(k_f-k_1)\phi_1}{k_1+(n_1-1)k_f+(k_f-k_s)\phi_1}$
Electrical conductivity (σ)	$\dfrac{\sigma_{nf}}{\sigma_f} = 1 + \dfrac{3(\sigma-1)\phi_1}{(\sigma+2)-(\sigma-1)\phi_1}$, where $\sigma = \dfrac{\sigma_s}{\sigma_f}$
Properties	Hybrid nanofluid (Au-$ZnO/C_2H_6O_2$)
Density (ρ)	$\rho_{hnf} = (1 - (\phi_1 + \phi_2))\rho_f + \phi_1\rho_{s_1} + \phi_2\rho_{s_2}$
Heat capacity (ρc_P)	$(\rho c_P)_{nf} = (1 - (\phi_1 + \phi_2))(\rho c_P)_f + \phi_1(\rho c_P)_{s_1} + \phi_2(\rho c_P)_{s_2}$
Dynamic viscosity (μ)	$\dfrac{\mu_{hnf}}{\mu_f} = \dfrac{1}{\left[1 - (\phi_1 + \phi_2)\right]^{2.5}}$
Thermal conductivity (k)	$\dfrac{k_{hnf}}{k_f} = \dfrac{k_2+(n_2-1)k_{nf}-(n_2-1)(k_{nf}-k_2)\phi_2}{k_2+(n_2-1)k_{nf}+(k_{nf}-k_2)\phi_2} \times \dfrac{k_1+(n_1-1)k_f-(n_1-1)(k_f-k_1)\phi_1}{k_1+(n_1-1)k_f+(k_f-k_1)\phi_1} \times k_f$
Electrical conductivity (σ_{hnf})	$\dfrac{\sigma_{hnf}}{\sigma_f} = 1 + \dfrac{3\left[\frac{\sigma_1\phi_1+\sigma_2\phi_2}{\sigma_f} - (\phi_1 + \phi_2)\right]}{2+\left[\frac{\sigma_1\phi_1+\sigma_2\phi_2}{(\phi_1+\phi_2)\sigma_f}\right] - \left[\frac{\sigma_1\phi_1+\sigma_2\phi_2}{\sigma_f} - (\phi_1 + \phi_2)\right]}$

Following transformations are used

$$f(\zeta)\Omega_1 r = u, \quad v = r\Omega_1 g(\zeta), \quad -2f(\zeta)H\Omega_1 = w, \quad \frac{-T_2 + T}{-T_2 + T_1} = \theta(\zeta), \quad \Omega_1\rho_f\nu_f\left(\frac{\epsilon r^2}{2H^2} + P(\zeta)\right) = P,$$

$$\varphi b_0 = b, \quad c = b_0\varphi_1, \quad \frac{z}{H} = \zeta, \quad (18)$$

where $\nu_f = \frac{\mu_f}{\rho_f}$ is the kinematic viscosity and ϵ is the pressure parameter.

Using the values from Equation (18) in Equations (4)–(11), the following eight Equations (19)–(26) are obtained

$$B_1 f''' + Re\left(2ff'' - f'^2 + g^2 - MB_2 f'\right) - \epsilon - k_2 Re B_1 f' - k_3 Re\frac{1}{\rho_{hnf}}(f')^2 = 0, \tag{19}$$

$$B_1 g'' + Re\left(2fg' - MB_2 g'\right) - k_2 Re B_1 g - k_3 Re\frac{1}{\rho_{hnf}}(g)^2 = 0, \tag{20}$$

$$P' = \frac{2}{k_2} - 4Reff' - f'', \tag{21}$$

$$B_3\frac{k_{hnf}}{k_f}\theta'' + \frac{1}{Rd}PrRe\left[2f\theta' + MEcB_4\left(g^2 + (f')^2\right)\right] = 0, \tag{22}$$

$$ScRe\left(2\varphi'f - k_4\varphi\varphi_1^2\right) + \varphi'' = 0, \tag{23}$$

$$\varphi_1'' + ScRe\left(2\varphi_1'f + k_4\varphi\varphi_1^2\right)\frac{1}{k_5} = 0, \tag{24}$$

$$f = 0, \quad f' = k_6, \quad g = 1, \quad \theta = 1, \quad \varphi' = k_7\varphi, \quad k_4\varphi_1' = -k_7\varphi, \quad P = 0 \quad at \quad \zeta = 0, \tag{25}$$

$$f = 0, \quad f' = k_8, \quad g = \Omega, \quad \theta = 0, \quad \varphi = 1, \quad \varphi_1 = 0 \quad at \quad \zeta = 1, \tag{26}$$

where $(')$ represents the derivative with respect to ζ. $B_1 = \left[1 - \dfrac{\frac{m_1}{\rho_1}}{\frac{m_1}{\rho_1} + \frac{m_2}{\rho_2} + \frac{m_f}{\rho_f}}\right]^{-2.5} \times$

$$\left[1 - \dfrac{\frac{m_1}{\rho_1}}{\frac{m_1}{\rho_1} + \frac{m_2}{\rho_2} + \frac{m_f}{\rho_f}} + \dfrac{\frac{m_1}{\rho_1}}{\frac{m_1}{\rho_1} + \frac{m_2}{\rho_2} + \frac{m_f}{\rho_f}} \dfrac{\rho_s}{\rho_f}\right]^{-1}, \qquad B_2 = 1 + \dfrac{3\left[\frac{\sigma_1\phi_1 + \sigma_2\phi_2}{\sigma_f} - (\phi_1 + \phi_2)\right]}{2 + \left[\frac{\sigma_1\phi_1 + \sigma_2\phi_2}{(\phi_1 + \phi_2)\sigma_f}\right] - \left[\frac{\sigma_1\phi_1 + \sigma_2\phi_2}{\sigma_f} - (\phi_1 + \phi_2)\right]},$$

$$B_3 = \dfrac{(\rho c_P)_f}{\left[\left[1 - \frac{\frac{m_1}{\rho_1}}{\frac{m_1}{\rho_1} + \frac{m_2}{\rho_2} + \frac{m_f}{\rho_f}}\right]\rho_f + \left[1 - \frac{\frac{m_1}{\rho_1}}{\frac{m_1}{\rho_1} + \frac{m_2}{\rho_2} + \frac{m_f}{\rho_f}}\right]\rho_s\right] \times \left[\left[1 - \frac{\frac{m_1}{\rho_1}}{\frac{m_1}{\rho_1} + \frac{m_2}{\rho_2} + \frac{m_f}{\rho_f}}\right](c_P)_f + \left[1 - \frac{\frac{m_1}{\rho_1}}{\frac{m_1}{\rho_1} + \frac{m_2}{\rho_2} + \frac{m_f}{\rho_f}}\right](c_P)_s\right]},$$

$B_4 = \dfrac{\sigma_{hnf}}{\rho_{hnf}}$. $k_2 = \dfrac{v_f}{S\Omega_1}$ is the porosity parameter, $k_3 = \dfrac{C_b}{S^{\frac{1}{2}}}$ is the inertial parameter due to Darcy Forchheimer

effect. The other non-dimensional parameters are $\Omega = \frac{\Omega_2}{\Omega_1}$, $Re = \frac{\Omega_1 H^2}{v_f}$, $M = \frac{\sigma_f B_0^2}{\rho_f \Omega_1}$, $Rd = \frac{16\sigma_1 T_1^3}{3k_f k_0}$, $Pr = \frac{(\rho c_P)_{hnf} v_f}{k_f}$, $Ec = \frac{r^2 \Omega_1^2}{c_P(T_1 - T_2)}$, $Sc = \frac{v_f}{D_B}$, $k_4 = \frac{k_c b_0^2}{\Omega_1}$, $k_5 = \frac{D_C}{D_B}$, $k_6 = \frac{a_1}{\Omega_1}$, $k_7 = \frac{k_s H}{D_B}$ and $k_8 = \frac{a_2}{\Omega_1}$ which are known as rotation parameter, Reynolds number, magnetic field parameter, thermal radiation parameter, Prandtl number, Eckert number, Schmidt number, homogeneous chemical reaction parameter, diffusion coefficient ratio, stretching parameter for lower disks, heterogeneous chemical reaction parameter and stretching parameter at upper disk, respectively.

Regarding the homogeneous-heterogeneous chemical reaction, the quantities B and C may be considered in a special case, i.e., if D_B is equal to D_C, then in such a case k_5 equals unity, which leads to

$$1 = \varphi_1(\zeta) + \varphi(\zeta). \tag{27}$$

Using Equation (27), Equations (23) and (24) generate

$$0 = ScRe\left[(1 - \varphi)^2 k_4\varphi + 2\varphi'f\right] + \varphi'', \tag{28}$$

whose corresponding boundary conditions become

$$\varphi' = k_7\varphi \quad for \quad \zeta = 0 \quad while \quad \varphi = 1 \quad for \quad \zeta = 1. \tag{29}$$

By taking derivative of Equation (19) with respect to ζ, it becomes

$$B_1 f''' + Re\left(2ff''' + 2gg' - MB_2 f''\right) - k_2 Re B_1 f'' - 2k_3 Re \frac{1}{\rho_{hnf}} ff'' = 0. \tag{30}$$

Considering Equation (21), Equations (25) and (26), the quantity ϵ is computed as

$$\epsilon = f''(0) - Re\left[-\left(g(0)\right)^2 + \left(f'(0)\right)^2 + MB_2 f'(0) + \frac{1}{B_1 k_2 f'(0)}\right]. \tag{31}$$

Integrating Equation (21) with respect to ζ by using the limit 0 to ζ for evaluating P as

$$P = -2\left[Re\left((f)^2 + \frac{1}{k_2}\int_0^\zeta f\right)\left(f' - f'(0)\right)\right]. \tag{32}$$

Skin Frictions and Nusselt Numbers

The important physical quantities are defined as

$$C_{f_1}(Local\ skin\ friction\ at\ lower\ disk) = \frac{\tau|_{z=0}}{\rho_{hnf}(r\Omega_1)^2},$$

$$C_{f_2}(Local\ skin\ friction\ at\ upper\ disk) = \frac{\tau|_{z=H}}{\rho_{hnf}(r\Omega_1)^2}, \tag{33}$$

where

$$\tau = \sqrt{(\tau_{zr})^2 + (\tau_{z\theta})^2},$$ (34)

denotes the sum of shear stress of tangential forces τ_{zr} and $\tau_{z\theta}$ along radial and tangential directions which are defined as

$$\tau_{zr} \ (Shear \ stress \ friction \ at \ lower \ disk) \ = \mu_{hnf}\frac{\partial u}{\partial z}|_{z=0} = \frac{\mu_{hnf}r\Omega_1 f''(0)}{H} \ and$$

$$\tau_{z\theta} = \mu_{hnf}\frac{\partial v}{\partial z}|_{z=0} = \frac{\mu_{hnf}r\Omega_1 g'(0)}{H}.$$ (35)

Using the information of Equations (34) and (35), Equation (33) proceeds to

$$C_{f_1} = \frac{1}{Re_r}\left[1 - \frac{\frac{m_1}{\rho_1}}{\frac{m_1}{\rho_1} + \frac{m_2}{\rho_2} + \frac{m_f}{\rho_f}}\right]^{-2.5}\left[\left(f''(0)\right)^2 + \left(g'(0)\right)^2\right]^{\frac{1}{2}},$$ (36)

$$C_{f_2} = \frac{1}{Re_r}\left[1 - \frac{\frac{m_1}{\rho_1}}{\frac{m_1}{\rho_1} + \frac{m_2}{\rho_2} + \frac{m_f}{\rho_f}}\right]^{-2.5}\left[\left(f''(1)\right)^2 + \left(g'(1)\right)^2\right]^{\frac{1}{2}},$$ (37)

where $Re_r = \frac{r\Omega_1 H}{\nu_{hnf}}$ is the Reynolds number.

Another important physical quantity is

$$Nu_{r_1} \ (Local \ Nusselt \ number \ at \ lower \ disk) \ = \frac{Hq_w}{k_f(T_1 - T_2)}|_{z=0},$$

$$Nu_{r_2} \ (Local \ Nusselt \ number \ at \ upper \ disk) \ = \frac{Hq_w}{k_f(T_1 - T_2)}|_{z=H},$$ (38)

where q_w is the surface temperature defined as

$$q_w \ (At \ lower \ disk) \ = -k_{hnf}\frac{\partial T}{\partial z}|_{z=0} = -k_{hnf}\frac{T_1 - T_2}{H}\theta'(0).$$ (39)

Taking information from Equation (39), Equation (38) becomes

$$Nu_{r_1} = -\frac{k_{hnf}}{k_f}\theta'(0), \quad Nu_{r_2} = -\frac{k_{hnf}}{k_f}\theta'(1).$$ (40)

3. Computational Methodology

Following the HAM, choosing the initial guesses and linear operators for the velocities, temperature and homogeneous-heterogeneous chemical concentration profiles as

$$f_0(\zeta) = \zeta^3(k_6 + k_8) - \zeta^2(2k_6 + k_8) + \zeta k_6, \quad g_0(\zeta) = \zeta\Omega + 1 - \zeta, \quad \theta_0(\zeta) = -\zeta + 1,$$

$$\varphi_0(\zeta) = \frac{\zeta k_7 + 1}{k_7 + 1},$$ (41)

$$\varphi'' = L_\varphi, \quad f''' = L_f, \quad g'' = L_g, \quad \theta'' = L_\theta,$$ (42)

characterizing

$$L_f\left[E_1 + E_2\zeta + E_3\zeta^2 + E_4\zeta^3\right] \ = \ 0, \quad L_g\left[E_5 + E_6\zeta\right] \ = \ 0, \quad L_\theta\left[E_7 + E_8\zeta\right] \ = \ 0, \quad L_\varphi\left[E_9 + E_{10}\zeta\right] \ = \ 0,$$ (43)

where $E_i(i = 1–10)$ are the arbitrary constants.

3.1. Zeroth Order Deformation Problems

Introducing the nonlinear operator $\aleph$ as

$$\aleph_f[f(\zeta,j),g(\zeta,j)] = B_1\frac{\partial^4 f(\zeta,j)}{\partial\zeta^4} + Re\left[2f(\zeta,j)\frac{\partial^3 f(\zeta,j)}{\partial\zeta^3} + 2g(\zeta,j)\frac{\partial g(\zeta,j)}{\partial\zeta} - MB_2\frac{\partial^2 f(\zeta,j)}{\partial\zeta^2}\right] -$$
$$k_2 ReB_1\frac{\partial^2 f(\zeta,j)}{\partial\zeta^2} - 2k_3 Re\frac{1}{\rho_{hnf}}\frac{\partial f(\zeta,j)}{\partial\zeta}\frac{\partial^2 f(\zeta,j)}{\partial\zeta^2}, \quad (44)$$

$$\aleph_g[f(\zeta,j),g(\zeta,j)] = B_1\frac{\partial^2 g(\zeta,j)}{\partial\zeta^2} + Re\left[2f(\zeta,j)\frac{\partial g(\zeta,j)}{\partial\zeta} - MB_2\frac{\partial g(\zeta,j)}{\partial\zeta}\right] - k_2 B_1 g(\zeta,j) - k_3\frac{1}{\rho_{hnf}}\left[g(\zeta,j)\right]^2, \quad (45)$$

$$\aleph_\theta[f(\zeta,j),g(\zeta,j),\theta(\zeta,j)] = B_3\frac{k_{hnf}}{k_f}\frac{\partial^2\theta(\zeta,j)}{\partial\zeta^2} + \frac{1}{Rd}PrRe\left[2f(\zeta,j)\frac{\partial\theta(\zeta,j)}{\partial\zeta} + MB_4 Ec\left(\frac{\partial f(\zeta,j)}{\partial\zeta}\right)^2 + \left(g(\zeta,j)\right)^2\right], \quad (46)$$

$$\aleph_\varphi[f(\zeta,j),\varphi(\zeta,j)] = \frac{\partial^2\varphi(\zeta,j)}{\partial\zeta^2} + ReSc\left[2f(\zeta,j)\frac{\partial\varphi(\zeta,j)}{\partial\zeta} + k_4\varphi(\zeta,j)(1-\varphi(\zeta,j))^2\right], \quad (47)$$

where j is the homotopy parameter such that $j \in [0, 1]$.

Moreover

$$(1-j)L_f[f(\zeta,j) - f_0(\zeta)] = j\hbar_f\aleph_f[f(\zeta,j),g(\zeta,j)], \quad (48)$$
$$(1-j)L_g[g(\zeta,j) - g_0(\zeta)] = j\hbar_g\aleph_g[f(\zeta,j),g(\zeta,j)], \quad (49)$$
$$(1-j)L_\theta[\theta(\zeta,j) - \theta_0(\zeta)] = j\hbar_\theta\aleph_\theta[f(\zeta,j),g(\zeta,j),\theta(\zeta,j)], \quad (50)$$
$$(1-j)L_\varphi[\varphi(\zeta,j) - \varphi_0(\zeta)] = j\hbar_\varphi\aleph_\varphi[f(\zeta,j),\varphi(\zeta,j)], \quad (51)$$

where $\hbar_\varphi, \hbar_f, \hbar_\theta$ and $\hbar_g$ are the convergence control parameters.

Boundary conditions of Equation (48) are

$$f(0,j) = 0, \quad f'(0,j) = k_6, \quad f(1,j) = 0, \quad f'(1,j) = k_8. \quad (52)$$

Boundary conditions of Equation (49) are

$$g(0,j) = 1, \quad g(1,j) = \Omega. \quad (53)$$

Boundary conditions of Equation (50) are

$$\theta(0,j) = 1, \qquad \theta(1,j) = 0. \quad (54)$$

Boundary conditions of Equation (51) are

$$\varphi'(0,j) = k_7\varphi(0,j), \qquad \varphi(1,j) = 1. \quad (55)$$

Characterizing $j = 0$ and $j = 1$, the calculations obtained as

$$j = 0 \Rightarrow f(\zeta,0) = f_0(\zeta) \quad and \quad j = 1 \Rightarrow f(\zeta,1) = f(\zeta), \quad (56)$$
$$j = 0 \Rightarrow g(\zeta,0) = g_0(\zeta) \quad and \quad j = 1 \Rightarrow g(\zeta,1) = g(\zeta), \quad (57)$$
$$j = 0 \Rightarrow \theta(\zeta,0) = \theta_0(\zeta) \quad and \quad j = 1 \Rightarrow \theta(\zeta,1) = \theta(\zeta), \quad (58)$$
$$j = 0 \Rightarrow \varphi(\zeta,0) = \varphi_0(\zeta) \quad and \quad j = 1 \Rightarrow \varphi(\zeta,1) = \varphi(\zeta). \quad (59)$$

$f(\zeta,j)$ becomes $f_0(\zeta)$ and $f(\zeta)$ as j assumes the values zero and one. $g(\zeta,j)$ becomes $g_0(\zeta)$ and $g(\zeta)$ as j assumes the values zero and one. $\theta(\zeta,j)$ becomes $\theta_0(\zeta)$ and $\theta(\zeta)$ as j assumes the values zero and one. Finally, $\varphi(\zeta,j)$ becomes $\varphi_0(\zeta)$ and $\varphi(\zeta)$ as j assumes the values zero and one.

Applying Taylor series expansion on the Equations (56)–(59), the results are obtained as

$$f(\zeta,j) = f_0(\zeta) + \sum_{m=1}^{\infty} f_m(\zeta)j^m, \quad f_m(\zeta) = \frac{1}{m!}\frac{\partial^m f(\zeta,j)}{\partial j^m}\Big|_{j=0}, \tag{60}$$

$$g(\zeta,j) = g_0(\zeta) + \sum_{m=1}^{\infty} g_m(\zeta)j^m, \quad g_m(\zeta) = \frac{1}{m!}\frac{\partial^m g(\zeta,j)}{\partial j^m}\Big|_{j=0}, \tag{61}$$

$$\theta(\zeta,j) = \theta_0(\zeta) + \sum_{m=1}^{\infty} \theta_m(\zeta)j^m, \quad \theta_m(\zeta) = \frac{1}{m!}\frac{\partial^m \theta(\zeta,j)}{\partial j^m}\Big|_{j=0}, \tag{62}$$

$$\varphi(\zeta,j) = \varphi_0(\zeta) + \sum_{m=1}^{\infty} \varphi_m(\zeta)j^m, \quad \varphi_m(\zeta) = \frac{1}{m!}\frac{\partial^m \varphi(\zeta,j)}{\partial j^m}\Big|_{j=0}. \tag{63}$$

$\hbar_\varphi$, $\hbar_f$, $\hbar_\theta$ and $\hbar_g$ are adjusted to obtain the convergence for the series in Equations (60)–(63) at $j = 1$, so Equations (60)–(63) transform to

$$f(\zeta) = f_0(\zeta) + \sum_{m=1}^{\infty} f_m(\zeta), \tag{64}$$

$$g(\zeta) = g_0(\zeta) + \sum_{m=1}^{\infty} g_m(\zeta), \tag{65}$$

$$\theta(\zeta) = \theta_0(\zeta) + \sum_{m=1}^{\infty} \theta_m(\zeta), \tag{66}$$

$$\varphi(\zeta) = \varphi_0(\zeta) + \sum_{m=1}^{\infty} \varphi_m(\zeta). \tag{67}$$

3.2. mth Order Deformation Problems

Considering Equations (48) and (52) for homotopy at *m*th order as

$$L_f[f_m(\zeta) - \chi_m f_{m-1}(\zeta)] = \hbar_f R_m^f(\zeta), \tag{68}$$

$$f_m(0) = 0, \quad f_m(1) = 0, \quad f'_m(0) = 0, \quad f'_m(1) = 0, \tag{69}$$

$$R_m^f(\zeta) = B_1 f'''''_{m-1} + Re\left[\sum_{k=0}^{m-1} f_{m-1-k}f'''_k + 2g_{m-1-k}g'_k - MB_2 f'''_{m-1}\right] - k_2 Re B_1 f''_{m-1} - $$
$$2k_3 Re \frac{1}{\rho_{hnf}} \sum_{k=0}^{m-1} f'_{m-1-k}f'''_k. \tag{70}$$

Considering Equations (49) and (53) for homotopy at *m*th order as

$$L_g[g_m(\zeta) - \chi_m g_{m-1}(\zeta)] = \hbar_g R_m^g(\zeta), \tag{71}$$

$$g_m(0) = 0, \quad g_m(1) = 0, \tag{72}$$

$$R_m^g(\zeta) = B_1 g''_{m-1} + Re\left[\sum_{k=0}^{m-1} 2f_{m-1-k}g'_k - MB_2 g'_{m-1}\right] - k_2 B_1 g_{m-1} - k_3 \frac{1}{\rho_{hnf}} \sum_{k=0}^{m-1} g_{m-1-k}g_k. \tag{73}$$

Considering Equations (50) and (54) for homotopy at *m*th order as

$$L_\theta[\theta_m(\zeta) - \chi_m \theta_{m-1}(\zeta)] = \hbar_\theta R_m^\theta(\zeta), \tag{74}$$

$$\theta_m(0) = 0, \quad \theta_m(1) = 0, \tag{75}$$

$$R_m^\theta(\zeta) = B_3 \frac{k_{hnf}}{k_f} \theta''_{m-1} + \frac{1}{Rd} PrRe \left[2 \sum_{k=o}^{m-1} f_{m-1-k}\theta'_k + MB_4 Ec \left[\sum_{k=o}^{m-1} f'_{m-1-k}f'_k + \sum_{k=o}^{m-1} g_{m-1-k}g_k \right] \right]. \tag{76}$$

Considering Equations (51) and (55) for homotopy at *m*th order as

$$L_\varphi[\varphi_m(\zeta) - \chi_m\varphi_{m-1}(\zeta)] = \hbar_\varphi R_m^\varphi(\zeta), \tag{77}$$

$$\varphi'_m(0) = 0, \qquad \varphi_m(1) = 0, \tag{78}$$

$$R_m^\varphi(\zeta) = \varphi''_{m-1} + ReSc \left[2 \sum_{k=o}^{m-1} f_{m-1-k}\varphi'_k + k_4 \left[\varphi_{m-1} + \varphi_{m-1-k} \sum_{l=o}^{k} \varphi_{k-l}\varphi_l - 2 \sum_{k=o}^{m-1} \varphi_{m-1-k}\varphi_k \right] \right], \tag{79}$$

$$\chi_m = \begin{cases} 0, & m \leq 1 \\ 1, & m > 1. \end{cases} \tag{80}$$

Adding the particular solutions $f_m^*(\zeta)$, $g_m^*(\zeta)$, $\theta_m^*(\zeta)$ and $\varphi_m^*(\zeta)$, Equations (68), (71), (74) and (77) yield the general solutions as

$$f_m(\zeta) = f_m^*(\zeta) + E_1 + E_2\zeta + E_3\zeta^2 + E_4\zeta^3, \tag{81}$$

$$g_m(\zeta) = g_m^*(\zeta) + E_5 + E_6\zeta, \tag{82}$$

$$\theta_m(\zeta) = \theta_m^*(\zeta) + E_7 + E_8\zeta, \tag{83}$$

$$\varphi_m(\zeta) = \varphi_m^*(\zeta) + E_9 + E_{10}\zeta. \tag{84}$$

4. Results and Discussion

Results and discussion provide the analysis of the problem through the impacts of all the relevant parameters. The non-dimensional Equations (20), (22), (28) and (30) with boundary conditions in Equations (25), (26) and (29) are analytically computed. The performances of different parameters on the velocity profiles with heat and concentration of homogeneous-heterogeneous chemical reactions are shown in the relevant graphs. The streamlines show the internal behaviors of flow. The physical representation of the problem is shown in Figure 1. Liao [61] introduced $\hbar$-curves for the convergence of the series solution to get the precise and convergent solutions of the problems. $\hbar$-curves are also called the convergence controlling parameters for solution in the homotopy analysis method (used for solution in the present case). These $\hbar$-curves specify the range of numerical values. These numerical values (optimum values) are selected from the valid region in straight line. These optimum values of $\hbar$-curves are selected from the straight lines parallel to the horizontal axis (please see carefully Figures 2–5) to control the convergence of problem solution. In the present case, the valid region of each profile $\hbar$-curve is specified. Therefore, the admissible $\hbar$-curves for $f(\zeta)$, $g(\zeta)$, $\theta(\zeta)$ and $\varphi(\zeta)$ are drawn in the ranges $-10.00 \leq \hbar_f \leq -4.00$, $-10.00 \leq \hbar_g \leq -5.00$, $-3.5 \leq \hbar_\theta \leq -2.50$ and $-1.50 \leq \hbar_\varphi \leq -0.50$ in Figures 2–5, respectively.

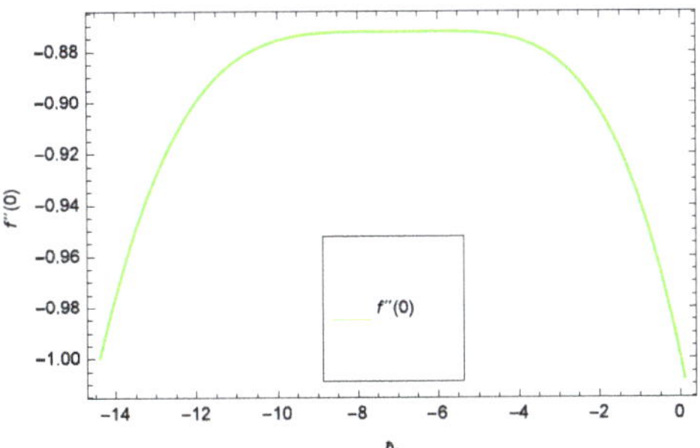

Figure 2. Illustration of the $\hbar_f$-curve of $f(\zeta)$.

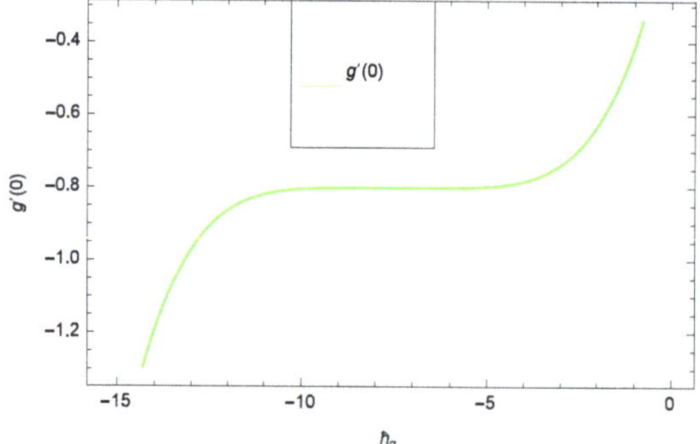

Figure 3. Illustration of the $\hbar_g$-curve of $g(\zeta)$.

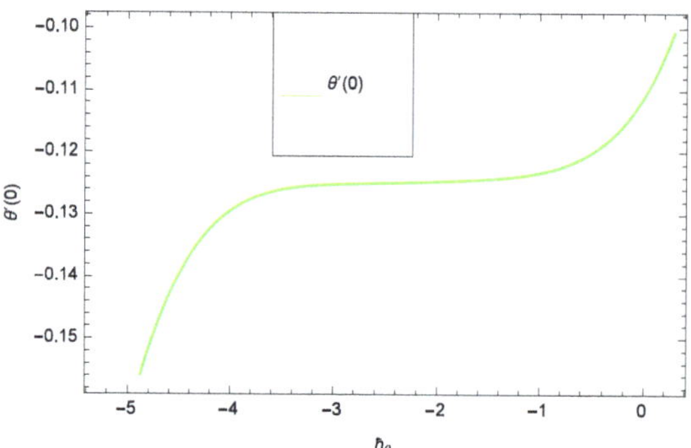

Figure 4. Illustration of the $\hbar_\theta$-curve of $\theta(\zeta)$.

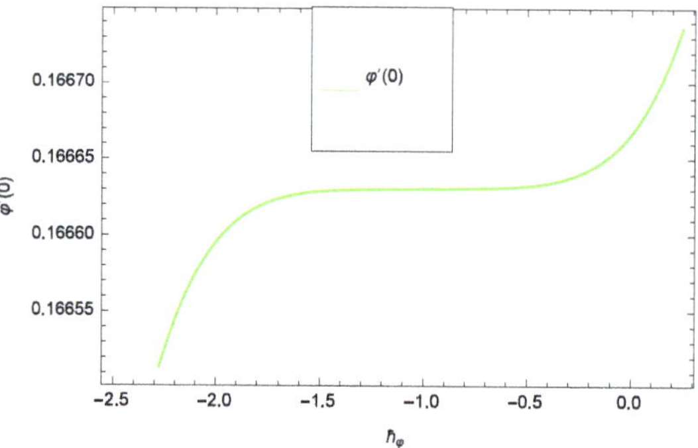

Figure 5. Illustration of the $\hbar_\varphi$-curve of $\varphi(\zeta)$.

4.1. Axial Velocity Profile

In the present study, two nanofluids namely ZnO-$C_2H_6O_2$ and Au-$ZnO/C_2H_6O_2$ are investigated whose behaviors are shown through the graphs under the effects of different parameters. In Figures 6–25, the green and magenta colors are used for ZnO-$C_2H_6O_2$ and Au-$ZnO/C_2H_6O_2$ while in Figures 24 and 25, the additional colors are also used. There are solid and dashed curves in Figures 6–23. The mechanism is that three positive increasing numerical values are given to one parameter in the HAM solution while all the remaining parameters are fixed to show the effect of that one parameter simultaneously on the two nanofluids namely ZnO-$C_2H_6O_2$ and Au-ZnO/$C_2H_6O_2$. When the solid lines locate below the dashed lines, then it shows the increasing effect and when the solid lines locate above the dashed lines, then it shows the decreasing effect. When the arrow head is from top to bottom, it shows the decreasing effect and when the arrow head is from bottom to top, it shows the increasing effect.

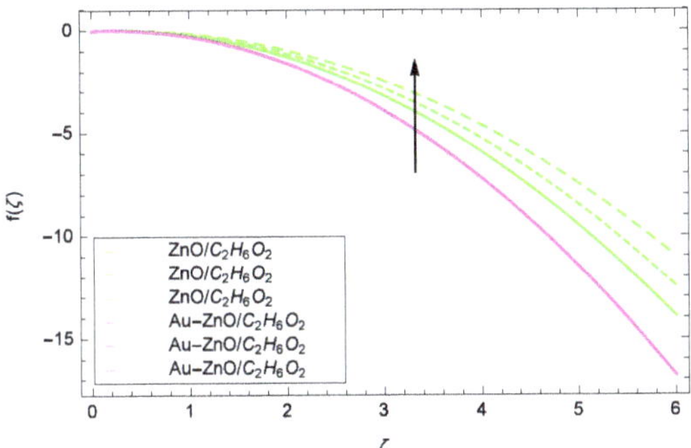

Figure 6. Illustration for the velocity $f(\zeta)$ and parameter Re = 1.00, 1.50, 2.00.

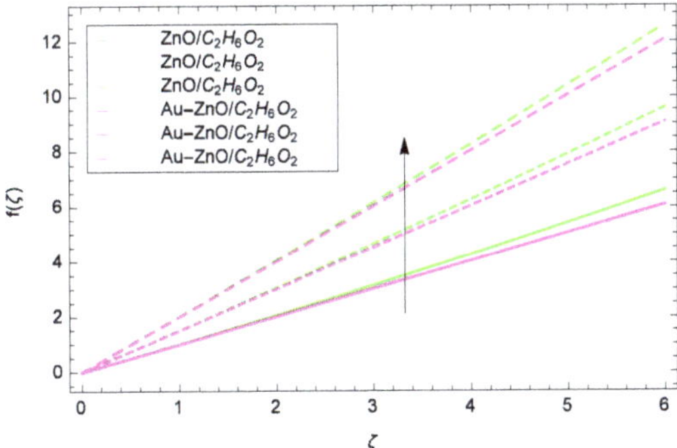

Figure 7. Illustration for the velocity $f(\zeta)$ and parameter k_6 = 1.00, 1.50, 2.00.

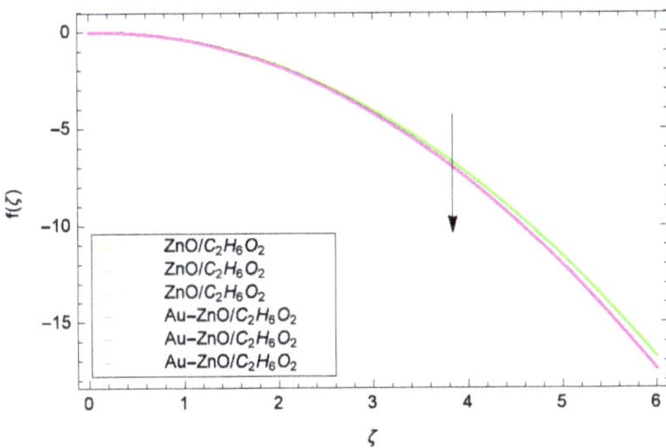

Figure 8. Illustration for the velocity $f(\zeta)$ and parameter M = 1.00, 1.50, 2.00.

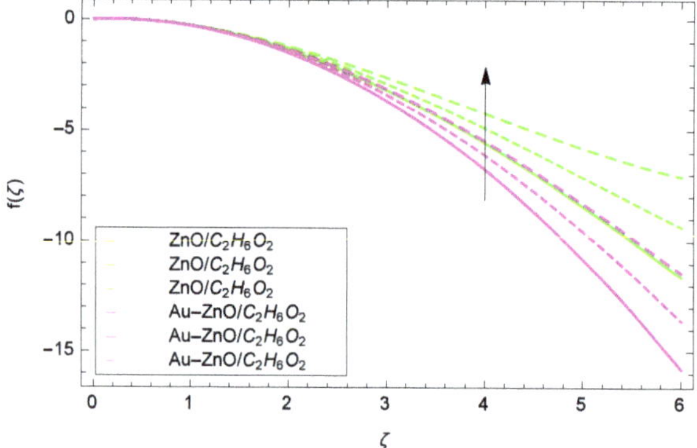

Figure 9. Illustration for the velocity $f(\zeta)$ and parameter $\Omega = 1.00, 1.50, 2.00$.

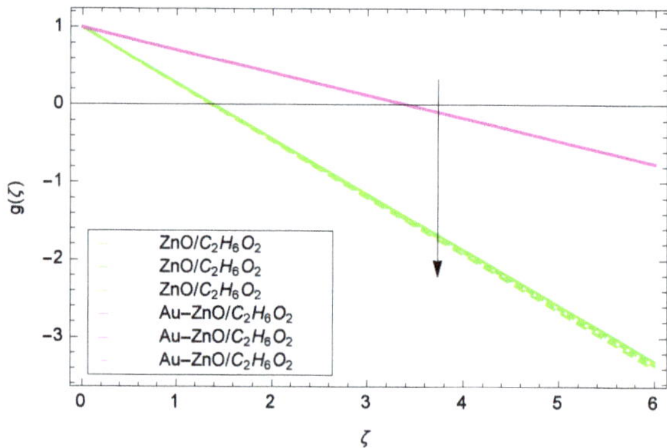

Figure 10. Illustration for the velocity $g(\zeta)$ and parameter $Re = 1.00, 10.50, 20.00$.

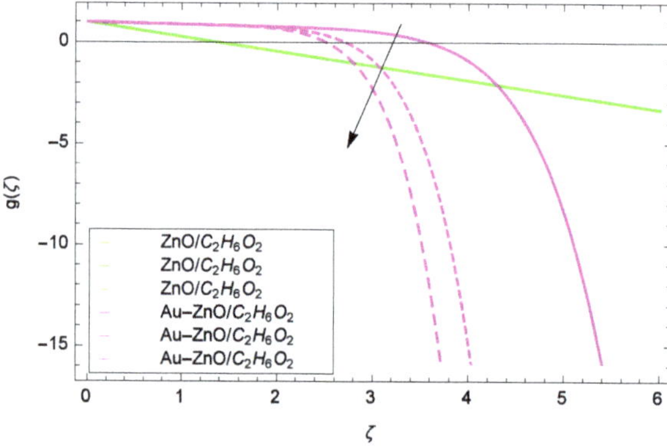

Figure 11. Illustration for the velocity $g(\zeta)$ and parameter $k_6 = 1.00, 10.50, 20.00$.

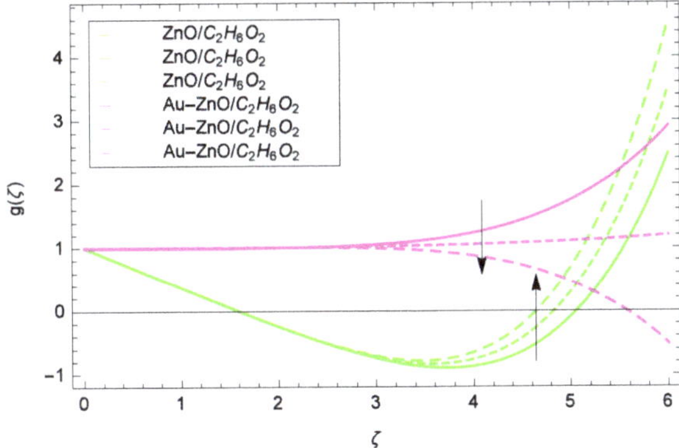

Figure 12. Illustration for the velocity $g(\zeta)$ and parameter $M = 1.00, 10.50, 20.00$.

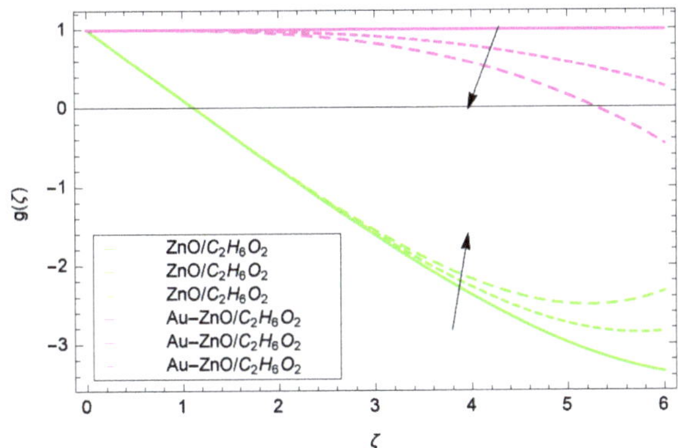

Figure 13. Illustration for the velocity $g(\zeta)$ and parameter $\Omega = 1.00, 1.50, 2.00$.

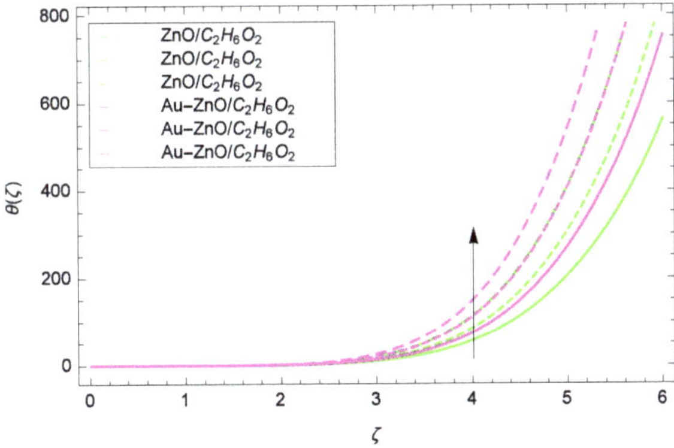

Figure 14. Illustration for the heat transfer $\theta(\zeta)$ and parameter $Re = 1.00, 1.50, 2.00$.

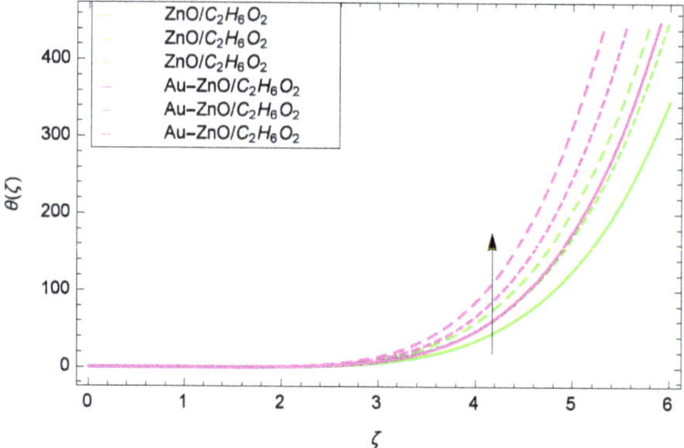

Figure 15. Illustration for the heat transfer $\theta(\zeta)$ and parameter $k_6 = 1.00, 1.50, 2.00$.

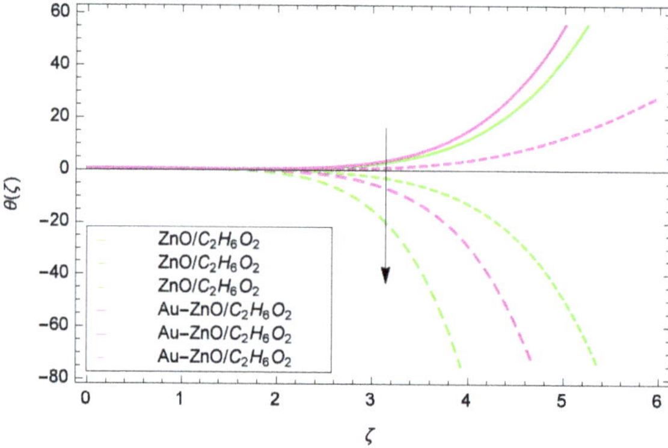

Figure 16. Illustration for the heat transfer $\theta(\zeta)$ and parameter $\Omega = 1.00, 5.50, 10.00$.

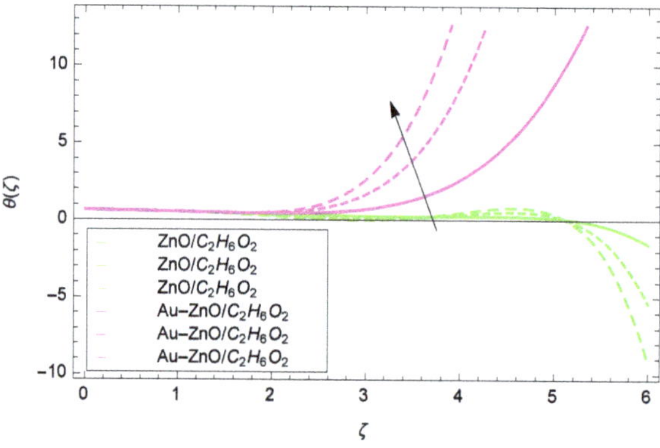

Figure 17. Illustration for the heat transfer $\theta(\zeta)$ and parameter $Pr = 1.00, 3.50, 6.00$.

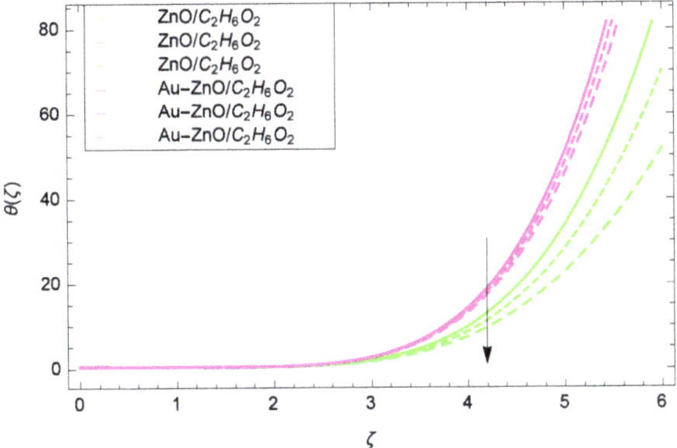

Figure 18. Illustration for the heat transfer $\theta(\zeta)$ and parameter $M = 1.00, 1.50, 2.00$.

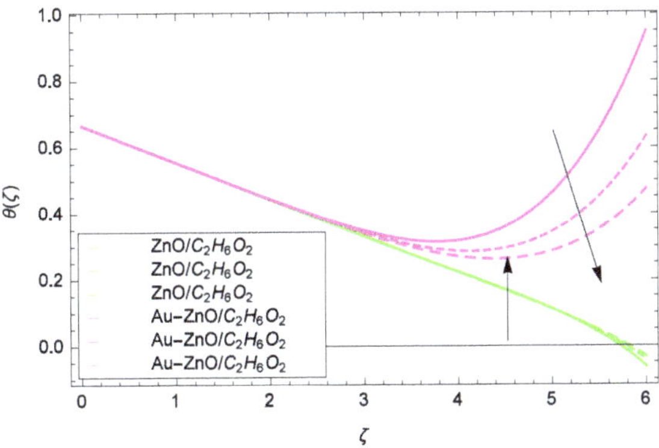

Figure 19. Illustration for the heat transfer $\theta(\zeta)$ and parameter $Rd = 1.00, 1.50, 2.00$.

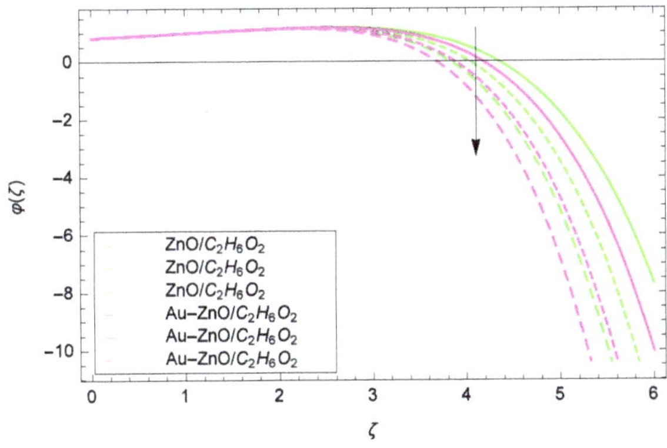

Figure 20. Illustration for the concentration $\varphi(\zeta)$ and parameter $Re = 1.00, 1.50, 2.00$.

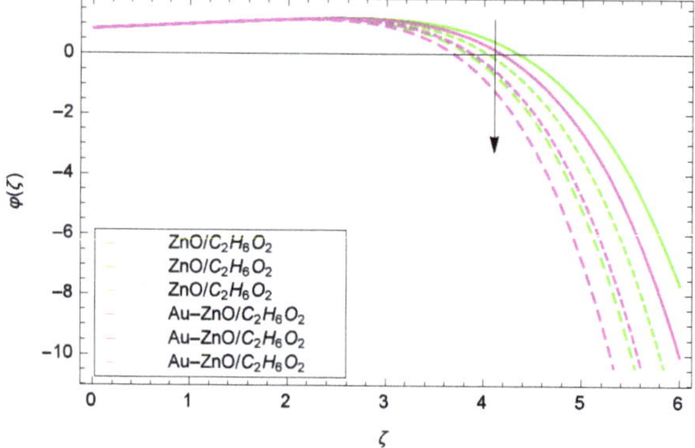

Figure 21. Illustration for the concentration $\varphi(\zeta)$ and parameter $k_4 = 1.00, 1.50, 2.00$.

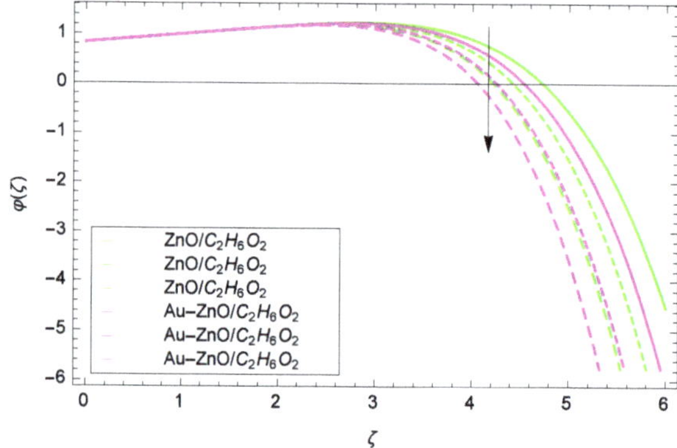

Figure 22. Illustration for the concentration $\varphi(\zeta)$ and parameter $k_6 = 1.00, 1.50, 2.00$.

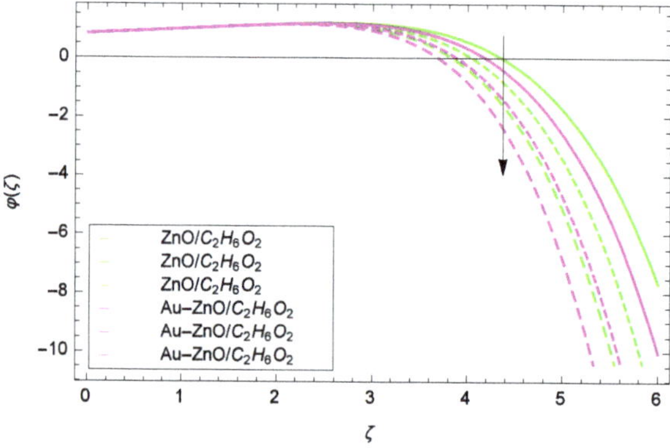

Figure 23. Illustration for the concentration $\varphi(\zeta)$ and parameter $Sc = 1.00, 1.50, 2.00$.

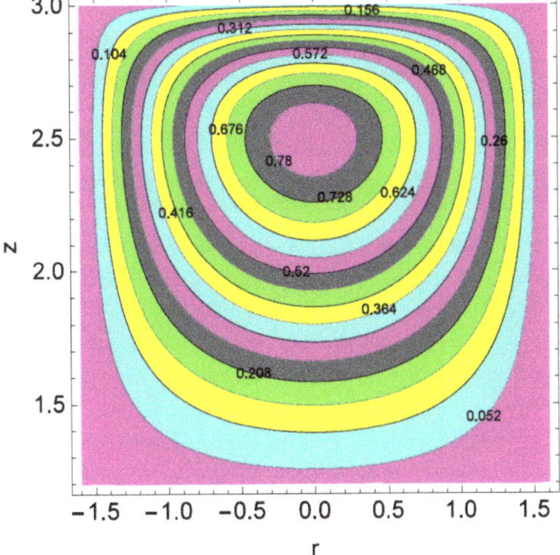

Figure 24. Illustration for the streamlines at upper disk and parameter $Re = 0.30$.

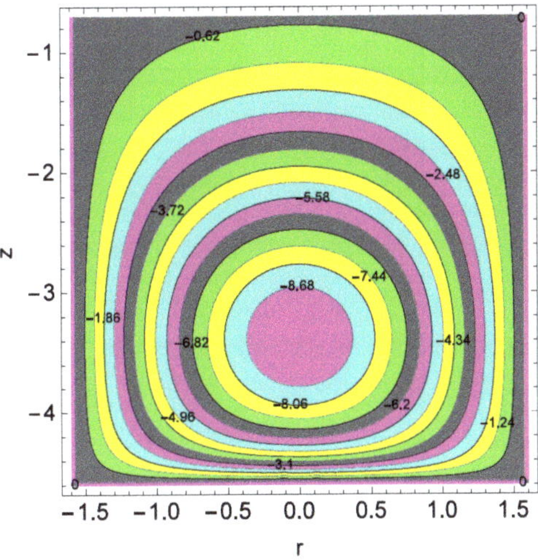

Figure 25. Illustration for the streamlines for lower disks and parameter $Re = 0.30$.

Figure 6 shows that for the different values of Reynolds number Re, the axial velocity $f(\zeta)$ is increased. In fact, the velocity of ZnO-$C_2H_6O_2$ and Au-ZnO/$C_2H_6O_2$ increase with increasing values of Reynolds number therefore overall motion is accelerated. Figure 7 shows the prominent role of stretching parameter k_6 due to lower disk in which the axial velocity $f(\zeta)$ increases. The present motion is due to stretching so if the stretching parameter is increased, the flow of fluids is also increased. In the mean time, porosity is responsible to decrease the axial flow. It shows that motion due to different nanofluids is reduced because the permeability at the edge of the accelerating surface increases. Surely, it is noted that excess of nanoparticles concentration is involved in decelerating the motion. It is worthy of notice that the axial velocity $f(\zeta)$ decreases against the inertia. Physically it means that the absorbency of the porous medium shows an increment in the thickness of the fluid. Figure 8 shows that magnetic field parameter resists the flow since due to magnetic field, the Lorentz forces are generated which resist the motion. The curves are shrink in response to the parameter effect. Figure 9 exhibits all the assigned values of Ω and axial velocity $f(\zeta)$ which offers opportunities to know about the rotating systems and shows that the flow of ZnO-$C_2H_6O_2$ and Au-ZnO/$C_2H_6O_2$ increase.

Some interesting results have been found in case of tangential velocity $g(\zeta)$. Figure 10 shows that as the Reynolds number Re increases, the opposite tendency has been observed in the motion of ZnO-$C_2H_6O_2$ and Au-ZnO/$C_2H_6O_2$. The flow of mono nanofluid ZnO-$C_2H_6O_2$ decreases while the flow of hybrid nanofluid Au-ZnO/$C_2H_6O_2$ shows no prominent change for increasing the Reynolds number Re. In Figure 11, the tangential velocity $f(\zeta)$ tends to decreasing. Tangential velocity assumes a likely downfall so the flow is not supported by stretching due to k_6. Figure 12 witnesses that the tangential velocity $g(\zeta)$ shifts to the effective decreasing for hybrid nanofluid Au-ZnO/$C_2H_6O_2$ and increases for ZnO-$C_2H_6O_2$ on behalf of the magnetic field parameter M. Figure 13 exhibits that rotation parameter Ω parameter resists the tangential flow of Au-ZnO/$C_2H_6O_2$ and enhances the tangential flow of ZnO-$C_2H_6O_2$.

4.2. Temperature Profile

Figure 14 shows the effect of Reynolds number Re on heat transfer. The larger values of Re increase the temperature of ZnO-$C_2H_6O_2$ and Au-ZnO/$C_2H_6O_2$. It has been observed in Figure 15 that as the stretching parameter k_6 increases, the temperature of ZnO-$C_2H_6O_2$ and Au-ZnO/$C_2H_6O_2$ increase. These observations indicate that the fluid temperature and its related layer are incremented for higher estimations of k_6. The rotation parameter Ω cannot generate an extra heating to the system as shown in Figure 16. Temperature $\theta(\zeta)$ is decreased on increasing the parameter Ω. The physical reason is that enhancement in Ω causes to improve the internal source of energy, that is why the fluid temperature is reduced. The system gets the parameter Pr for the designated values 1.00, 3.50, and 6.00 during the process and increases the temperature shown through Figure 17. The direct relation of Pr and thermal conductivity increases the thickness of thermal boundary layer. Larger values of Pr generate the high diffusion of heat transfer. The temperature $\theta(\zeta)$ is changed to lowest level after the exchange of high values of magnetic field parameter M as shown in Figure 18. The reason is that strong Lorentz forces resist the flow of nanoparticles, so causing no high collision among the nanoparticles, consequently, the temperature is decreased. Figure 19 depicts that with the increasing values of thermal radiation parameter Rd, the temperature $\theta(\zeta)$ of ZnO-$C_2H_6O_2$ increases while the temperature of hybrid nanofluid Au-ZnO-$C_2H_6O_2$ decreases. The reason is that radiation enhances more heat in the working fluids.

4.3. Concentration of Homogeneous-Heterogeneous Chemical Reactions

Looking at the non-dimensional Equation (28), the suitable values of Re, Sc and k_4 are the basic quantities for generating a cubic autocatalysis chemical reaction. The concentration of chemical reaction $\varphi(\zeta)$ is low with the Reynolds number Re as shown in Figure 20. Figure 21 shows that for the homogeneous chemical reaction parameter k_4, the concentration of chemical reaction is decreased. From Equation (28), it is witnessed that the homogeneous chemical reaction parameter k_4 is a part of performance with the multiple solutions. Enhancement in k_4 makes dominant the concentration. In Figure 22, the stretching parameter k_6 upgrades concentration of chemical reaction with low level by performing active role in the rotating motion. The stretching parameter k_6 makes compact the homogeneous reaction and hence the concentration profile $\varphi(\zeta)$. Figure 23 stands for the outcomes of Schmidt number Sc and concentration $\varphi(\zeta)$. Momentum diffusivity to mass diffusivity is known as Schmidt number. The parameter Sc causes to make low the homogeneous chemical reaction.

4.4. Streamlines

Figure 24 shows the streamlines at upper disks. The size of the streamlines increases at upper disk compared to that of lower disk. Both mono nanofluid and hybrid nanofluid proceed towards the edges of disks. Figure 25 shows the streamlines for the Reynolds number Re at lower disks. The compression of streamlines are clear from Figure 25. The plumes power is strong for lower disks.

4.5. Authentication of the Present Work

The important physical quantities introduced in Section 2 are evaluated to compare the validity of the solution with the published work [8]. Table 4 shows the tabulations to the several values for the parameter Re. There exists a nice agreement with the published work [8]. Similarly in Table 5, the values of heat transfer rate are computed for the volume fraction ϕ = 0.10, 0.20, 0.30, and 0.40. These values also have the close agreement with the published work [8].

Table 4. Comparison of the present and published work.

Reynolds Number (Re)	$f''(1)$ [8]	$f''(1)$ (Present)	$g'(1)$ [8]	$g'(1)$ (Present)
0.10	0.292991	0.292993	0.284684	0.284683
0.20	0.237792	0.237791	0.224995	0.224994
0.30	0.208284	0.208283	0.197046	0.197044
0.40	0.206995	0.206994	0.203117	0.203115

Table 5. Comparison of the present and published work.

Volume Fraction (ϕ)	$-\theta'(0)$ [8]	$-\theta'(0)$ (Present)	$-\theta'(1)$ [8]	$-\theta'(1)$ (Present)
0.10	3.677172	3.677170	2.26814	2.26813
0.20	4.53192	4.53190	3.128083	3.128081
0.30	2.983936	2.983935	1.648859	1.648857
0.40	3.00208	3.00207	1.624995	1.624993

5. Conclusions

A significant modification in the mathematical model for hybrid nanofluid has been made for the analysis of flow, heat and mass transfer. Chemical species reactions are shown in hybrid nanofluid. The problem is modeled in rotating systems for the nanoparticles ZnO and Au with base fluid ethylene glycol and solved through HAM. In ethylene glycol-based fluid ($C_2H_6O_2$), two types of nanoparticles, namely ZnO (zinc oxide) and Au (gold), with volume fractions ϕ_1 = 0.03 and ϕ_2 = 0.04 are investigated, respectively. It is noted that for ϕ_1 = 0.00 and ϕ_2 = 0.00, the problem becomes about viscous fluid with the absence of nanoparticles volume fractions. If ϕ_1 = 0.00, $Ag/C_2H_6O_2$ is obtained and if ϕ_2 = 0.00, $ZnO_2/C_2H_6O_2$ is constructed. Achieving better comprehension, the competencies of active parameters on flow, heat transfer and concentration of heterogeneous-homogeneous chemical reactions are noted. There exists a nice agreement between the present and published work in Tables 4 and 5. The problem has potential for renewable energy system and researchers to investigate the thermal conductivity of nanoparticles like silver, aluminum, copper etc. with different base fluids like water, benzene, engine oil etc. The results for flow, heat transfer and concentration of homogeneous-heterogeneous chemical reactions are summarized as following.

(1) Axial velocity $f(\zeta)$ increases for ZnO-$C_2H_6O_2$ and Au-ZnO/$C_2H_6O_2$ with the increasing values of Reynolds number Re, stretching parameter k_6 and rotation parameter Ω while axial velocity $f(\zeta)$ decreases for ZnO-$C_2H_6O_2$ and Au-ZnO/$C_2H_6O_2$ with the increasing values of magnetic field parameter M.

(2) Tangential velocity $g(\zeta)$ increases for ZnO-$C_2H_6O_2$ with the increasing values of magnetic field parameter M and rotation parameter Ω while the same velocity decreases for Au-ZnO/$C_2H_6O_2$ with the increasing values of magnetic field parameter M and rotation parameter Ω. Moreover, tangential velocity $g(\zeta)$ decreases for ZnO-$C_2H_6O_2$ and Au-ZnO/$C_2H_6O_2$ with the increasing values of Reynolds number Re and stretching parameter k_6.

(3) Heat transfer $\theta(\zeta)$ increases for ZnO-$C_2H_6O_2$ and Au-ZnO/$C_2H_6O_2$ with the increasing values of Reynolds number Re, stretching parameter k_6. Similarly, heat transfer $\theta(\zeta)$ increases for ZnO-$C_2H_6O_2$ with increasing values of thermal radiation parameter Rd while it is decreased for ZnO-$C_2H_6O_2$ and Au-ZnO/$C_2H_6O_2$ with the increasing values of rotation parameter Ω, magnetic

field parameter M. In case of Au-ZnO/$C_2H_6O_2$, heat transfer $\theta(\zeta)$ also decreases with increasing values of thermal radiation parameter Rd.

(4) The concentration of homogeneous-heterogeneous chemical reactions $\varphi(\zeta)$ decreases for ZnO-$C_2H_6O_2$ and Au-ZnO/$C_2H_6O_2$ with the increasing values of Reynolds number Re, stretching parameter k_6 and Schmidt number Sc.

(5) Streamlines are compressed at the upper portion of upper disk while these are compressed at the lower portion of lower disk when the Reynolds number Re assumes the value 0.30.

(6) Tables 4 and 5 show an excellent agreement of the present work with published work.

Author Contributions: Conceptualization, N.S.K.; methodology, N.S.K.; software, N.S.K.; validation, N.S.K.; formal analysis, P.K.; investigation, N.S.K.; resources, P.T.; data curation, P.T.; writing—original draft preparation, N.S.K.; writing—review and editing, N.S.K.; visualization, P.K.; supervision, N.S.K.; project administration, P.K.; funding acquisition, P.K. All authors have read and agreed to the revised version of the manuscript.

Funding: This research is funded by the Center of Excellence in Theoretical and Computational Science (TaCS-CoE), KMUTT.

Acknowledgments: This work was partially supported by the International Research Partnerships: Electrical Engineering Thai-French Research Center (EE-TFRC) between King Mongkut's University of Technology North Bangkok and Universite' de Lorraine under Grant KMUTNB-BasicR-64-17. The authors are cordially thankful to the honorable reviewers for their constructive comments to improve the quality of the paper. This research is supported by the Postdoctoral Fellowship from King Mongkut's University of Technology Thonburi (KMUTT), Thailand. This project is supported by the Theoretical and Computational Science (TaCS) Center under Computational and Applied Science for Smart Innovation Research Cluster (CLASSIC), Faculty of Science, KMUTT.

Conflicts of Interest: The authors declare no conflict of interest.

References

1. Kasaeian, A.; Eshghi, A.T.; Sameti, M. A review on the applications of nanofluids in solar energy systems. *Renew. Sustain. Energy Rev.* **2015**, *43*, 584–598. [CrossRef]

2. Ettefaghi, E.; Ghobadian, B.; Rashidi, S.; Najafi, G.; Khoshtaghaza, M.H.; Rashtchi, M.; Sadeghian, S. A novel bio-nano emulsion fuel based on biodegradable nanoparticles to improve diesel engines performance and reduce exhaust emissions. *Renew. Energy* **2018**, *125*, 64–72. [CrossRef]

3. Gunjo, D.G.; Jena, S.R.; Mahanta, P.; Robi, P.S. Melting enhancement of a latent heat storage with dispersed Cu, CuO and Al_2O_3 nanoparticles for solar thermal application. *Renew. Energy* **2018**, *121*, 652–665. [CrossRef]

4. Khanafer, K.; Vafi, K. A review on the aplications of nanofluids in solar energy field. *Renew. Energy* **2018**, *123*, 398–406. [CrossRef]

5. Caglar, M.; Ilican, S., Caglar, Y.; Yakuphanoglu, F. Electrical conductivity and optical properties of Zno nanostructured thin film. *Appl. Surf. Sci.* **2009**, *225*, 4491–4496. [CrossRef]

6. Islam, R.I.; Shabani, B.; Rosengarten, G. Electrical and thermal conductivities of water-ethylene glycol based TiO_2 nanofluids to be used as coolants in PEM fuel cells. *Energy Procedia* **2017**, *110*, 101–108. [CrossRef]

7. Fal, J.; Sidorowicz, A.; Zyla, G. Electrical conductivity of ethylene glycol based nanofluids with different types of thulium oxide nanoparticles. *Acta Phys. Pol. A* **2017**, *132*, 146–148. [CrossRef]

8. Rout, B.C.; Mishra, S.R.; Nayak, B. Semi analytical solution of axisymmetric flows of Cu- and Ag-water nanofluids between two rotating disks. *Heat Transf. Asian Res.* **2019**, *132*, 1–25. [CrossRef]

9. Hamilton, R.L.; Crosser, O.K. Thermal conductivity of heterogeneous two component systems. *Ind. Eng. Chem. Fundam.* **1962**, *1*, 187–191. [CrossRef]

10. Choi, S.U.S.; Eastman, J.A. Enhancing thermal conductivity of fluids with nanoparticles. In Proceedings of the 1995 International Mechanical Engineering Congress and Exhibition, San Francisco, CA, USA, 12–17 November 1995; Volume 231, pp. 99–106.

11. Vallejo, J.P.; Zyla, G.; Fernandez-Seara, J.; Lugo, L. Influence of six carbon-based nanomaterials on the rheological properties of nanofluids. *Nanomaterials* **2019**, *9*, 146. [CrossRef]

12. Alihosseini, S.; Jafari, A. The effect of porous medium configuration on nanofluid heat transfer. *Appl. Nanosci.* **2020**, *10*, 895–906. [CrossRef]

13. Sheikholeslami, M.; Shah, Z.; Tassaddiq, A.; Shafee, A.; Khan, I. Application of electric field for augmentation of Ferrofluid heat transfer in an enclosure including double moving walls. *IEEE Access* **2019**, *7*, 21048–21056. [CrossRef]

14. Al-Kouz, W.; Al-Muhtady, A.; Owhaib, W.; Al-Dahidi, S.; Hadar, M.; Abu-Alghanam, R. Entropy generation optimization for rarified nanofluid flows in a square cavity with two fins at the hot wall. *Entropy* **2019**, *21*, 103. [CrossRef]

15. Atta, A.M.; Abdullah, M.M.S.; Al-Lohedan, H.A.; Mohamed, N.H. Novel superhydrophobic sand and polyurethane sponge coated with silica/modified asphaltene nanoparticles for rapid oil spill cleanup. *Nanomaterials* **2019**, *9*, 187. [CrossRef]

16. Rout, A.; Boltaev, G.S.; Ganeev, R.A.; Fu, Y.; Maurya, S.K.; Kim, V.V.; Rao, K.S.; Guo, C. Nonlinear optical studies of gold nanoparticles films. *Nanomaterials* **2019**, *9*, 291. [CrossRef] [PubMed]

17. Alvarez-Regueiro, E.; Vallejo, J.P.; Fernandez-Seara, J.; Fernandez, J.; Lugo, L. Experimental convection heat transfer analysis of a nano-enhanced industrial coolant. *Nanomaterials* **2019**, *9*, 267. [CrossRef]

18. Alsagri, A.S.; Nasir, S.; Gul, T.; Islam, S.; Nisar, K.S.; Shah, Z.; Khan, I. MHD thin film flow and thermal analysis of blood with CNTs nanofluid. *Coatings* **2019**, *9*, 175. [CrossRef]

19. Mishra, S.K.; Chandra, H.; Arora, A. Effect of velocity and rheology of nanofluid on heat transfer of laminar vibrational flow through a pipe under constant heat flux. *Int. Nano Lett.* **2019**, *9*, 245–256. [CrossRef]

20. Abbas, S.Z.; Khan, W.A.; Sun, H.; Ali, M.; Irfan, M.; Shahzed, M.; Sultan, F. Mathematical modeling and analysis of cross nanofluid flow subjected to entropy generation. *Appl. Nanosci.* **2019**, *10*, 3149–3160. [CrossRef]

21. Ali, M.; Khan, W.A.; Irfan, M.; Sultan, F.; Shahzed, M.; Khan, M. Computational analysis of entropy generation for cross nanofluid flow. *Appl. Nanosci.* **2019**, *10*, 3045–3055. [CrossRef]

22. Sharma, R.P.; Seshadri, R.; Mishra, S.R.; Munjam, S.R. Effect of thermal radiation on magnetohydrodynamic three-dimensional motion of a nanofluid past a shrinking surface under the influence of a heat source. *Heat Transf. Asian Res.* **2019**, *48*, 2105–2121. [CrossRef]

23. Jahan, S.; Sakidin, H.; Nazar, R.; Pop, I. Analysis of heat transfer in nanofluid past a convectively heated permeable stretching/shrinking sheet with regression and stability analyses. *Results Phys.* **2018**, *10*, 395–405. [CrossRef]

24. Hossinzadeh, K.; Asadi, A.; Mogharrebi, A.R.; Khalesi, J.; Mousavisani, S.; Ganji, D.D. Entropy generation analysis of $(CH_2OH)_2$ containing CNTs nanofluid flow under effect of MHD and thermal radiation. *Case Stud. Therm. Eng.* **2019**, *14*, 100482. [CrossRef]

25. Khan, N.S. Bioconvection in second grade nanofluid flow containing nanoparticles and gyrotactic microorganisms. *Braz. J. Phys.* **2018**, *43*, 227–241. [CrossRef]

26. Khan, N.S.; Gul, T.; Khan, M.A.; Bonyah, E.; Islam, S. Mixed convection in gravity-driven thin film non-Newtonian nanofluids flow with gyrotactic microorganisms. *Results Phys.* **2017**, *7*, 4033–4049. [CrossRef]

27. Khan, N.S.; Gul, T.; Islam, S.; Khan, I.; Alqahtani, A.M.; Alshomrani, A.S. Magnetohydrodynamic nanoliquid thin film sprayed on a stretching cylinder with heat transfer. *J. Appl. Sci.* **2017**, *7*, 271. [CrossRef]

28. Zuhra, S.; Khan, N.S.; Khan, M.A.; Islam, S.; Khan, W.; Bonyah, E. Flow and heat transfer in water based liquid film fluids dispensed with graphene nanoparticles. *Results Phys.* **2018**, *8*, 1143–1157. [CrossRef]

29. Khan, N.S.; Gul, T.; Islam, S.; Khan, W. Thermophoresis and thermal radiation with heat and mass transfer in a magnetohydrodynamic thin film second-grade fluid of variable properties past a stretching sheet. *Eur. Phys. J. Plus* **2017**, *132*, 11. [CrossRef]

30. Palwasha, Z.; Khan, N.S.; Shah, Z.; Islam, S.; Bonyah, E. Study of two dimensional boundary layer thin film fluid flow with variable thermo-physical properties in three dimensions space. *AIP Adv.* **2018**, *8*, 105318. [CrossRef]

31. Khan, N.S.; Gul, T.; Islam, S.; Khan, A.; Shah, Z. Brownian motion and thermophoresis effects on MHD mixed convective thin film second-grade nanofluid flow with Hall effect and heat transfer past a stretching sheet. *J. Nanofluids* **2017**, *6*, 812–829. [CrossRef]

32. Khan, N.S.; Zuhra, S.; Shah, Z.; Bonyah, E.; Khan, W.; Islam, S. Slip flow of Eyring-Powell nanoliquid film containing graphene nanoparticles. *AIP Adv.* **2019**, *8*, 115302. [CrossRef]

33. Khan, N.S.; Gul, T.; Kumam, P.; Shah, Z.; Islam, S.; Khan, W.; Zuhra, S.; Sohail, A. Influence of inclined magnetic field on Carreau nanoliquid thin film flow and heat transfer with graphene nanoparticles. *Energies* **2019**, *12*, 1459. [CrossRef]

34. Khan, N.S. Study of two dimensional boundary layer flow of a thin film second grade fluid with variable thermo-physical properties in three dimensions space. *Filomat* **2019**, *33*, 5387–5405. [CrossRef]
35. Khan, N.S.; Zuhra, S. Boundary layer unsteady flow and heat transfer in a second grade thin film nanoliquid embedded with graphene nanoparticles past a stretching sheet. *Adv. Mech. Eng.* **2019**, *11*, 1–11. [CrossRef]
36. Khan, N.S.; Gul, T.; Islam, S.; Khan, W.; Khan, I.; Ali, L. Thin film flow of a second-grade fluid in a porous medium past a stretching sheet with heat transfer. *Alex. Eng. J.* **2017**, *57*, 1019–1031. [CrossRef]
37. Zuhra, S.; Khan, N.S.; Alam, A.; Islam, S.; Khan, A. Buoyancy effects on nanoliquids film flow through a porous medium with gyrotactic microorganisms and cubic autocatalysis chemical reaction. *Adv. Mech. Eng.* **2020**, *12*, 1–17. [CrossRef]
38. Palwasha, Z.; Islam, S.; Khan, N.S.; Ayaz, H. Non-Newtonian nanoliquids thin film flow through a porous medium with magnetotactic microorganisms. *Appl. Nanosci.* **2018**, *8*, 1523–1544. [CrossRef]
39. Khan, N.S. Mixed convection in MHD second grade nanofluid flow through a porous medium containing nanoparticles and gyrotactic microorganisms with chemical reaction. *Filomat* **2019**, *33*, 4627–4653. [CrossRef]
40. Zuhra, S.; Khan, N.S.; Shah, Z.; Islam, Z.; Bonyah, E. Simulation of bioconvection in the suspension of second grade nanofluid containing nanoparticles and gyrotactic microorganisms. *AIP Adv.* **2018**, *8*, 105210. [CrossRef]
41. Khan, N.S.; Shah, Z.; Shutaywi, M.; Kumam, P.; Thounthong, P. A comprehensive study to the assessment of Arrhenius activation energy and binary chemical reaction in swirling flow. *Sci. Rep.* **2020**, *10*, 7868. [CrossRef]
42. Zuhra, S.; Khan, N.S.; Islam, S. Magnetohydrodynamic second grade nanofluid flow containing nanoparticles and gyrotactic microorganisms. *Comput. Appl. Math.* **2018**, *37*, 6332–6358. [CrossRef]
43. Zuhra, S.; Khan, N.S.; Islam, S.; Nawaz, R. Complexiton solutions for complex KdV equation by optimal homotopy asymptotic method. *Filomat* **2020**, *33*, 6195–6211. [CrossRef]
44. Zahra, A.; Mahanthesh, B.; Basir, M.F.M.; Imtiaz, M.; Mackolil, J.; Khan, N.S.; Nabwey, H.A.; Tlili, I. Mixed radiated magneto Casson fluid flow with Arrhenius activation energy and Newtonian heating effects: Flow and sensitivity analysis. *Alex. Eng. J.* **2020**, *57*, 1019–1031.
45. Liaqat, A.; Asifa, T.; Ali, R.; Islam, S.; Gul, T.; Kumam, P.; Mukhtar, S.; Khan, N.S.; Thounthong, P. A new analytical approach for the research of thin-film flow of magneto hydrodynamic fluid in the presence of thermal conductivity and variable viscosity. *ZAMM J. Appl. Math. Mech. Z. Angewwandte Math. Mech.* **2020**, 1–13. [CrossRef]
46. Khan, N.S.; Zuhra, S.; Shah, Q. Entropy generation in two phase model for simulating flow and heat transfer of carbon nanotubes between rotating stretchable disks with cubic autocatalysis chemical reaction. *Appl. Nanosci.* **2019**, *9*, 1797–1822. [CrossRef]
47. Khan, N.S.; Shah, Z.; Islam, S.; Khan, I.; Alkanhal, T.A.; Tlili, I. Entropy generation in MHD mixed convection non-Newtonian second-grade nanoliquid thin film flow through a porous medium with chemical reaction and stratification. *Entropy* **2019**, *21*, 139. [CrossRef]
48. Khan, N.S.; Zuhra, S.; Shah, Z.; Bonyah, E.; Khan, W.; Islam, S.; Khan, A. Hall current and thermophoresis effects on magnetohydrodynamic mixed convective heat and mass transfer thin film flow. *J. Phys. Commun.* **2019**, *3*, 035009. [CrossRef]
49. Khan, N.S.; Kumam, P.; Thounthong, P. Renewable energy technology for the sustainable development of thermal system with entropy measures. *Int. J. Heat Mass Transf.* **2019**, *145*, 118713. [CrossRef]
50. Khan, N.S.; Kumam, P.; Thounthong, P. Second law analysis with effects of Arrhenius activation energy and binary chemical reaction on nanofluid flow. *Sci. Rep.* **2020**, *10*, 1226. [CrossRef]
51. Khan, N.S.; Shah, Q.; Bhaumik, A.; Kumam, P.; Thounthong, P.; Amiri, I. Entropy generation in bioconvection nanofluid flow between two stretchable rotating disks. *Sci. Rep.* **2020**, *10*, 4448. [CrossRef]
52. Khan, N.S.; Shah, Q.; Sohail, A. Dynamics with Cattaneo-Christov heat and mass flux theory of bioconvection Oldroyd-B nanofluid. *Adv. Mech. Eng.* **2020**. [CrossRef]
53. Khan, N.S.; Shah, Q.; Sohail, A.; Kumam, P.; Thounthong, P.; Bhaumik, A.; Ullah, Z. Lorentz forces effects on the interactions of nanoparticles in emerging mechanisms with innovative approach. *Symmetry* **2020**, *5*, 1700. [CrossRef]
54. Liaqat, A.; Khan, N.S.; Ali, R.; Islam, S.; Kumam, P.; Thounthong, P. Novel insights through the computational techniques in unsteady MHD second grade fluid dynamics with oscillatory boundary conditions. *Heat Transf.* **2020**. [CrossRef]

55. Khan, N.S.; Ali, L.; Ali, R.; Kumam, P.; Thounthong, P. A novel algorithm for the computation of systems containing different types of integral and integro-differential equations. *Heat Transf.* **2020**. [CrossRef]

56. Ahmad, S.; Nadeem, S.; Ullah, N. Entropy generation and temperature-dependent viscosity in the study of SWCNT-MWCNT hybrid nanofluid. *Appl. Nanosci.* **2020**. [CrossRef]

57. Dinarvand, S.; Rostami, M.N. An innovative mass-based model of aqueous zinc oxide-gold hybrid nanofluid for von Karman's swirling flow. *J. Therm. Anal. Calorim.* **2019**. [CrossRef]

58. Ahmed, S.; Xu, H. Mixed convection in gravity driven thin nano-liquid film flow with homogeneous-heterogeneous reactions. *Phys. Fluids* **2020**, *32*, 023604. [CrossRef]

59. Hayat, T.; Haider, F.; Muhammad, T.; Ahmad, B. Darcy-Forchheimer flow of carbon nanotubes due to a convectively heated rotating disk with homogeneous-heterogeneous chemical reactions. *J. Therm. Anal. Calorim.* **2019**, *137*, 1939–1949. [CrossRef]

60. Suleman, M.; Ramzan, M.; Ahmad, S.; Lu, D.; Muhammad, T.; Chung, J.D. A numerical simulation of silver-water nanofluid flow with impacts of Newtonian heating and homogeneous-heterogeneous reactions past a nonlinear stretched cylinder. *Symmetry* **2019**, *11*, 295. [CrossRef]

61. Liao, S.J. *Homotopy Analysis Method in Non-Linear Differential Equations*; Higher Education Press: Beijing, China; Springer: Berlin/Heidelberg, Germany, 2012.

Publisher's Note: MDPI stays neutral with regard to jurisdictional claims in published maps and institutional affiliations.

Article

Cationically Modified Nanocrystalline Cellulose/Carboxyl-Functionalized Graphene Quantum Dots Nanocomposite Thin Film: Characterization and Potential Sensing Application

Najwa Norimanina Muhammad Rosddi [1], Yap Wing Fen [1,2,*] , Nur Ain Asyiqin Anas [2,3], Nur Alia Sheh Omar [2], Nur Syahira Md Ramdzan [1] and Wan Mohd Ebtisyam Mustaqim Mohd Daniyal [2]

[1] Department of Physics, Faculty of Science, Universiti Putra Malaysia, UPM Serdang 43400, Selangor, Malaysia; najwanorimanina@gmail.com (N.N.M.R.); nursyahira.upm@gmail.com (N.S.M.R.)

[2] Functional Devices Laboratory, Institute of Advanced Technology, Universiti Putra Malaysia, UPM Serdang 43400, Selangor, Malaysia; nurainanas.upm@gmail.com (N.A.A.A.); nuralia.upm@gmail.com (N.A.S.O.); wanmdsyam@gmail.com (W.M.E.M.M.D.)

[3] Physics Unit, Centre of Foundation Studies for Agricultural Science, Universiti Putra Malaysia, UPM Serdang 43400, Selangor, Malaysia

* Correspondence: yapwingfen@gmail.com; Tel.: +60-39769-6689

Received: 5 August 2020; Accepted: 5 September 2020; Published: 27 September 2020

Abstract: In this study, highly functional cationically modified nanocrystalline cellulose (NCC)/carboxyl-functionalized graphene quantum dots (CGQD) has been described. The surface of NCC was first modified with hexadecyltrimethylammonium bromide (CTA) before combining with CGQD. The CGQD, CTA-NCC and CTA-NCC/CGQD nanocomposites thin films were prepared using spin coating technique. The obtained nanocomposite thin films were then characterized by using the Fourier transform infrared spectroscopy (FTIR) which confirmed the existence of hydroxyl groups, carboxyl groups and alkyl groups in CTA-NCC/CGQD. The optical properties of the thin films were characterized using UV–Vis spectroscopy. The absorption of CTA-NCC/CGQD was high with an optical band gap of 4.127 eV. On the other hand, the CTA-NCC/CGQD nanocomposite thin film showed positive responses towards glucose solution of different concentration using an optical method based on surface plasmon resonance phenomenon. This work suggests that the novel nanocomposite thin film has potential for a sensing application in glucose detection.

Keywords: nanocrystalline cellulose; graphene quantum dots; thin film; optical; sensing; glucose; surface plasmon resonance

---

1. Introduction

In recent years, there has been an interest in the production of nanocrystalline cellulose (NCC) from cellulosic material because of its biodegradability, renewability, abundance and excellent mechanical properties [1]. In this world, cellulose is one of the most numerous natural renewable and biodegradable polysaccharides. NCC is the nano-scaled of needle or rod-shaped crystalline which has hundreds of nanometers in length and 1–10 nm in width [2,3]. NCC is obtained when cellulose undergoes acid hydrolysis with conditions where the amorphous regions are selectively hydrolyzed [4]. Mineral acids including hydrochloric acid and sulfuric acid are used in the mixture of hydrolysis of cellulose to prepare NCC [5]. Thus, NCC is constitutively acidic and exhibits a lyotropic phase behavior depending on the concentration. NCC has the potential in various applications as a rheology modifier such as drilling fluids, consumer products, drug delivery, artificial tissue formation and injectable hydrogels [6–8].

To enhance the NCC properties, the hydroxyl functional group in NCC can be modified by using several methods [9–12]. In this present work, NCC has been cationically modified using hexadecyltrimethylammonium bromide (CTA). CTA can enhance the absorption by improving hyperchromicity and sensitization of NCC [13–17]. To further increase the performance, this modified NCC was chosen as a matrix for carboxyl functionalized graphene quantum dots (CGQD) as CGQD has beneficial and unique properties including hydrophilicity, strong photoluminescence and photo-stability [18–20]. Due to its outstanding properties, CGQD provides unprecedented opportunities for different fields of application such as optical sensing, catalysis and bioimaging [21–25].

As far as we know, the optical properties of the CTA cationically modified NCC/CGQD (CTA-NCC/CGQD) nanocomposite thin film and its potential application for detection of glucose using surface plasmon resonance technique (SPR) have yet to be reported. SPR is known as a simple optical method for surface studies of thin films and can act as a very sensitive spectroscopy for detection of a variety of targets [26–42]. Hence in this study, the fabrication of the CTA-NCC/CGQD nanocomposite thin film, its characterization and potential sensing application were explored.

2. Materials and Methods

2.1. Reagent and Materials

Hexadecyltrimethyl ammonium bromide (CTA) and nanocrystalline cellulose (NCC) were purchased from Sigma Aldrich (St. Louis, MO, USA). Carboxyl-graphene quantum dots (CGQD) solution (0.1 wt%) was purchased from ACS Material (Pasadena, CA, USA) and glucose was purchased from R&M Marketing (Essex, UK).

2.2. Preparation of Chemicals

To prepare NCC solution, 1 g of NCC was diluted in 100 mL deionized water. Then, 0.2 g of CTA was diluted in 20 mL of deionized water to obtain CTA solution. NCC solution was then dropped into CTA solution drop by drop while heat stirred for 24 h. The CTA-NCC solution was centrifuged at 3000 rpm for 15 min. Then, CTA-NCC/CGQD solution (0.05 wt%) was obtained by dispersing 1 mL of CGQD into 1 mL of CTA-NCC. The glucose solution was prepared by dissolving 9.91 mg of glucose with 100 mL of deionized water to produce 10 µM of glucose solution. To prepare glucose solution with various concentration, the 10 µM of glucose solution was diluted with deionized water based on the formula $M_1V_1 = M_2V_2$ to obtain 0.005, 0.01, 0.03, 0.05 and 0.1 µM of glucose [43–45].

2.3. Preparation of CTA-NCC/CGQD Nanocomposite Thin Film

Glass cover slips (24 mm × 24 mm × 0.1 mm) were used as the substrates. The glass slip was first sputtered with gold (SC7640 sputter coater machine) for 67 seconds to obtain 50 nm of gold thin film [46–48]. Then, spin coating technique was used to deposit the CTA-NCC/CGQD solution homogenously on the gold surface. About 1000 µL of CTA-NCC/CGQD solution was added on a gold coated glass slip and was spun at 3000 rev/min for 30 seconds using spin coater P-6708D to obtain around 12–15 nm thickness of the CTA-NCC/CGQD layer. The summarized flow chart for the preparation of CTA-NCC/CGQD is shown in Figure 1.

Figure 1. Preparation of hexadecyltrimethylammonium bromide (CTA)-nanocrystalline cellulose (NCC)/carboxyl-functionalized graphene quantum dots (CGQD) thin film.

2.4. Characterization Instrument

The Fourier transform infrared (FTIR) spectrum of CGQD, CTA-NCC and CTA-NCC/CGQD solutions were analyzed using the Fourier Transform Infrared Spectrometer model spectrum 100 (PerkinElmer, Waltham, MA, USA), with wavelength set from 400 to 4000 cm^{-1} which is to determine the functional groups and the chemical interaction of the composites. Other than that, the purity of the compound can be obtained from the collection of the absorption band from the spectrum. For optical properties, the absorption of all samples with wavelength range from 220 nm to 500 nm was investigated using UV–Vis-NIR spectrometer (UV-3600 Shimadzu, Kyoto, Japan). The absorbance coated thin film was measured at room temperature. The energy band gap was determined by analyzing the graph of absorption peak against wavelength obtained using UV–Vis spectrometer.

2.5. Surface Plasmon Resonance

Surface plasmon resonance (SPR) is used to identify the potential of CTA-NCC/CGQD nanocomposite thin film for glucose detection. SPR is an optical process in which light satisfying resonance conditions excite a charge-density wave propagating along the interface between a dielectric material and metal by p-polarized and monochromatic light beam [49]. The reflected light intensity is reduced at a specific incident angle producing a sharp shadow due to the resonance occurs between surface plasmon wave and incident beam [50]. The SPR measurement was carried out by determining the reflected He-Ne laser beam (532.8 nm, 5 mW) [51]. Figure 2 shows the setup of SPR sensor. The SPR setup consisted of an He-Ne laser, a light attenuator, a polarizer and optical chopper (SR 540) and an optical stage driven by a stepper motor MM 3000 with a resolution of 0.001° (Newport, CA, USA). The reflected beam was detected by photodiode and then processed by the lock-in-amplifier (SR530) [52–55].

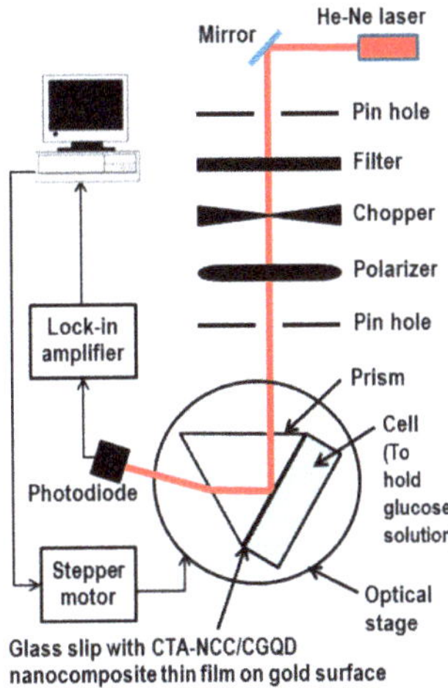

Figure 2. Experimental setup of surface plasmon resonance (SPR) sensor.

3. Results

3.1. FTIR Analysis

The FTIR spectrum of CGQD, CTA-NCC and CTA-NCC/CGQD solutions are shown in Figure 3. From the spectrum of the CGQD solution, the peak present at 3310 cm^{-1} represented O–H stretching. The peak at 2891 cm^{-1} was attributed to the C–H stretching. The characteristic band appearing at 1625 cm^{-1} corresponded to the stretching of C=O of the carboxylic group in graphene quantum dots and the peak at 1037 cm^{-1} represented the C–O stretching [18].

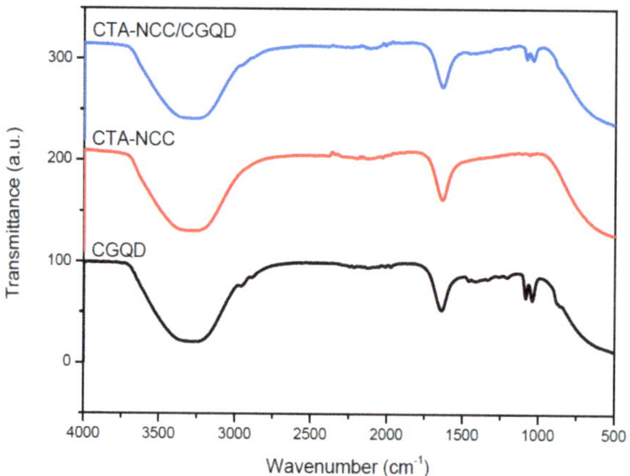

Figure 3. FTIR spectrum of CGQD, CTA-NCC and CTA-NCC/CGQD solutions.

Next, in the spectrum of CTA-NCC solution, the peak at 3332 cm^{-1} corresponded to the O–H stretching. The peak at 1617 cm^{-1} was attributed to the stretching vibration of C–O and the peak at 1056 cm^{-1} corresponded to C–O stretching [56].

The spectra of CTA-NCC/CGQD solution displayed the properties similar to CGQD and CTA-NCC thin film where there was a broad absorption peak at 3277 cm^{-1} that was attributed to the O–H stretching vibration. The peak at 2885 cm^{-1} corresponded to C–H stretching. The peak at 1637 cm^{-1} can be assigned to C=O stretching and is similar to the peak for both spectrums of CGQD and CTA-NCC. The characteristics band that appeared at 1032 cm^{-1} corresponded to the stretching of C–O. From the results, it is successfully confirmed that the functional groups of O–H, C–H, C=O and C–O existed in the composite solution. CGQD that are rich with oxygen-containing groups might interact with the hydroxyl groups and oxygen atoms in CTA-NCC through the hydrogen bonding. Furthermore, the possible structure of the composite is also presented in Figure 4.

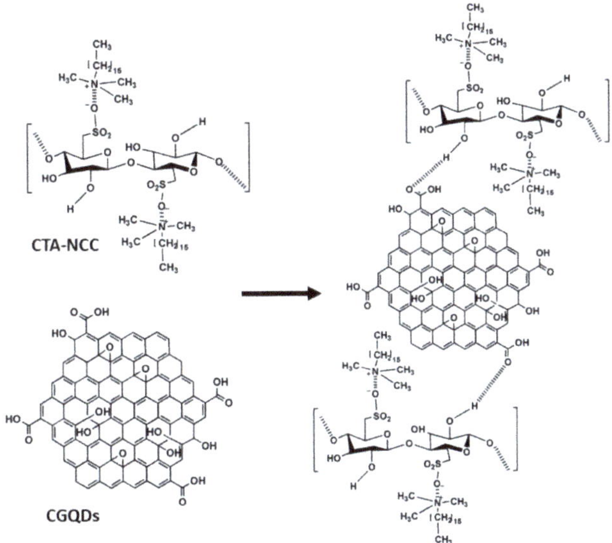

Figure 4. The possible structure of CTA-NCC/CGQD nanocomposite.

3.2. Optical Studies

The optical properties were analyzed using the absorbance spectrum of the thin film with wavelength range 220–500 nm. The absorbance curves for CGQD, CTA-NCC and CTA-NCC/CGQD thin film are presented in Figure 5.

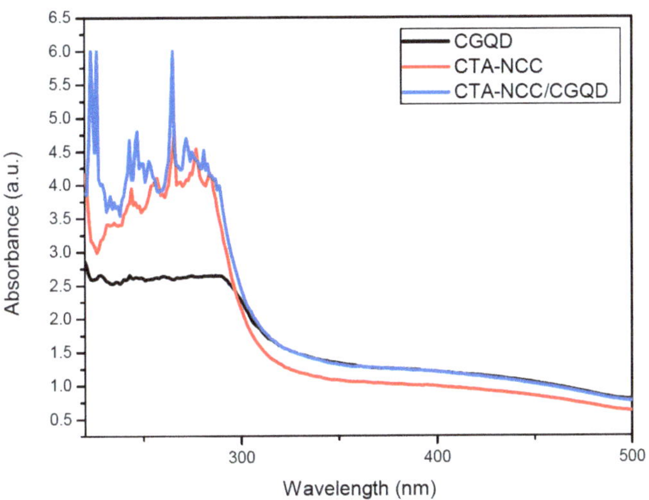

Figure 5. Absorbance spectrum of CGQD, CTA-NCC and CTA-NCC/CGQD thin film.

As shown in Figure 5, the absorbance curve for the thin film is different. From the figure, the CTA-NCC/CGQD thin film shows the highest absorption spectra with the absorption peaks at 222.9 nm, 225.9 nm and 264.9 nm. The highest absorption was contributed by the modification of CGQD with CTA-NCC, which confirmed the presence plasmon resonance in carbonaceous material [57]. On the other hand, the lowest absorbance belongs to CGQD thin film. The characteristic peaks that appeared in the nanocomposite thin film can be attributed to the presence of $\pi \rightarrow \pi^*$ bond transitions of the carbonyl groups [58]. In addition, it can be observed that the maximum absorption length can be determined from 263.04 nm to 266.63 nm. The results obtained are in the range of the absorption peaks for sulfur doped graphene quantum dots which are at 216–464 nm [59].

To proceed with the determination of the optical band gap, the relationship between the absorbance and the intensities of the monochromatic light was used [60]. The absorbance, A of samples can be related with the ratio of the initial light intensity on the detector I_0 to the light intensity with the presence of the sample I_t.

$$A = \log_{10} \frac{I_0}{I_t} \tag{1}$$

The absorbance coefficient is another quantity that can be measured. It is a very useful quantity which is used to compare samples of a varying thickness. The absorbance coefficient, α can be expressed as

$$\alpha = 2.303 \frac{A}{t} \tag{2}$$

where t is the thin film thickness in meters and the α is in units of m^{-1}. The energy band gap of these composites has been figured out with the help of the absorption coefficient. To obtain the optical band gap from the absorption spectra, the Tauc relation is used:

$$\alpha = \frac{k \left(h\upsilon - E_g \right)^n}{h\upsilon} \tag{3}$$

where k is a constant, h is the Plank's constant, υ is the frequency of the incident photon, the multiplication of h and υ, $h\upsilon$ represents the incident photon energy, E_g is the optical band gap and n is the state of transition. In this study we use $n = 1/2$ for direct transitions. Rearranging Equation (3) gives

$$(\alpha h\upsilon)^2 = k \left(h\upsilon - E_g \right) \tag{4}$$

Based on Equation (4), a graph of $(\alpha h\upsilon)^2$ against $h\upsilon$ can be plotted using linear fitting techniques and the optical band gap of the thin films can be determined [61–63]. According to Abdulla and Abbo (2012), the intersection of the straight line on the x-axis is taken as the value of the optical band gap [64]. The graph of $(\alpha h\upsilon)^2$ versus $h\upsilon$ for CGQD thin film, CTA-NCC thin film and CTA-NCC/CGQD thin film are shown in Figures 6–8, respectively.

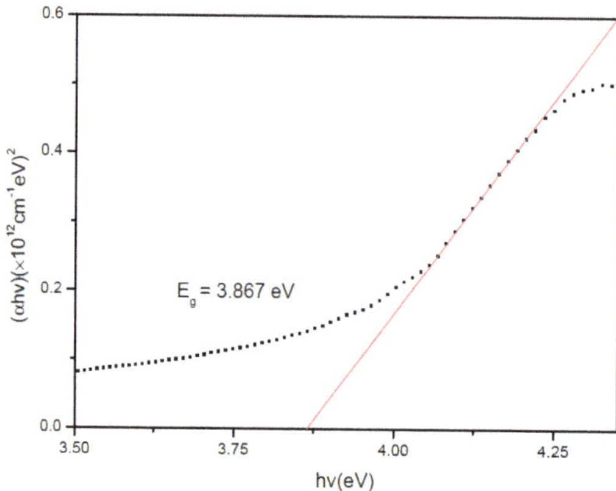

Figure 6. Optical band gap for CGQD thin film.

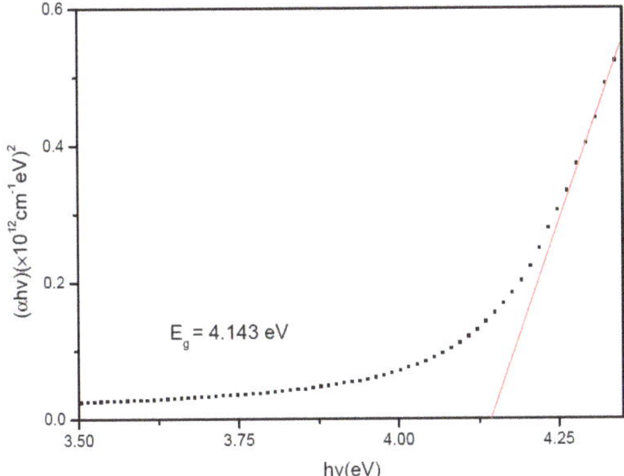

Figure 7. Optical band gap for CTA-NCC thin film.

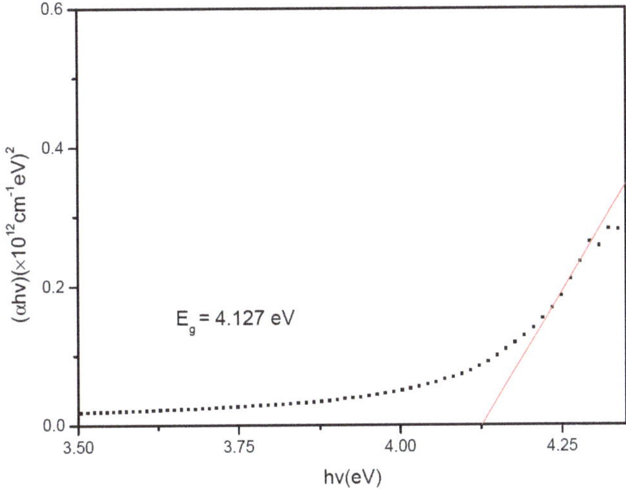

Figure 8. Optical band gap for CTA-NCC/CGQD thin film.

As can be seen from the figures, the intersection of the linear fitted line on the x-axis gives the value of the optical band gap. The term band gap is denoting the energy difference between the top of the valence band to the bottom of the conduction band where electrons can jump from one band to another. It necessitates a specific minimum extent of energy for the transition to permit an electron to jump from a valence band to a conduction band and this energy is called as the band gap energy. The optical band gap energies of CGQD, CTA-NCC and CTA-NCC/CGQD are 3.867 eV, 4.143 eV and 4.127 eV, respectively. Based on the result, CTA-NCC had the highest band gap energy among energy band gap results of all three thin films. The variations of optical band gap for the composite thin films, probably due to the presence of CTA-NCC solution as the band gap of CTA-NCC/CGQD, is higher than CGQD. The CTA-NCC/CGQD has a higher energy band gap as compared to CGQD thin film, which is in good agreement with the work reported by Daniyal et al. (2018), i.e., CTA-NCC increased the optical band gap of the composite [56].

3.3. Potential Sensing Analysis

The SPR experiment was first conducted using gold CTA-NCC/CGQD nanocomposite thin film in contact with deionized water (or 0 µM of glucose). The resonance angle for the first part of this experiment was obtained as 54.400°, where this value was used to compare the resonance angle for different concentrations of glucose solution. The SPR experiment was then continued for different concentrations of glucose solution that ranged from 0.005 µM to 0.1 µM. The glucose solution was injected into the cell one after another [65]. The reflectance as a function of incident angle of CTA-NCC/CGQD thin film, in contact with different concentrations of glucose solutions is shown in Figure 7. From the curves, the resonance angle can be obtained for the glucose concentration of 0.005, 0.01, 0.03, 0.05 and 0.1 µM.

From Figure 9, it can be observed that the resonance angle was shifted to the right and it further increased with the increase of glucose concentration [66,67]. The resonance angles obtained for 0.005, 0.01, 0.03, 0.05 and 0.1 µM were 54.571°, 54.732°, 54.755°, 54.767° and 54.769°, respectively. The shift of the SPR curves and resonance angle change demonstrated that the CTA-NCC/CGQD nanocomposite thin film has affinity towards glucose, where its incorporation with SPR can be a potential sensor for glucose.

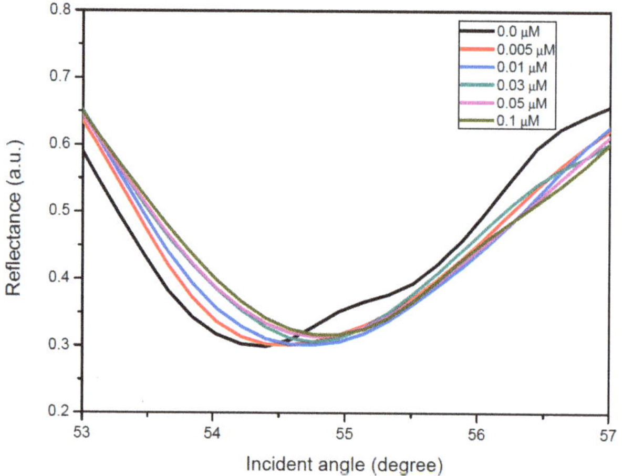

Figure 9. SPR curves for CTA-NCC/CGQD nanocomposite thin film for glucose solution with different concentrations (0.0–0.1 µM).

4. Conclusions

In this study, the CTA cationically modified NCC/CGQD nanocomposite thin film has been successfully fabricated. The functional groups that existed in the thin film were confirmed from the FTIR results. The absorbance value of CTA-NCC/CGQD was the highest with energy band gap of 4.127 eV. The studies of the CTA-NCC/CGQD nanocomposite thin film using the SPR technique have successfully shown that the novel thin film can detect various concentrations of glucose with the lowest detection of 5 nM. This study gives an important idea that the CTA-NCC/CGQD nanocomposite thin film has high potential as an application in sensing glucose when incorporated with the SPR technique and can be further investigated in future studies.

Author Contributions: Conceptualization, methodology, writing—original draft preparation, N.N.M.R.; validation, supervision, writing—review and editing, funding acquisition, Y.W.F.; investigation, formal analysis, N.A.A.A.; software, resources, N.A.S.O.; visualization, N.S.M.R. and W.M.E.M.M.D. All authors have read and agreed to the published version of the manuscript.

Funding: This research was funded and supported by the Ministry of Education Malaysia through the Fundamental Research Grant Scheme (FRGS) (FRGS/1/2019/STG02/UPM/02/1) and Putra Grant Universiti Putra Malaysia.

Acknowledgments: The authors acknowledged the laboratory facilities provided by the Institute of Advanced Technology, Department of Physics, and Department of Chemistry, Universiti Putra Malaysia.

Conflicts of Interest: The authors declare no conflict of interest.

References

1. Azrinaa, Z.A.Z.; Beg, M.D.H.; Rosli, M.Y.; Ramli, R.; Junadi, N.; Alam, A.K.M.M. Spherical nanocrystalline cellulose (NCC) from oil palm empty fruitbunch pulp via ultrasound assisted hydrolysis. *Carbohydr. Polym.* **2017**, *162*, 115–120. [CrossRef] [PubMed]
2. Daniyal, W.M.E.M.M.; Fen, Y.W.; Abdullah, J.; Sadrolhosseini, A.R.; Saleviter, S.; Omar, N.A.S. Exploration of surface plasmon resonance for sensing copper ion based on nanocrystalline cellulose-modified thin film. *Opt. Express* **2018**, *26*, 34880–34893. [CrossRef] [PubMed]
3. Junior, E.A.P.; Dávila, J.L.; d'Ávila, M.A. Rheological studies on nanocrystalline cellulose/alginate suspensions. *J. Mol. Liq.* **2019**, *277*, 418–423.
4. Kaboorani, A.; Riedl, B.; Blanchet, P.; Fellin, M.; Hosseinaei, O.; Wang, S. Nanocrystalline cellulose (NCC): A renewable nanomaterial for polyvinyl acetate (PVA) adhesive. *Eur. Polym. J.* **2012**, *48*, 1829–1837. [CrossRef]
5. Sun, Y.; Liu, P.; Liu, Z. Catalytic conversion of carbohydrates to 5-hydroxymethylfurfuralfrom the waste liquid of acid hydrolysis NCC. *Carbohydr. Polym.* **2016**, *142*, 177–182. [CrossRef] [PubMed]
6. Li, M.C.; Wu, Q.; Song, K.; Qing, Y.; Wu, Y. Cellulose nanoparticles as modifiers for rheology and fluid loss in bentonite water-based fluids. *ACS Appl. Mater. Interfaces* **2015**, *7*, 5006–5016. [CrossRef]
7. Taheri, A.; Mohammadi, M. The use of cellulose nanocrystals for potential application in topical delivery of hydroquinone. *Chem. Biol. Drug Des.* **2015**, *86*, 102–106. [CrossRef]
8. Yang, Y.; Chen, Z.; Zhang, J.; Wang, G.; Zhang, R.; Suo, D. Preparation and Applications of the Cellulose Nanocrystal. *Int. J. Polym. Sci.* **2019**, *2019*, 1767028. [CrossRef]
9. Daniyal, W.M.E.M.M.; Fen, Y.W.; Anas, N.A.A.; Omar, N.A.S.; Ramdzan, N.S.M.; Nakajima, H.; Mahdi, M.A. Enhancing the sensitivity of a surface plasmon resonance-based optical sensor for zinc ion detection by the modification of a gold thin film. *RSC Adv.* **2019**, *9*, 41729–41736. [CrossRef]
10. Salajková, M.; Berglund, L.A.; Zhou, Q. Hydrophobic cellulose nanocrystals modified with quaternary ammonium salts. *J. Mater. Chem.* **2012**, *22*, 19798–19805. [CrossRef]
11. Grunert, M.; Winter, W.T. Nanocomposites of cellulose acetate butyrate reinforced with cellulose nanocrystals. *J. Polym. Environ.* **2002**, *10*, 27–30. [CrossRef]
12. Abitbol, T.; Marway, H.; Cranston, E.D. Surface modification of cellulose nanocrystals with cetyltrimethylammonium bromide. *Nord. Pulp Pap. Res. J.* **2014**, *29*, 7764–7779. [CrossRef]
13. Mao, Y.; Fan, Q.; Li, J.; Yu, L.; Qu, L.-B. A novel and green CTAB-functionalized graphene nanosheets electrochemical sensor for Sudan I determination. *Sens. Actuators B Chem.* **2014**, *203*, 759–765. [CrossRef]
14. Selvi, S.S.T.; Linet, J.M.; Sagadevan, S. Influence of CTAB surfactant on structural and optical properties of CuS and CdS nanoparticles by hydrothermal route. *J. Exp. Nanosci.* **2018**, *13*, 130–143. [CrossRef]
15. Yao, P.J.; Wang, J.; Du, H.Y.; Zhao, L.; Yan, W.P. Effect of CTAB concentration on synthesis and sensing properties of perovskite oxide via a hydrothermal process. *Key Eng. Mater.* **2013**, *543*, 330–333. [CrossRef]
16. Jin, Y.; Liu, F.; Tong, M.; Hou, Y. Removal of arsenate by cetyltrimethylammonium bromide modified magnetic nanoparticles. *J. Hazard. Mater.* **2012**, *227*, 461–468. [CrossRef]
17. Leng, Y.; Qian, S.; Wang, Y.; Lu, C.; Ji, X.; Lu, Z.; Lin, H. Single-indicator-based multidimensional sensing: Detection and identification of heavy metal ions and understanding the foundations from experiment to simulation. *Sci. Rep.* **2016**, *6*, 25354. [CrossRef]
18. Ramdzan, N.S.M.; Fen, Y.W.; Omar, N.A.S.; Saleviter, S.; Zainudin, A.A. Optical and surface plasmon resonance sensing properties for chitosan/carboxyl-functionalized graphene quantum dots thin film. *Optik* **2019**, *178*, 802–812. [CrossRef]
19. Shen, S.; Wang, J.; Wu, Z.; Du, Z.; Tang, Z.; Wu, X. Graphene quantum dots with high yield and high quality synthesized from low cost precursor of aphanitic graphite. *Nanomaterials* **2020**, *10*, 375. [CrossRef]
20. Kaciulis, S.; Mezzi, A.; Soltani, P.; Pizzoferrato, R.; Ciotta, E.; Prosposito, P. Graphene quantum dots obtained by unfolding fullerene. *Thin Solid Films* **2019**, *673*, 19–25. [CrossRef]
21. Li, L.; Li, L.; Wang, C.; Liu, K.; Zhu, R.; Qiang, H.; Lin, Y. Synthesis of nitrogen-doped and amino acid-functionalized graphene quantum dots from glycine, and their application to the fluorometric determination of ferric ion. *Microchim. Acta* **2014**, *182*, 763–770. [CrossRef]

22. Tang, Y.; Li, J.; Guo, Q.; Nie, G. An ultrasensitive electrochemiluminescence assay for Hg^{2+} through graphene quantum dots and poly(5-formylindole) nanocomposite. *Sens. Actuators B Chem.* **2019**, *282*, 824–830. [CrossRef]

23. Tian, P.; Tang, L.; Teng, K.S.; Lau, S.P. Graphene quantum dots from chemistry to applications. *Mater. Today Chem.* **2018**, *10*, 221–258. [CrossRef]

24. Li, Y.; Hu, Y.; Zhao, Y.; Shi, G.; Deng, L.; Hou, Y.; Qu, L. An electrochemical avenue to green-luminescent graphene quantum dots as potential electron-acceptors for photovoltaics. *Adv. Mater.* **2010**, *23*, 776–780. [CrossRef] [PubMed]

25. Kováčová, M.; Špitalská, E.; Markovic, Z.; Špitálský, Z. Carbon quantum dots as antibacterial photosensitizers and their polymer nanocomposite applications. *Part. Part. Syst. Charact.* **2020**, *37*, 1900348. [CrossRef]

26. Fen, Y.W.; Yunus, W.M.M. Surface plasmon resonance spectroscopy as an alternative for sensing heavy metal ions: A review. *Sens. Rev.* **2013**, *33*, 305–314.

27. Omar, N.A.S.; Fen, Y.W.; Saleviter, S.; Daniyal, W.M.E.M.M.; Anas, N.A.A.; Ramdzan, N.S.M.; Roshidi, M.D.A. Development of a graphene-based surface plasmon resonance optical sensor chip for potential biomedical application. *Materials* **2019**, *12*, 1928. [CrossRef]

28. Anas, N.A.A.; Fen, Y.W.; Omar, N.A.S.; Daniyal, W.M.E.M.M.; Ramdzan, N.S.M.; Saleviter, S. Development of graphene quantum dots-based optical sensor for toxic metal ion detection. *Sensors* **2019**, *19*, 3850. [CrossRef]

29. Eddin, F.B.K.; Fen, Y.W. The principle of nanomaterials based surface plasmon resonance biosensors and its potential for dopamine detection. *Molecules* **2020**, *25*, 2769. [CrossRef] [PubMed]

30. Hashim, H.S.; Fen, Y.W.; Omar, N.A.S.; Omar, N.A.S.; Abdullah, J.; Daniyal, W.M.E.M.M.; Saleviter, S. Detection of phenol by incorporation of gold modified-enzyme based graphene oxide thin film with surface plasmon resonance technique. *Opt. Express* **2020**, *28*, 9738–9752. [CrossRef] [PubMed]

31. Daniyal, W.M.E.M.M.; Saleviter, S.; Fen, Y.W. Development of surface plasmon resonance spectroscopy for metal ion detection. *Sens. Mater.* **2018**, *30*, 2023–2038. [CrossRef]

32. Jia, Y.; Peng, Y.; Bai, J.; Zhang, X.; Cui, Y.; Ning, B.; Cui, J.; Gao, Z. Magnetic nanoparticle enhanced surface plasmon resonance sensor for estradiol analysis. *Sens. Actuators B Chem.* **2017**, *254*, 629–635. [CrossRef]

33. Fen, Y.W.; Yunus, W.M.M.; Yusof, N.A. Surface plasmon resonance optical sensor for detection of essential heavy metal ions with potential for toxicity: Copper, zinc and manganese ions. *Sens. Lett.* **2011**, *9*, 1704–1711. [CrossRef]

34. Fen, Y.W.; Yunus, W.M.M.; Yusof, N.A. Detection of mercury and copper ions using surface plasmon resonance optical sensor. *Sens. Mater.* **2011**, *23*, 325–334.

35. Zainudin, A.A.; Fen, Y.W.; Yusof, N.A.; Al-Rekabi, S.H.; Mahdi, M.A.; Omar, N.A.S. Incorporation of surface plasmon resonance with novel valinomycin doped chitosan-graphene oxide thin film for sensing potassium ion. *Spectrochim. Acta A Mol. Biomol. Spectrosc.* **2018**, *191*, 111–115. [CrossRef]

36. Omar, N.A.S.; Fen, Y.W. Recent development of SPR spectroscopy as potential method for diagnosis of dengue virus E-protein. *Sens. Rev.* **2018**, *38*, 106–116. [CrossRef]

37. Saleviter, S.; Fen, Y.W.; Omar, N.A.S.; Zainudin, A.A.; Yusof, N.A. Development of optical sensor for determination of Co(II) based on surface plasmon resonance phenomenon. *Sens. Lett.* **2017**, *15*, 862–867. [CrossRef]

38. Fen, Y.W.; Yunus, W.M.M.; Talib, Z.A. Analysis of Pb(II) ion sensing by crosslinked chitosan thin film using surface plasmon resonance spectroscopy. *Optik* **2013**, *124*, 126–133. [CrossRef]

39. Ramdzan, N.S.M.; Fen, Y.W.; Anas, N.A.A.; Omar, N.A.S.; Saleviter, S. Development of biopolymer and conducting polymer-based optical sensors for heavy metal ion detection. *Molecules* **2020**, *25*, 2548. [CrossRef]

40. Anas, N.A.A.; Fen, Y.W.; Omar, N.A.S.; Ramdzan, N.S.M.; Daniyal, W.M.E.M.M.; Saleviter, S.; Zainudin, A.A. Optical properties of chitosan/hydroxyl-functionalized graphene quantum dots thin film for potential optical detection of ferric(III) ion. *Opt. Laser Technol.* **2019**, *120*, 105724. [CrossRef]

41. Al-Rekabi, S.H.; Kamil, Y.M.; Bakar, M.H.A.; Fen, Y.W.; Lim, H.N.; Kanagesan, S.; Mahdi, M.A. Hydrous ferric oxide-magnetite-reduced graphene oxide nanocomposite for optical detection of arsenic using surface plasmon resonance. *Opt. Laser Technol.* **2019**, *111*, 417–423. [CrossRef]

42. Fen, Y.W.; Yunus, W.M.M.; Talib, Z.A.; Yusof, N.A. Development of surface plasmon resonance sensor for determining zinc ion using novel active nanolayers as probe. *Spectrochim. Acta A Mol. Biomol. Spectrosc.* **2015**, *134*, 48–52. [CrossRef]

43. Saleviter, S.; Fen, Y.W.; Daniyal, W.M.E.M.M.; Abdullah, J.; Sadrolhosseini, A.R.; Omar, N.A.S. Design and analysis of surface plasmon resonance optical sensor for determining cobalt ion based on chitosan-graphene oxide decorated quantum dots-modified gold active layer. *Opt. Express* **2019**, *27*, 32294–32307. [CrossRef] [PubMed]

44. Omar, N.A.S.; Fen, Y.W.; Abdullah, J.; Sadrolhosseini, A.R.; Kamil, Y.M.; Fauzi, N.I.F.; Hashim, H.S.; Mahdi, M.A. Quantitative and selective surface plasmon resonance response based on a reduced graphene oxide-polyamidoamine nanocomposite for detection of dengue virus e-proteins. *Nanomaterials* **2020**, *10*, 569. [CrossRef] [PubMed]

45. Fen, Y.W.; Yunus, W.M.M.; Yusof, N.A. Optical properties of crosslinked chitosan thin film as copper ion detection using surface plasmon resonance technique. *Opt. Appl.* **2011**, *41*, 999–1013.

46. Roshidi, D.A.; Fen, Y.W.; Daniyal, W.M.E.M.M.; Omar, N.A.S.; Zulholinda, M. Structural and optical properties of chitosan–poly(amidoamine) dendrimer composite thin film for potential sensing Pb^{2+} using an optical spectroscopy. *Optik* **2019**, *185*, 351–358. [CrossRef]

47. Omar, N.A.S.; Fen, Y.W.; Abdullah, J.; Zaid, M.H.M.; Daniyal, W.M.E.M.M.; Mahdi, M.A. Sensitive surface plasmon resonance performance of cadmium sulfide quantum dots-amine functionalized graphene oxide based thin film towards dengue virus E-protein. *Opt. Laser Technol.* **2019**, *114*, 204–208. [CrossRef]

48. Anas, N.A.A.; Fen, Y.W.; Yusof, N.A.; Omar, N.A.S.; Ramdzan, N.S.M.; Daniyal, W.M.E.M.M. Investigating the properties of cetyltrimethylammonium bromide/hydroxylated graphene quantum dots thin film for potential optical detection of heavy metal ions. *Materials* **2020**, *13*, 2591. [CrossRef]

49. Homola, J. Surface plasmon resonance sensors for detection of chemical and biological species. *Chem. Rev.* **2008**, *108*, 462–493. [CrossRef]

50. Fen, Y.W.; Yunus, W.M.M. Utilization of chitosan-based sensor thin films for the detection of lead ion by surface plasmon resonance optical sensor. *IEEE Sens. J.* **2013**, *13*, 1413–1418. [CrossRef]

51. Fen, Y.W.; Yunus, W.M.M.; Yusof, N.A. Surface plasmon resonance optical sensor for detection of Pb^{2+} based on immobilized p-tert-butylcalix[4]arene-tetrakis in chitosan thin film as an active layer. *Sens. Actuators B Chem.* **2012**, *171–172*, 287–293. [CrossRef]

52. Saleviter, S.; Fen, Y.W.; Omar, N.A.S.; Daniyal, W.M.E.M.M.; Abdullah, J.; Zaid, M.H.M. Structural and optical studies of cadmium sulfide quantum dot-graphene oxide-chitosan nanocomposite thin film as a novel SPR spectroscopy active layer. *J. Nanomater.* **2018**, *2018*, 4324072. [CrossRef]

53. Zainudin, A.A.; Fen, Y.W.; Yusof, N.A.; Omar, N.A.S. Structural, optical and sensing properties of ionophore doped graphene based bionanocomposite thin film. *Optik* **2017**, *144*, 308–315. [CrossRef]

54. Fen, Y.W.; Yunus, W.M.M.; Moksin, M.M.; Talib, Z.A.; Yusof, N.A. Surface plasmon resonance optical sensor for mercury ion detection by crosslinked chitosan thin film. *J. Optoelectron. Adv. Mater.* **2011**, *13*, 279–285.

55. Fauzi, N.I.M.; Fen, Y.W.; Omar, N.A.S.; Saleviter, S.; Daniyal, W.M.E.M.M.; Hashim, H.S.; Nasrullah, M. Nanostructured chitosan/maghemite composites thin film for potential optical detection of mercury ion by surface plasmon resonance investigation. *Polymers* **2020**, *12*, 1497. [CrossRef]

56. Daniyal, W.M.E.M.M.; Fen, Y.W.; Abdullah, J.; Saleviter, S.; Omar, N.A.S. Preparation and characterization of hexadecyltrimethyl-ammonium bromide modified nanocrystalline cellulose/graphene oxide composite thin film and its potential in sensing copper ion using surface plasmon resonance technique. *Optik* **2018**, *173*, 71–77. [CrossRef]

57. Zamiranvari, A.; Solati, E.; Dorranian, D. Effect of CTAB concentration on the properties nanosheet produced by laser ablation. *Opt. Laser Technol.* **2017**, *97*, 209–218. [CrossRef]

58. Rattana, T.; Chaiyakun, S.; Witit-Anun, N.; Nuntawong, N.; Chindaudom, P.; Oaew, S.; Kedkeaw, C.; Limsuwan, P. Preparation and characterization of graphene oxide nanosheets. *Procedia Eng.* **2012**, *32*, 759–764. [CrossRef]

59. Li, X.; Lao, S.P.; Tang, L.; Ji, R.; Yang, P. Sulphur doping: A facile approach to tune the electronic structure and optical properties of graphene quantum dots. *Nanoscale* **2014**, *6*, 5323–5328. [CrossRef]

60. Fen, Y.W.; Yunus, W.M.M.; Yusof, N.A.; Ishak, N.S.; Omar, N.A.S.; Zainudin, A.A. Preparation, characterization and optical properties of ionophore doped chitosan biopolymer thin film and its potential application for sensing metal ion. *Optik* **2015**, *126*, 4688–4692. [CrossRef]

61. Saleviter, S.; Fen, Y.W.; Omar, N.A.; Zainudin, A.A.; Daniyal, W.M.E.M.M. Optical and structural characterization of immobilized 4-(2-pyridylazo) resorcinol in chitosan-graphene oxide composite thin film and its potential for Co^{2+} sensing using surface plasmon resonance technique. *Results Phys.* **2018**, *11*, 118–122. [CrossRef]

62. Roshidi, M.D.A.; Fen, Y.W.; Omar, N.A.S.; Saleviter, S.; Daniyal, W.M.E.M.M. Optical studies of graphene oxide/poly(amidoamine) dendrimer composite thin film and its potential for sensing Hg^{2+} using surface plasmon resonance spectroscopy. *Sens. Mater.* **2018**, *31*, 1147–1156. [CrossRef]

63. Hashim, H.S.; Fen, Y.W.; Omar, N.A.S.; Daniyal, W.M.E.M.M.; Saleviter, S.; Abdullah, J. Structural, optical and potential sensing properties of tyrosinase immobilized graphene oxide thin film on gold surface. *Optik* **2020**, *212*, 164786. [CrossRef]

64. Abdulla, H.S.; Abbo, A.I. Optical and electrical properties of thin films of polyaniline and polypyrrole. *Int. J. Electrochem. Sci.* **2012**, *7*, 10666–10678.

65. Daniyal, W.M.E.M.M.; Fen, Y.W.; Abdullah, J.; Sadrolhosseini, A.R.; Saleviter, S.; Omar, N.A.S. Label-free optical spectroscopy for characterizing binding properties of highly sensitive nanocrystalline cellulose-graphene oxide based nanocomposite towards nickel ion. *Spectrochim. Acta Part A Mol. Biomol. Spectrosc.* **2019**, *212*, 25–31. [CrossRef]

66. Omar, N.A.S.; Fen, Y.W.; Saleviter, S.; Kamil, Y.M.; Daniyal, W.M.E.M.M.; Abdullah, J.; Mahdi, M.A. Experimental evaluation on surface plasmon resonance sensor performance based on sensitive hyperbranched polymer nanocomposite thin films. *Sens. Actuators A Phys.* **2020**, *303*, 111830. [CrossRef]

67. Sadrolhosseini, A.; Noor, A.; Bahrami, A.; Lim, H.N.; Talib, Z.A.; Mahdi, M.A. Application of polypyrrole multi-walled carbon nanotube composite layer for detection of mercury, lead and iron ions using surface plasmon resonance technique. *PLoS ONE* **2014**, *9*, e93962. [CrossRef]

 *crystals*

Article

Mixed Convective Radiative Flow through a Slender Revolution Bodies Containing Molybdenum-Disulfide Graphene Oxide along with Generalized Hybrid Nanoparticles in Porous Media

Umair Khan [1], Aurang Zaib [2,3,*], Mohsen Sheikholeslami [4], Abderrahim Wakif [5] and Dumitru Baleanu [6,7,8]

[1] Department of Mathematics and Social Sciences, Sukkur IBA University, Sukkur 65200, Sindh, Pakistan; umairkhan@iba-suk.edu.pk

[2] Department of Natural Sciences, The Begum Nusrat Bhutto Women University, Sukkur 65170, Pakistan

[3] Department of Mathematical Sciences, Federal Urdu University of Arts, Science & Technology, Gulshan-e-Iqbal, Karachi 75300, Pakistan

[4] Department of Mechanical Engineering, Babol Noshiravni University of Technology, P.O. Box 484, Babol, Mazandaran, Iran; mohsen.sheikholeslami@yahoo.com

[5] Laboratory of Mechanics, Faculty of Sciences Aïn Chock, Hassan II University, B.P. 5366 Mâarif, Casablanca 20100, Morocco; wakif.abderrahim@gmail.com

[6] Department of Mathematics, Cankaya University, Ankara 06790, Turkey; dumitru.baleanu@gmail.com

[7] Institute of Space Sciences, 077125 Magurele, Romania

[8] Department of Medical Research, China Medical University, Taichung 40447, Taiwan

* Correspondence: aurangzaib@fuuast.edu.pk

Received: 29 July 2020; Accepted: 28 August 2020; Published: 31 August 2020

Abstract: The current framework tackles the buoyancy flow via a slender revolution bodies comprising Molybdenum-Disulfide Graphene Oxide generalized hybrid nanofluid embedded in a porous medium. The impact of radiation is also provoked. The outcomes are presented in this analysis to examine the behavior of hybrid nanofluid flow (HNANF) through the cone, the paraboloid, and the cylinder-shaped bodies. The opposing flow (OPPF) as well as the assisting flow (ASSF) is discussed. The leading flow equations of generalized hybrid nanoliquid are worked out numerically by utilizing bvp4c solver. This sort of the problem may meet in the automatic industries connected to geothermal and geophysical applications where the sheet heat transport occurs. The impacts of engaging controlled parameters of the transmuted system on the drag force and the velocity profile are presented through the graphs and tables. The achieved outcomes suggest that the velocity upsurges due to the dimensionless radius of the slender body parameter in case of the assisting flow and declines in the opposing flow. Additionally, an increment is observed owing to the shaped bodies as well as in type A nanofluid and type B hybrid nanofluid.

Keywords: hybrid nanofluid; slender body revolution; porous media; radiation effect; mixed convection

---

1. Introduction

The research regarding the convective-flow entrenched in porous media widely has been utilized owing to its vast engineering applications as solar collectors, heat exchangers, post-accidental heat exclusion in nuclear reactors, building construction, drying processes, oil recovery and geothermal, ground water pollution, etc. Nield [1] analyzed the liquid flow of stability ensuing through a vertical mass and thermal gradients via a horizontal-layer immersed in porous media. Bejan and Khair [2] explored the marvel of mass and heat transfer through a vertical sheet entrenched in a porous medium and they have taken unvarying concentration as well as temperature. The impact of mixed as well as free convective flows with heat transport through a slender revolution of the body in a porous medium was examined by Lai et al. [3] and they concluded that the temperature gradient shrinks due to dimensional radius. The thermal and mass diffusion through a cone embedded in porous medium was scrutinized by Yih [4]. Bano and Singh [5] explored the radiation influence on mass and heat transport from a radiated thin needle in a saturated porous medium by utilizing a technique of Von Karman-integral. Singh and Chandarki [6] inspected the free convective flow with mass and heat transfer through a vertical cylinder occupied in porous media. The mixed convective flow through a vertical flat surface in a porous medium with nanoliquid was examined by Ahmad and Pop [7]. Talebizadeh et al. [8] scrutinized the impact of radiation on natural convective flow through a porous vertical surface and found numerical as well as exact solutions. Moghimi et al. [9] discussed the MHD (magnetohydrodynamic) influence on natural convective flow via a sphere saturated in a porous medium. Moghimi et al. [10] obtained an exact solution of flow through a flat surface with constant heat flux and slip effect by utilizing DQM (differential quadrature method) and HAM (homotopy analysis method). Raju and Sandeep [11] inspected the magnetic influence on flow of mass and heat transport containing Casson fluid through a rotated vertical cone/plate in porous media with micro-organisms. They perceived that the mass and heat transfer from a cone is superior compared to flow over a plate. Raju et al. [12] applied the Buongiorno model to inspect the phenomenon of mass and heat transport through a radiated revolution slender body in porous media. They perceived that the behavior of temperature shrinks in a flow revolution over a cone compared to flow revolution through the cylinder and paraboloid.

Several recent explorations divulged that nanofluids have superior capability of heat transport than convectional fluids. Thus, it is likely to swap conventional heat transport fluids via nanofluids in the numerous designs of heat transport like heat exchanger, heat generators, and cooling systems. Choi [13] observed that, by scattering metallic nanometer sized particles in regular heat transport liquids, the ensuing nanoliquids hold greater thermal conductivity than those of presently utilized ones. Further, Eastman et al. [14] discovered that the shape of the particle has a stronger effect on nanofluid effective thermal conductivity than the size of particle or thermal conductivity of the particle. To augment the nanofluids heat transport owing to nanoparticles migration and the resulting boundary-layer disturbance was experimentally examined by Wen and Ding [15]. The Boltzmann technique to inspect the magnetic impact on the natural convective flow comprising nanoliquid through a cylindrical annulus was studied by Ashorynejad et al. [16]. The impact of distinct shapes of nanoscale particles in EG (entropy generation) based aqueous solution was inspected by Ellahi et al. [17]. Inspired by the importance of nanofluids, several researchers recently were engaged with the debate of flow with heat transport to nanoliquids via different perspective [18–21].

Recently, a novel type of fluid, suggested hybrid nanofluids has been utilized to augment the heat transport in applications of thermal [22–24]. Hybrid nanoliquids consist of two or more different nanoparticles in either mixture of non-composite forms. Hybrid nanofluids envisages in the fields of heat transport as electronic and generator cooling, thermal storage, biomedical, cooling of transformer, lubrication, solar heating, spacecraft and aircraft, welding, protection, refrigeration, and heat pump. Minea [25] inspected the estimations of distinct viscosity of hybrid nanoliquid by scattering the water-based TiO_2, Al_2O_3, and SiO_2 nanomaterials. The impact of nonlinear radiation on the magneto flow of micropolar hybrid (Cu-Al_2O_3) dusty nanofluid from a stretched sheet was examined by Ghadikolaei et al. [26]. Sheikoleslami et al. [27] scrutinized the modeling of the porous domain with Lorentz forces and radiation impact and obtained the solution by CVFEM (control volume based finite element method). They explored that temperature gradient has greater influence due to the greater buoyancy parameter. Gholinia et al. [28] inspected the steady magneto flow of hybrid CNTs (carbon nanotubes) nanofluid over a permeable stretched cylinder. Recently, Khan et al. [29]

discussed the influence of magnetic function comprising ethylene glycol-based hybrid nanofluid from a stretched/shrinking wedge with mixed convection and stability analysis performed.

The earlier review literature reveals that the mixed convective generalized hybrid (MoS_2-Go) nanofluid flow through a slender revolution bodies in porous media has been highlighted as a mostly unknown area. The radiation impact with generalized hybrid single phase model is discussed here, which is not considered in the earlier published results [7,11]. The leading PDEs (partial differential equations) are altered into ODEs (ordinary differential equation) through suitable transformations and then tackled via three stage Lobatto formula. Impacts of the pertinent variables are portrayed and investigated through the graphs.

2. Problem Formulation

In the current exploration, the buoyancy flow through a slender revolution bodies containing Molybdenum-Disulfide Graphene Oxide along with the generalized hybrid nanofluid entrenched in a saturated porous medium is shown schematically in Figure 1. To optimize the heat transport, the impact of radiation in the occurrence of the opposing and the assisting flows is analyzed. It is assumed that the ambient velocity is considered as $u_e(x) = U_\infty x^m$ with constant U_∞, while T_∞ the constant free stream temperature. The temperature of the slender revolution body is taken as $T_w(x)$ with $T_w(x) > T_\infty$ utilizes for (ASSF) and $T_w(x) < T_\infty$ uses for (OPPF). The system of coordinate from the slender body vortex is at origin, while the distance along the revolution body and normal to the slender revolution body is presented by the cylindrical coordinate (x, r). Applying the single-phase model suggested by the Tiwari and Das [30] with the approximations of Boussinesq and boundary layer scaling, the leading equations are

$$\frac{\partial(ru)}{\partial x} + \frac{\partial(vr)}{\partial r} = 0 \tag{1}$$

$$\frac{\mu_{HNF}}{\mu_F}\frac{\partial u}{\partial r} = \frac{Kg(\rho\beta)_F}{\mu_F}\frac{(\rho\beta)_{HNF}}{(\rho\beta)_F}\frac{\partial T}{\partial r} \tag{2}$$

$$u\frac{\partial T}{\partial x} + v\frac{\partial T}{\partial r} = \frac{k_{HNF}}{(\rho c_p)_{HNF}}\frac{1}{r}\frac{\partial}{\partial r}\left(r\frac{\partial T}{\partial r}\right) - \frac{1}{(\rho c_p)_{HNF}}\frac{\partial}{\partial r}(rq_r) \tag{3}$$

with the corresponding boundary conditions

$$\begin{aligned} v = 0, \ T = T_w(x) = T_\infty + Ax^\lambda \text{ at } r = R(x), \\ u = u_e(x) = U_\infty x^m, \ T = T_\infty \text{ as } r \to \infty. \end{aligned} \tag{4}$$

Integrating Equation (2) and utilizing Equation (4), it becomes

$$\frac{\mu_{HNF}}{\mu_F}u = \frac{\mu_{HNF}}{\mu_F}u_e + \frac{Kg\beta_F}{\nu_F}\frac{(\rho\beta)_{HNF}}{(\rho\beta)_F}(T - T_\infty) \tag{5}$$

Here v, u signify Darcy's law velocity components in (r, x) directions, T temperature of the hybrid nanofluid, g acceleration owing to gravity, μ_{HNF} hybrid viscosity, μ_F base fluid viscosity, K permeability of the porous medium, k_{HNF} hybrid nanofluid thermal conductivity, $(\rho c_p)_{HNF}$ hybrid nanofluid heat capacitance, $(\rho\beta)_{HNF}$ hybrid nanofluid thermal expansion, and $R(x)$ surface shape of the axisymmetric body.

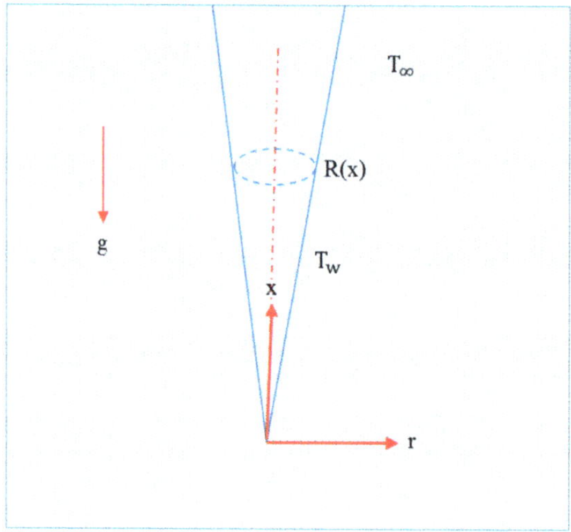

Figure 1. Physical diagram of the problem.

For radiation effect, the Rosseland approximation is illustrated as

$$q_r = -\frac{4\gamma_1}{3k_1}\left(\frac{\partial T^4}{\partial r}\right) \tag{6}$$

where k_1 and γ_1 signify coefficient of mean absorption and Stefan–Boltzmann, respectively. Applying Taylor series to expand T^4 about T_∞ and prohibiting the terms involving higher-order, one gets

$$T^4 \cong 4TT_\infty^3 - 3T_\infty^4 \tag{7}$$

The similarity variables for further analysis are introduced as

$$\psi = \alpha_F x F(\eta), \quad \eta = Pe_x \frac{r^2}{x^2} = \frac{U_\infty r^2 x^{m-1}}{\alpha_F}, \quad \theta(\eta) = (T - T_\infty)/(T_w - T_\infty),$$
$$Pe_x = \frac{U_\infty x^{m+1}}{\alpha_F}, \quad u = 2u_e F', \quad v = \frac{\alpha_F}{r}\eta(1-m)F' - \frac{\alpha_F}{r}F, \quad R_d = \frac{4\gamma_1 T_\infty^3}{k_1 k_F}. \tag{8}$$

Equation (1) is identically true and Equation (3) and Equation (5) are transformed to

$$\frac{\mu_{HNF}}{\mu_F}(2F'-1) - \frac{Kg\beta_F x^{\lambda-m}}{\nu_F U_\infty}\frac{(\rho\beta)_{HNF}}{(\rho\beta)_F}\theta = 0 \tag{9}$$

$$\left(\frac{k_{HNF}}{k_F} + \frac{4}{3}R_d\right)(2\eta\theta'' + 2\theta') + \frac{(\rho c_p)_{HNF}}{(\rho c_p)_F}(\theta'F - \lambda F'\theta) = 0 \tag{10}$$

It is perceptible that Equation (9) and Equation (10) will consent the similarity solutions if the power of x in Equation (9) disappears, i.e.,:

$$m = \lambda \tag{11}$$

With this classified condition, Equation (9) can be rewritten as

$$\frac{\mu_{HNF}}{\mu_F}(2F'-1) - \xi\frac{(\rho\beta)_{HNF}}{(\rho\beta)_F}\theta = 0 \tag{12}$$

where the dimensionless constraint involved in the aforementioned equations is mathematically expressed as

$$\xi = \frac{Ra}{Pe_x} = \frac{Kg\beta_F(T_w - T_\infty)x}{\nu_F\alpha_F}\frac{\alpha_F}{U_\infty x^{m+1}} = \frac{Kg\beta_F A}{\nu_F U_\infty}, \quad Ra = \frac{Kg\beta_F(T_w - T_\infty)x}{\nu_F\alpha_F},$$

$$Pe_x = \frac{U_\infty x^{m+1}}{\alpha_F}, \quad Pr = \frac{\nu_F}{\alpha_F}, \quad R_d = \frac{4\gamma_1 T_\infty^3}{k_1 k_F}.$$

and the interpretation of these constraints are the mixed convection parameter, the local Rayleigh number for a porous medium, the Peclet number, the Prandtl number and the radiation parameter, respectively.

Placing $\eta = b$, where b is constant and utilized for a slender body, it is numerically small. Equation (9) stipulated the body size as well as body shape with surface is defined via

$$R(x) = \left(\frac{\nu_F\alpha_F bRa}{Pe_x Kg\beta_F A}\right)^{\frac{1}{2}} x^{(\frac{1-\lambda}{2})} \quad \text{OR} \quad \left(\frac{b}{Pe_x}\right)^{\frac{1}{2}} x \tag{13}$$

The problems concerning the realistic interest, the amount of $\lambda \leq 1$. For instance, $\lambda = 1, \lambda = 0$ and $\lambda = -1$ represent the cylinder, paraboloid, and cone shape bodies.

The boundary restrictions are

$$b(1-m)F' - F = 0, \ \theta = 1 \text{ at } \eta = b, \tag{14}$$
$$F' \to 0.5, \ \theta \to 0 \text{ as } \eta \to \infty.$$

If Equation (10) and Equation (12) are merged, one gets

$$\left(\frac{k_{HNF}}{k_F} + \frac{4}{3}R_d\right)(4\eta F''' + 4F'') + \frac{(\rho c_p)_{HNF}}{(\rho c_p)_F}\left(2FF'' - 2\lambda F'^2 + \lambda F'\right) = 0 \tag{15}$$

along with the modified boundary restrictions

$$b(1-m)F'(b) - F(b) = 0, \quad \frac{\mu_{HNF}}{\mu_F}(2F'(b)-1) = \xi\frac{(\rho\beta)_{HNF}}{(\rho\beta)_F}, \quad F'(\infty) \to 0.5. \tag{16}$$

The quantities of practical interest and to measure the liquid behaviors is the skin friction which is explained as

$$C_F = \frac{\tau_w}{\rho u_e^2} = \frac{\mu_{HNF}\frac{\partial u}{\partial r}\big|_{r=R(x)}}{\rho u_e^2} \tag{17}$$

The skin friction in the dimensional form is

$$\frac{Pe_x^{0.5}}{Pr}C_F = 4\frac{\mu_{HNF}}{\mu_F}b^{\frac{1}{2}}F''(b). \tag{18}$$

3. Model of Generalized Hybrid Nanoliquid

In numerical and experimental investigations on the behaviors of nanofluid, modeling their physical quantities using condensed mathematical relationships between solid particles and regular liquid is a common procedure. Numerous experiments have been performed to validate such terms for nanofluids dilute in scattering of a single sort of solid material [31] and mixtures of two kinds of particles (Suresh et al. [32]). Devi and Devi [33] recommended a collection of correlation for hybrid nanofluid physical-quantities. They approached the liquid involving a single sort of nanoparticle as the regular liquid and the other sort of nanoparticle as the individual particle. The relationship of thermal conductivity and viscosity matched Suresh et al.'s [32] experimental outcomes. In the approach of Devi and Devi [33], there exists the non-linear terms owing to the communication of two sorts of distinct nanoparticles. However, in dilute mixtures where the volumetric fractions of nanoparticle are generally tiny, the impacts of these non-linear conditions may not be important. Thus, it realistically ignores the non-linear terms in the model of Devi and Devi. The hybrid nanofluid model in simplified form and the models of normal nanofluid and Devi and Devi are scheduled in Table 1, where Type A signifies the conventional nanoliquid model, Types B and C, respectively; indicate the hybrid nanoliquid model of Devi and Devi and the model of hybrid nanoliquid in simplified form. Devi and Devi [33] used the approach of the recurrence formulae to signify the density, viscosity, thermal conductivity, and specific heat of the hybrid nanoliquid related generally to the N-th type of nanoparticles as

$$\rho_{HNF} \equiv \rho_{HNF_N} = (1 - \phi_N)\rho_{HNF_{N-1}} + \phi_N \rho_{S_N} \tag{19}$$

$$\mu_{HNF} \equiv \mu_{HNF_N} = \frac{\rho_{HNF_{N-1}}}{(1 - \phi_N)^{2.5}} \tag{20}$$

$$k_{HNF} \equiv k_{HNF_N} = \frac{k_{S_N} + k_{HNF_{N-1}}(M-1) - \phi_N(M-1)\left(k_{HNF_{N-1}} - k_{S_N}\right)}{k_{S_N} + k_{HNF_{N-1}}(M-1) + \phi_N\left(k_{HNF_{N-1}} - k_{S_N}\right)} \tag{21}$$

$$\left(\rho c_p\right)_{HNF} \equiv \left(\rho c_p\right)_{HNF_N} = \phi_N\left(\rho c_p\right)_{S_N} + (1 - \phi_N)\left(\rho c_p\right)_{HNF_{N-1}} \tag{22}$$

$$\beta_{HNF} \equiv \beta_{HNF_N} = (1 - \phi_N)\beta_{HNF_{N-1}} + \phi_N \beta_{S_N} \tag{23}$$

From the above correlations, ignoring the non-linear terms, one obtains

$$\rho_{HNF} = 1 - \rho_F \sum_{a=1}^{N} \phi_a + \sum_{a=1}^{N} \phi_a \rho_{S_a} \tag{24}$$

$$\mu_{HNF} = \frac{\mu_F}{\left(1 - \sum\limits_{a=1}^{N} \phi_a\right)^{2.5}} \tag{25}$$

$$\left(\rho c_p\right)_{HNF} = 1 - \left(\rho c_p\right)_F \sum_{a=1}^{N} \phi_a + \sum_{a=1}^{N} \phi_a \left(\rho c_p\right)_{S_a} \tag{26}$$

$$\beta_{HNF} = 1 - \beta_F \sum_{a=1}^{N} \phi_a + \sum_{a=1}^{N} \phi_a \beta_{S_a} \tag{27}$$

It is worth mentioning that for k_{HNF}, we remain the recurrence Formula (21) because of the connections amid dissimilar particles can barely be uttered by the Maxwell equations. Additionally, throughout the research the values of $M = 3$ is taken which implies that the shape of the particle is spherical.

Table 1. Thermo-physical models of nanofluid and hybrid nanofluid.

Property	Types	Correlation
Density	A	$\rho_{NF} = (1-\phi)\rho_F + \phi\rho_S$
	B	$\rho_{HNF} = (1-\phi_2)\left[(1-\phi_1)\rho_F + \phi_1\rho_{S_1}\right] + \phi_2\rho_{S_2}$
	C	$\rho_{HNF} = (1-\phi_1-\phi_2)\rho_F + \phi_1\rho_{S_1} + \phi_2\rho_{S_2}$
Viscosity	A	$\mu_{NF} = \dfrac{\mu_F}{(1-\phi)^{2.5}}$
	B	$\mu_{HNF} = \dfrac{\mu_F}{(1-\phi_1)^{2.5}(1-\phi_2)^{2.5}}$
	C	$\mu_{HNF} = \dfrac{\mu_F}{(1-\phi_1-\phi_2)^{2.5}}$
Heat Capacity	A	$\left(\rho c_p\right)_{NF} = (1-\phi)\left(\rho c_p\right)_F + \phi\left(\rho c_p\right)_S$
	B	$\left(\rho c_p\right)_{HNF} =$ $(1-\phi_2)\left[(1-\phi_1)\left(\rho c_p\right)_F + \phi_1\left(\rho c_p\right)_{S_1}\right] + \phi_2\left(\rho c_p\right)_{S_2}$
	C	$\left(\rho c_p\right)_{HNF} = (1-\phi_1-\phi_2)\left(\rho c_p\right)_F + \phi_1\left(\rho c_p\right)_{S_1} + \phi_2\left(\rho c_p\right)_{S_2}$
Thermal conductivity	A	$k_{NF} = \dfrac{k_S + (M-1)k_F - (M-1)\phi(k_F - k_S)}{k_S + (M-1)k_F + \phi(k_F - k_S)}$
	B & C	$k_{HNF} = \dfrac{k_{S_2} + (M-1)k_{MF} - (M-1)\phi_2\left(k_{MF} - k_{S_2}\right)}{k_{S_2} + (M-1)k_{MF} + \phi_2\left(k_{MF} - k_{S_2}\right)}$ where $k_{MF} = \dfrac{k_{S_1} + (M-1)k_F - (M-1)\phi_1\left(k_F - k_{S_1}\right)}{k_{S_1} + (M-1)k_F + \phi_1\left(k_F - k_{S_1}\right)}k_F$
Thermal expansion	A	$\beta_{NF} = (1-\phi)\beta_F + \phi\beta_S$
	B	$\beta_{HNF} = (1-\phi_2)\left[(1-\phi_1)\beta_F + \phi_1\beta_{S_1}\right] + \phi_2\beta_{S_2}$
	C	$\beta_{HNF} = (1-\phi_1-\phi_2)\beta_F + \phi_1\beta_{S_1} + \phi_2\beta_{S_2}$
Pr	6.2	-

4. Results and Discussion

The primary intention is to scrutinize the characteristics of generalized hybrid nanoliquid for the mixed convective flow comprising MoS_2-Go nanoparticles through slender revolution bodies. The non-linear Equation (15) with boundary restrictions (Equation (16)) has been worked out numerically via Lobatto IIIa formula. The thickness of the boundary layer is considered as 30 for convergence of profiles asymptotically which is essential for this type of problem. Table 2 represents the thermo-physical characteristics of the base and nanofluids. The outcomes of the sundry parameters in the presence of the water-based fluid and the type B hybrid nanofluid on the field of velocity distribution and the skin friction have been examined in Figures 2–13. For validation, the results of the current problem have been compared with the outcomes of Ahmad et al. [34] and Saleh et al. [35] as shown in Table 3. An excellent harmony is seen. Whereas the numerical values of types B and C hybrid nanofluid for (ASSF) as well as for (OPPF) are displayed in Tables 4 and 5.

Table 2. Thermo-physical properties of the base fluid and hybrid nanoparticles.

Characteristic Properties	H_2O	MoS_2	GO
ρ	997.1	5060	1800
c_p	4179	397.21	717
k	0.613	904.4	5000
β	21	2.8424×10^{-5}	2.84×10^{-4}

Table 3. Comparison values of $F''(b)$ when $m = 1$, $\lambda = 0$, $\xi = -1$, $R_d = \phi = 0$, $\mathrm{Pr} = 1$ for the distinct values of b.

b	Ahmed et al. [34]	Saleh et al. [35]	Current
0.01	8.4924360	8.4924452	8.4924456
0.1	1.2888171	1.2888299	1.28883009

Table 4. Computation of the $\mathrm{Pr}^{-1}Pe_x^{0.5}C_F$ for the assisting flow $\xi = 1.1$ utilizing the two different types of models varying ϕ_2 while the other fixed parameters are $m = \lambda = 1$, $b = 0.1$, $R_d = 0.5$.

ϕ_1	ϕ_2	Type B	Type C
	0.025	−2.3995	−2.3973
0.025	0.030	−2.3760	−2.3733
	0.035	−2.3527	−2.3496
	0.025	−2.3754	−2.3727
0.030	0.030	−2.3522	−2.3490
	0.035	−2.3292	−2.3255
	0.025	−2.3516	−2.3485
0.035	0.030	−2.3287	−2.3250
	0.035	−2.3060	−2.3017

Table 5. Computation of the $\mathrm{Pr}^{-1}Pe_x^{0.5}C_F$ for the opposing flow $\xi = -1.1$ utilizing the two different types of models varying ϕ_2 while the other fixed parameters are $m = \lambda = 1$, $b = 0.1$, $R_d = 0.5$.

ϕ_1	ϕ_2	Type B	Type C
	0.025	1.9037	1.9031
0.025	0.030	1.8954	1.8946
	0.035	1.8869	1.8859
	0.025	1.8951	1.8943
0.030	0.030	1.8866	1.8856
	0.035	1.8779	1.8767
	0.025	1.8862	1.8853
0.035	0.030	1.8776	1.8764
	0.035	1.8688	1.8673

The outcomes, indicating from these tables the type B hybrid nanofluid, are superior to the type C hybrid nanofluid. Moreover, in case of the assisting flow, the skin friction increases by 0.979% for the type B hybrid nanofluid whereas the skin friction in the type C hybrid nanofluid augments by 1.001%. As the value of the nanoparticle volumetric fraction increases, the skin friction decreases continuously for both types of hybrid nanofluid. In contrast, the skin friction decreases up to 0.435%

for the type B hybrid nanofluid due to the fixed value of the parameter $\phi_2 = 0.025$ in the example of opposing flow while for the type C hybrid nanofluid, it is decreased by 0.446%. Due to very small negligible differences in the outcomes of both the types of hybrid nanofluid, therefore the computation throughout the paper is done only for the type B hybrid nanofluid.

Figures 2 and 7 are set to inspect the impact of the dimensionless radius of the slender body parameter b and the volume fraction of nanoparticle ϕ_2 on the velocity gradient against the similarity variable η for the three different phenomena such as the assisting and opposing flows, shape bodies and the normal nanofluid as well for the hybrid nanofluid. In the example of assisting flow, the velocity distribution and the momentum boundary-layer flow (MBLF) increase with increasing the dimensionless radius of the slender body parameter b, while in the phenomenon of the opposing flow, the behavior of the motion of the fluid behaves in the contrary direction as shown in Figure 2. It is transparent to observe from the outcomes that the gap between the curves is initially more significant in the ASSF as well as in the OPPF, while as we upsurge, the value of the dimensionless radius of the slender body parameter b the gap between the solution curves is reduced.

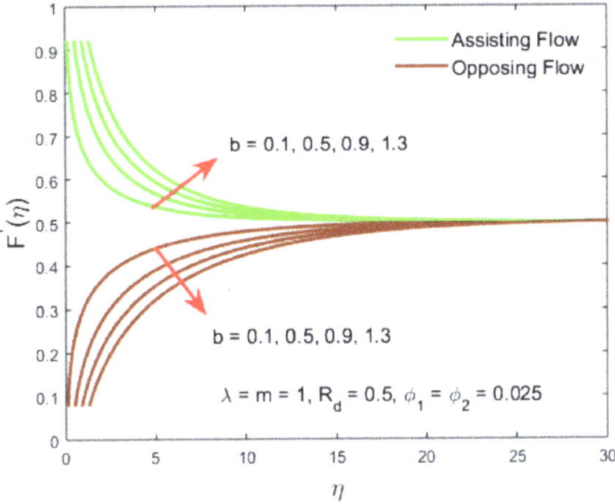

Figure 2. The variation of the velocity profile $F'(\eta)$ for the case of assisting and opposing flow versus the similarity variable η for the distinct values of the dimensionless radius of the slender body parameter b.

On the other hand, the velocity distribution is enhancing the function of the shape bodies as well as for the type B hybrid nanofluid and the type A nanofluid for the higher values of b as shown in Figures 3 and 4, respectively. It is perceived from Figure 3 that the velocity field is superior in the flow through a cone compared to paraboloid and cylindrical type's bodies. Moreover, it is perfectly visible from the graph that the liquid flow accelerates more for the MoS_2/water nanoparticle or type A nanofluid as compared to the type B hybrid nanoparticles as highlighted in Figure 4. The gap between the solution curves in the shape bodies is more when compared to the solution curves like Figure 4. Since Figure 5, it has been noticed that for the (ASSF), the velocity distribution decreases with escalating ϕ_2, whereas for the opposing flow, it is augmented.

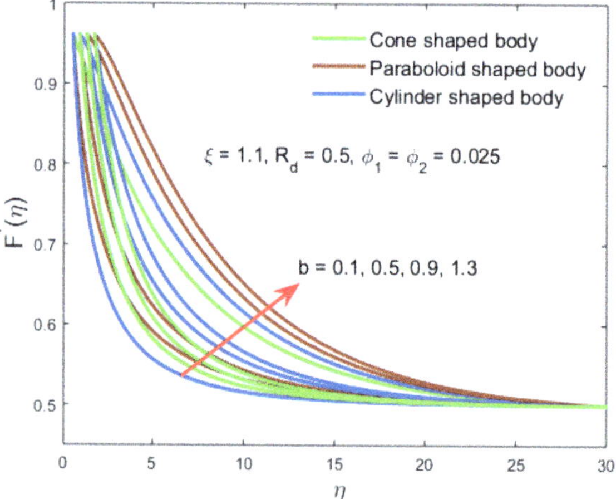

Figure 3. The variation of the velocity profile $F'(\eta)$ for the three important cases of shape bodies such as Cone, paraboloid, and cylinder versus the similarity variable η for the distinct values of the dimensionless radius of the slender body parameter b.

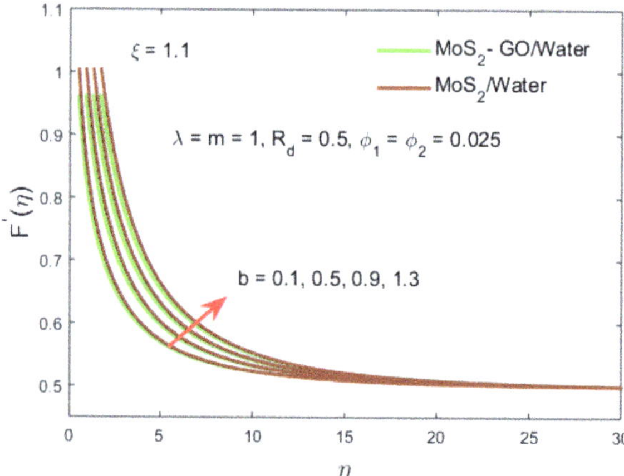

Figure 4. The variation of the velocity profile $F'(\eta)$ for the type A normal nanofluid as well as for the type B hybrid nanofluid versus the similarity variable η for the distinct values of the dimensionless radius of the slender body parameter b.

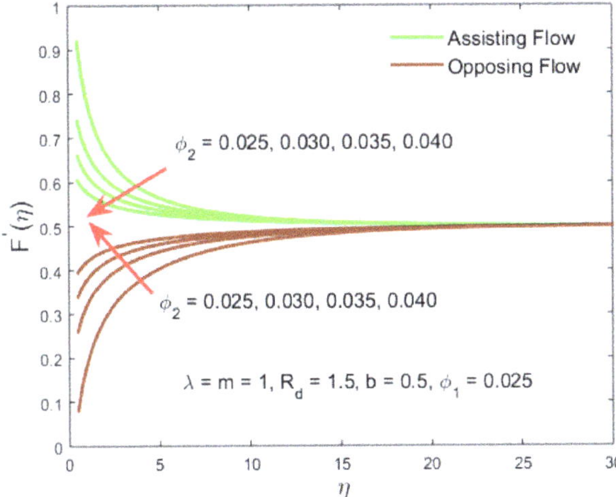

Figure 5. Shows the variation of the velocity profile $F'(\eta)$ for the case of assisting and opposing flow versus the similarity variable η for the distinct values of the volume fraction of nanoparticle ϕ_2.

Additionally, it explains that the thickness of the velocity and the (MBLF) declines with ϕ_2 for the ASSF and augments for the OPPF. The outcome of the velocity gradient in Figure 5 is showing a contrary behavior in both the cases of ASSF and OPPF as we compare with the solution curves of Figure 2. From Figure 6, it is transparent that due to the shape bodies, the velocity field is decelerated for the higher impact of ϕ_2. In comparison, the flow of the velocity field over the cone shape body is finer than the rest of the two-shape body. Generally, the velocity upsurges due to the fact that type B hybrid nanofluid dynamic viscosity has an inverse relationship to nanoparticle volume fraction. Hence, an augmenting in ϕ_2 guides to the decline's viscosity of base liquid and therefore speeds up the motion of the liquid flow.

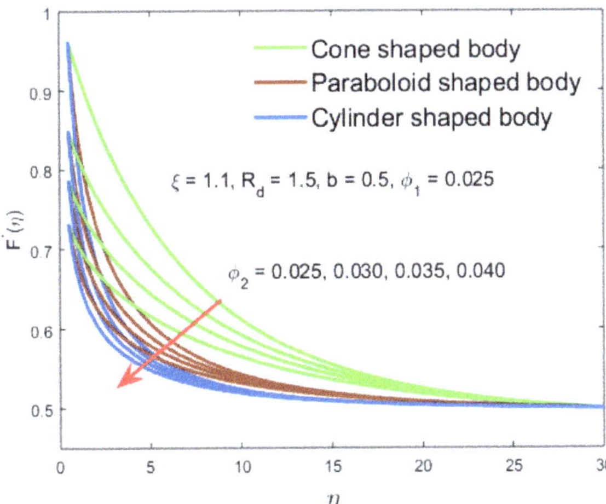

Figure 6. Shows the variation of the velocity profile $F'(\eta)$ for the three important cases of shape bodies such as Cone, paraboloid, and cylinder versus the similarity variable η for the distinct values of the volume fraction of nanoparticle ϕ_2.

Moreover, the velocity profile reduces due to ϕ_2 for the type B hybrid nanofluid and type A nanofluid as portrayed in Figure 7. The impact between the curves is very minor for the parameter ϕ_2 while the outcome for the type B hybrid nanofluid is higher than the type A nanofluid.

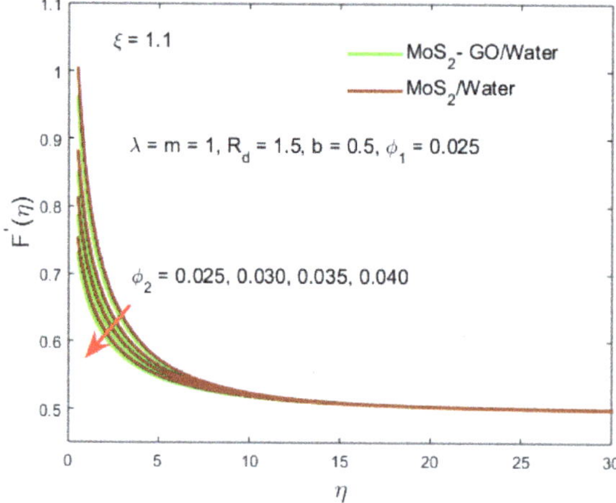

Figure 7. The variation of the velocity profile $F'(\eta)$ for the type A normal nanofluid as well as for the type B hybrid nanofluid versus the similarity variable η for the distinct values of the volume fraction of nanoparticle ϕ_2.

Figures 8 and 9 highlight the stimulus of radiation parameter R_d on the velocity distribution against the similarity variable of the two distinct phenomena such as the assisting and opposing flows as well as for the type A nanofluid and type B hybrid nanofluid. It is clear from Figure 8 as depicted graphically that the velocity distribution augments for the ASSF and declines for the OPPF as the value of the radiation parameter R_d upsurges. The solution behavior is similarly observed like Figure 2 while the contrary behavior of the outcomes in Figure 5 is seen for both the cases ASSF and as well as in the OPPF, while in comparison the influence between the curves is less. Additionally, R_d is exploited to drop the molecules of liquid into hydrogen. Figure 9 displays that the velocity upsurges are owed to magnifying the radiation parameter for the type A nanofluid as well as for the type B hybrid nanofluid. The motion of the fluid flow for the type A nanofluid is superior as compare to the type B hybrid nanofluid. In addition, the solution impact in the curves for the velocity gradient is more owing to the impact of radiation parameter as compared to the significant impact of the other two parameters which is exercised in Figures 4 and 7, respectively.

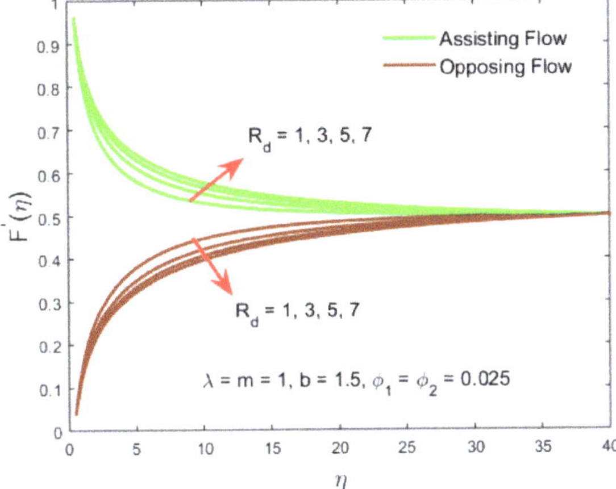

Figure 8. Shows the variation of the velocity profile $F'(\eta)$ for the case of assisting and opposing flow versus the similarity variable η for the distinct values of the radiation parameter R_d.

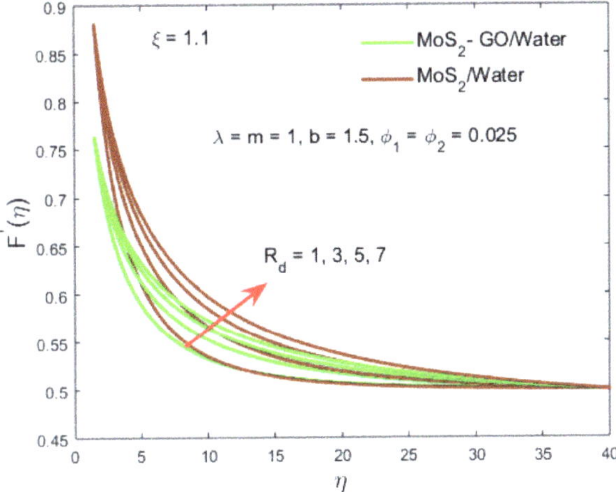

Figure 9. The variation of the velocity profile $F'(\eta)$ for the type A normal nanofluid as well as for the type B hybrid nanofluid versus the similarity variable η for the distinct values of the radiation parameter R_d.

Figures 10–13 show the variation of the dimensionless radius of the slender body parameter b and the nanoparticle volume fraction ϕ_2 on the skin friction against the mixed convection parameter ξ for the two distinct cases such as the flow over the shape bodies and the corresponding type A nanofluid and the type B hybrid nanofluid. The skin friction is reduced with enhanced b along the horizontal axis of the range $(-\infty < \xi \leq 0)$ and augments in the spectrum $(0 \leq \xi < \infty)$ for the flow over the shape bodies as well as for the types A and B nanofluid and hybrid nanofluid as highlighted in Figures 10 and 11, respectively. The skin friction is higher in the range $(0 \leq \xi < \infty)$ for the flow over the cone shape body and also for the type B hybrid nanofluid as compared to the flow on the paraboloid and the cylindrical shape bodies as well as for the type A nanofluid, while the opposite trend is observed in the range $(-\infty < \xi \leq 0)$ as shown in Figures 10 and 11. The range in both graphs is taken from $-0.5 \leq \xi \leq 0.5$ and it is not the fixed one we can vary from this range as a real number $(-\infty < \xi < \infty)$ but due to this small range in the solution, the collision is more effectively significant. A similar behavior is detected (Figures 10 and 11) for the skin friction owing to the nanoparticle volume fraction ϕ_2 as described in Figures 12 and 13, respectively. Hence, in Figures 12 and 13 illustrate more transparently the impact of the curves of the flow over the shape bodies against the mixed convection parameter owing to b is greater compared to ϕ_2 on the skin friction while this gap for the type B hybrid nanofluid and type A nanofluid owing the parameter ϕ_2 is coarsened as compared to the dimensionless radius of the slender body parameter b.

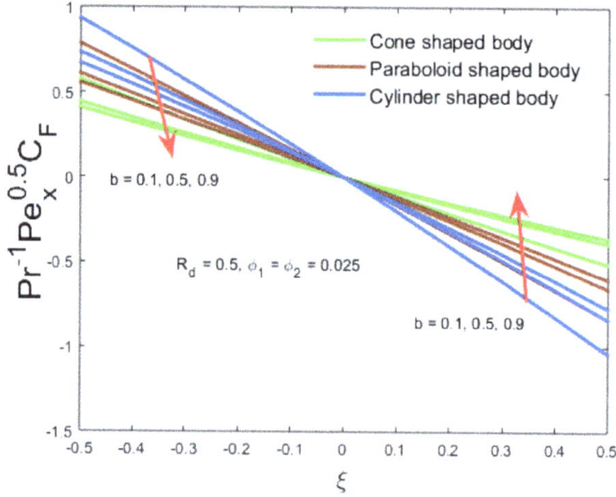

Figure 10. Deviation of the skin friction versus the mixed convection parameter ξ for the distinct values of the dimensionless radius of the slender body parameter b.

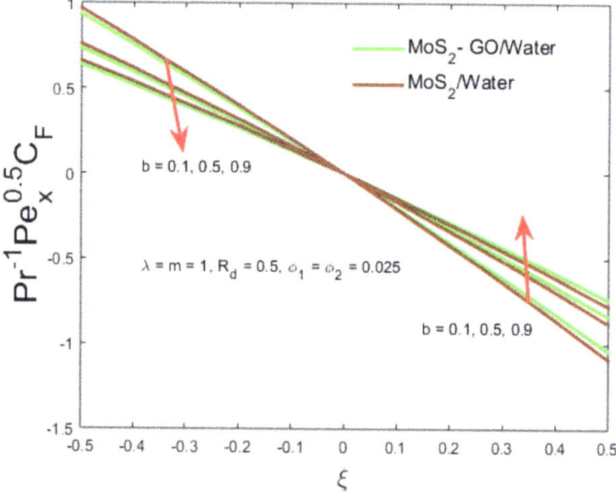

Figure 11. Deviation of the skin friction versus the mixed convection parameter ξ for the distinct values of the dimensionless radius of the slender body parameter b.

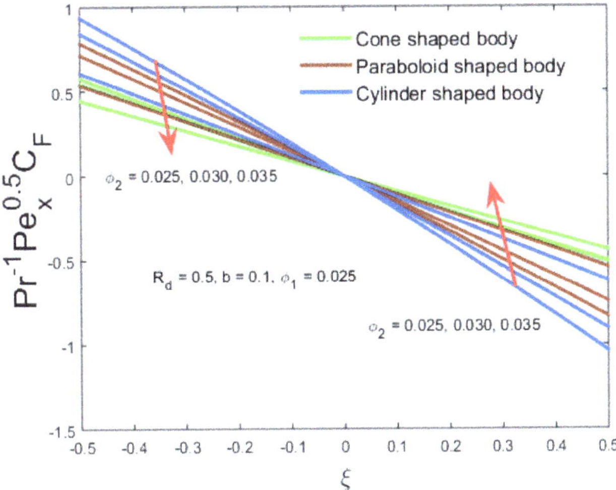

Figure 12. Deviation of the skin friction versus the mixed convection parameter ξ for distinct values of the volume fraction of nanoparticle ϕ_2.

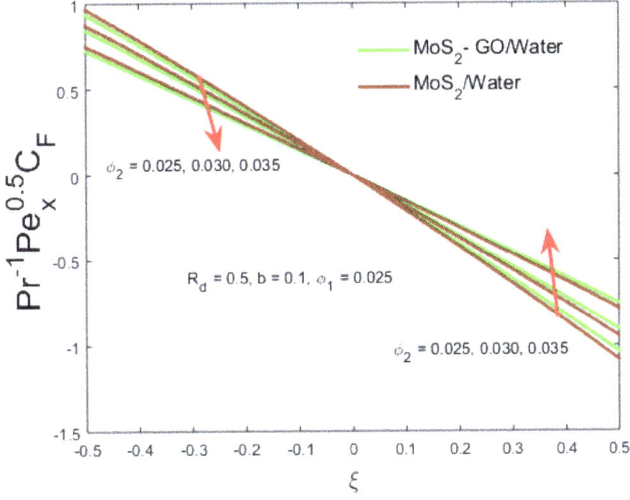

Figure 13. Deviation of the skin friction versus the mixed convection parameter ξ for distinct values of the volume fraction of nanoparticle ϕ_2.

5. Concluding Remarks

The key points of the current study are summarized as:

- The velocity augments for the (ASSF) and declines for the (OPPF) owing to magnifying the dimensionless radius of the slender body parameter b while the change behavior is detected in response of the higher nanoparticle volume fraction ϕ_2.
- The persistent effect of b, the velocity upsurges for the flow of the shape bodies like paraboloid shape body ($\lambda = 0$), cylindrical shape body ($\lambda = 1$) as well as cone shape body ($\lambda = -1$) while the same behavior of the velocity is seen in the type A nanofluid and type B hybrid nanofluid.

- The velocity field decreases for the type A nanofluid as well as for the type B hybrid nanofluid and also for the flow over the different shape bodies owing to ϕ_2.

- Due to R_d, the velocity upsurges for the type A nanofluid as well as for the type B hybrid nanofluid while for the type B hybrid nanofluid, the velocity is lesser relative to the type A nanofluid.

- The skin friction under the distinct shape bodies and along the type A nanofluid and type B hybrid nanofluid are augmented for both parameter b and ϕ_2 along the x-axis of the slender sheet in the range $0 \leq \xi < \infty$ while vice versa in the range $-\infty < \xi \leq 0$.

This sort of problem may be useful in geothermal and geophysical applications. Thus, the outcomes of the problem will be obliging in the evaluation and assessment of resources in geothermal energy.

Author Contributions: Conceptualization, A.Z.; Data curation, A.W.; Formal analysis, M.S.; Investigation, A.W.; Methodology, U.K.; Resources, A.Z.; Software, U.K. and A.Z.; Supervision, D.B. All authors have read and agreed to the published version of the manuscript.

Funding: This research received no external funding.

Conflicts of Interest: The authors declare no conflict of interest.

References

1. Nield, D.A. Onset of Thermohaline Convection in a Porous Medium. *Water Resour. Res.* **1968**, *4*, 553–560. [CrossRef]

2. Bejan, A.; Khair, K.R. Heat and mass transfer by natural convection in a porous medium. *Int. J. Heat Mass Transf.* **1985**, *28*, 909–918. [CrossRef]

3. Lai, F.; Pop, I.; Kulack, F. Free and mixed convection from slender bodies of revolution in porous media. *Int. J. Heat Mass Transf.* **1990**, *33*, 1767–1769. [CrossRef]

4. Yih, K.A. Coupled heat and mass transfer by free convection over a truncated cone in porous media: VWT/VWC or VHF/VMF. *Acta Mech.* **1999**, *137*, 83–97. [CrossRef]

5. Bano, N.; Singh, B.B. An integral treatment for coupled heat and mass transfer by natural convection from a radiating vertical thin needle in a porous medium. *Int. Commun. Heat Mass Transf.* **2017**, *84*, 41–48. [CrossRef]

6. Singh, B.; Chandarki, I. Integral treatment of coupled heat and mass transfer by natural convection from a cylinder in porous media. *Int. Commun. Heat Mass Transf.* **2009**, *36*, 269–273. [CrossRef]

7. Ahmad, S.; Pop, I. Mixed convection boundary layer flow from a vertical flat plate embedded in a porous medium filled with nanofluids. *Int. Commun. Heat Mass Transf.* **2010**, *37*, 987–991. [CrossRef]

8. Talebizadeh, P.; Moghimi, M.A.; Kimiaeifar, A.; Ameri, M. Numerical and analytical solutions for natural convection flow with thermal radiation and mass transfer past a moving vertical porous plate by dqm and ham. *Int. J. Comput. Methods* **2011**, *8*, 611–631. [CrossRef]

9. Moghimi, M.A.; Talebizadeh, P.; Mehrabian, M.A. Heat generation/absorption effects on magnetohydrodynamic natural convection flow over a sphere in a non-Darcian porous medium. *Proc. Inst. Mech. Eng. Part E J. Process. Mech. Eng.* **2010**, *225*, 29–39. [CrossRef]

10. Moghimi, M.A.; Tabaei, H.; Kimiaeifar, A. HAM and DQM solutions for slip flow over a flat plate in the presence of constant heat flux. *Math. Comput. Model.* **2013**, *58*, 1704–1713. [CrossRef]

11. Raju, C.S.K.; Sandeep, N. Heat and mass transfer in MHD non-Newtonian bio-convection flow over a rotating cone/plate with cross diffusion. *J. Mol. Liq.* **2016**, *215*, 115–126. [CrossRef]

12. Raju, C.S.K.; Saleem, S.; Upadhya, S.M.; Hussain, I. Heat and mass transport phenomena of radiated slender body of three revolutions with saturated porous: Buongiorno's model. *Int. J. Therm. Sci.* **2018**, *132*, 309–315. [CrossRef]

13. Choi, S.U.S. Enhancing thermal conductivity of fluids with nanoparticles. In Proceedings of the 1995 International Mechanical Engineering Congress and Exhibition, San Francisco, CA, USA, 12–17 November 1995.

14. Eastman, J.A.; Choi, S.; Li, S.; Yu, W.; Thompson, L.J. Anomalously increased effective thermal conductivities of ethylene glycol-based nanofluids containing copper nanoparticles. *Appl. Phys. Lett.* **2001**, *78*, 718–720. [CrossRef]

15. Wen, D.; Ding, Y. Experimental investigation into convective heat transfer of nanofluids at the entrance region under laminar flow conditions. *Int. J. Heat Mass Transf.* **2004**, *47*, 5181–5188. [CrossRef]

16. Ashorynejad, H.R.; Mohamad, A.A.; Sheikholeslami, M. Magnetic field effects on natural convection flow of a nanofluid in a horizontal cylindrical annulus using Lattice Boltzmann method. *Int. J. Therm. Sci.* **2013**, *64*, 240–250. [CrossRef]

17. Ellahi, R.; Zeeshan, A.; Hassan, M. Particle shape effects on Marangoni convection boundary layer flow of a nanofluid. *Int. J. Numer. Methods Heat Fluid Flow* **2016**, *26*, 2160–2174. [CrossRef]

18. Soomro, F.A.; Zaib, A.; Haq, R.U.; Sheikholeslami, M. Dual nature solution of water functionalized copper nanoparticles along a permeable shrinking cylinder: FDM approach. *Int. J. Heat Mass Transf.* **2019**, *129*, 1242–1249. [CrossRef]

19. Tabassum, R.; Mehmood, R. Crosswise stream of methanol–iron oxide ($CH3OH–Fe_3O_4$) with temperature-dependent viscosity and suction/injection effects. *Proc. Inst. Mech. Eng. Part E: J. Process. Mech. Eng.* **2019**, *233*, 1013–1023. [CrossRef]

20. Zaib, A.; Khan, U.; Wakif, A.; Zaydan, M. Numerical Entropic Analysis of Mixed MHD Convective Flows from a Non-Isothermal Vertical Flat Plate for Radiative Tangent Hyperbolic Blood Biofluids Conveying Magnetite Ferroparticles: Dual Similarity Solutions. *Arab. J. Sci. Eng.* **2020**, *45*, 5311–5330. [CrossRef]

21. Sayed, A.Y.; Abdel-Wahed, M.S. Entropy analysis for an MHD nanofluid with a microrotation boundary layer over a moving permeable plate. *Eur. Phys. J. Plus* **2020**, *135*, 1–17. [CrossRef]

22. Alsarraf, J.; Rahmani, R.; Shahsavar, A.; Afrand, M.; Wongwises, S.; Tran, M.D. Effect of magnetic field on laminar forced convective heat transfer of MWCNT–Fe_3O_4/water hybrid nanofluid in a heated tube. *J. Therm. Anal. Calorim.* **2019**, *137*, 1809–1825. [CrossRef]

23. Shahsavar, A.; Sardari, P.T.; Toghraie, D. Free convection heat transfer and entropy generation analysis of water–Fe_3O_4/CNT hybrid nanofluid in a concentric annulus. *Int. J. Numer. Methods Heat Fluid Flow* **2019**, *29*, 915–934. [CrossRef]

24. Khan, U.; Shafiq, A.; Zaib, A.; Baleanu, D. Hybrid nanofluid on mixed convective radiative flow from an irregular variably thick moving surface with convex and concave effects. *Case Stud. Therm. Eng.* **2020**, *21*, 100660. [CrossRef]

25. Minea, A.A. Hybrid nanofluids based on Al_2O_3, TiO_2 and SiO_2: Numerical evaluation of different approaches. *Int. J. Heat Mass Transf.* **2017**, *104*, 852–860. [CrossRef]

26. Ghadikolaei, S.; Hosseinzadeh, K.; Hatami, M.; Ganji, D. MHD boundary layer analysis for micropolar dusty fluid containing Hybrid nanoparticles (Cu-Al2O3) over a porous medium. *J. Mol. Liq.* **2018**, *268*, 813–823. [CrossRef]

27. Sheikholeslami, M.; Gerdroodbary, M.B.; Shafee, A.; Tlili, I. Hybrid nanoparticles dispersion into water inside a porous wavy tank involving magnetic force. *J. Therm. Anal. Calorim.* **2019**, 1–7. [CrossRef]

28. Gholinia, M.; Armin, M.; Ranjbar, A.; Ganji, D. Numerical thermal study on CNTs/ $C_2H_6O_2$– H_2O hybrid base nanofluid upon a porous stretching cylinder under impact of magnetic source. *Case Stud. Therm. Eng.* **2019**, *14*, 100490. [CrossRef]

29. Khan, U.; Zaib, A.; Mebarek-Oudina, F. Mixed Convective Magneto Flow of SiO_2–MoS_2/$C_2H_6O_2$ Hybrid Nanoliquids Through a Vertical Stretching/Shrinking Wedge: Stability Analysis. *Arab. J. Sci. Eng.* **2020**, 1–13. [CrossRef]

30. Tiwari, R.K.; Das, M.K. Heat transfer augmentation in a two-sided lid-driven differentially heated square cavity utilizing nanofluids. *Int. J. Heat Mass Transf.* **2007**, *50*, 2002–2018. [CrossRef]

31. Pak, B.C.; Cho, Y.I. Hydrodynamic and heat transfer study of dispersed fluids with submicron metallic oxide particles. *Exp. Heat Transf.* **1998**, *11*, 151–170. [CrossRef]

32. Suresh, S.; Venkitaraj, K.; Selvakumar, P.; Chandrasekar, M. Synthesis of Al_2O_3–Cu/water hybrid nanofluids using two step method and its thermo physical properties. *Colloids Surfaces A: Physicochem. Eng. Asp.* **2011**, *388*, 41–48. [CrossRef]

33. Devi, S.S.U.; Devi, S.A. Numerical investigation of three-dimensional hybrid Cu–Al_2O_3/water nanofluid flow over a stretching sheet with effecting Lorentz force subject to Newtonian heating. *Can. J. Phys.* **2016**, *94*, 490–496. [CrossRef]

34. Ahmad, R.; Mustafa, M.; Hina, S. Buongiorno's model for fluid flow around a moving thin needle in a flowing nanofluid: A numerical study. *Chin. J. Phys.* **2017**, *55*, 1264–1274. [CrossRef]

35. Salleh, S.N.A.; Bachok, N.; Arifin, N.M.; Ali, F.M.; Pop, I. Magnetohydrodynamics Flow Past a Moving Vertical Thin Needle in a Nanofluid with Stability Analysis. *Energies* **2018**, *11*, 3297. [CrossRef]

Article

Triple Solutions and Stability Analysis of Micropolar Fluid Flow on an Exponentially Shrinking Surface

Liaquat Ali Lund [1,2], **Zurni Omar** [1], **Ilyas Khan** [3,*], **Dumitru Baleanu** [4,5,6] **and Kottakkaran Sooppy Nisar** [7]

1   School of Quantitative Sciences, Universiti Utara Malaysia, Sintok 06010, Kedah, Malaysia; balochliaqatali@gmail.com
2   KCAET Khairpur Mir's, Sindh Agriculture University, Tandojam Sindh 70060, Pakistan; zurni@uum.edu.my
3   Faculty of Mathematics and Statistics, Ton Duc Thang University, Ho Chi Minh City 72915, Vietnam
4   Department of Mathematics, Cankaya University, 06790 Ankara, Turkey; dumitru@cankaya.edu.tr or Baleanu@mail.cmuh.org.tw
5   Institute of Space Sciences, 077125 Magurele, Romania
6   Department of Medical Research, China Medical University Hospital, China Medical University, Taichung 40447, Taiwan
7   Department of Mathematics, College of Arts and Sciences, Prince Sattam bin Abdulaziz University, Wadi Aldawaser 11991, Saudi Arabia; n.sooppy@psau.edu.sa
*   Correspondence: ilyaskhan@tdtu.edu.vn

Received: 11 March 2020; Accepted: 27 March 2020; Published: 9 April 2020

Abstract: In this article, we reconsidered the problem of Aurangzaib et al., and reproduced the results for triple solutions. The system of governing equations has been transformed into the system of non-linear ordinary differential equations (ODEs) by using exponential similarity transformation. The system of ODEs is reduced to initial value problems (IVPs) by employing the shooting method before solving IVPs by the Runge Kutta method. The results reveal that there are ranges of multiple solutions, triple solutions, and a single solution. However, Aurangzaib et al., only found dual solutions. The effect of the micropolar parameter, suction parameter, and Prandtl number on velocity, angular velocity, and temperature profiles have been taken into account. Stability analysis of triple solutions is performed and found that a physically possible stable solution is the first one, while all leftover solutions are not stable and cannot be experimentally seen.

Keywords: similarity solution; triple solutions; stability analysis; shooting method; three-stage Lobatto III-A formula

1. Introduction

Generally, the investigation of the non-Newtonian fluid flow in two-dimensional problems is a hard task when multiple solutions are attempted to find because equations are related to high nonlinear terms. In spite of these challenges, the researchers are making efforts to tackle these problems for multiple solutions due to their wide range of applications in various science and industrial fields. The category of non-Newtonian fluids which pacts with the suspended micro-rotational particles, is known as the micropolar fluid. Eringen introduced the theory of micropolar [1,2]. He explained the impacts of couple stresses and local rotational inertia that cannot be described by the standard equations of Navier–Stokes. The micropolar equations are mathematically described for the theory of porous media and the theory of lubrication in the books by Lukaszewicz [3] and Eringen [4]. There are various applications of the micropolar fluids, for example, liquid crystals, particle suspension, animal blood, lubrication, turbulent shear flows, and paints. Lok et al., [5] considered the stagnation point flow of micropolar nanofluid and then succeeded to find the dual solution. It is also stated that only the stable solution is the upper branch by performing an analysis of stability. Sheremet et al., [6] examined micropolar fluid with a convectional effect in a wavy triangular cavity. Bhattacharyya et al., [7] studied the micropolar fluid with thermal radiation effect and discovered that there exist dual solutions, while they did not perform the stability analysis. The dual solution has been discovered in the opposing flow case for the micropolar fluid [8], whereas stability analysis has not been performed in their examination of the multiple solutions. Ramzan et al., [9] investigated the nanofluid of micropolar non-Newtonian kind of fluid on the stretching sheet and claimed that velocity and angular velocity have an inverse relation with slip parameter. Turkyilmazoglu [10] analytically examined the Magnetohydrodynamic (MHD) flow of micropolar fluid. Shah et al., [11] examined micropolar nanofluid with the effect of the Casson parameter in the channel and stated that the thermal boundary layer becomes thicker for the higher values of the Brownian motion parameter. Some papers have recently shown some development on the micropolar fluid [12–16].

Nowadays, multiple solutions of fluid flow have received extensive consideration. Generally, many researchers stated in their studies that non-uniqueness of solutions, depending on the non-linearity in the fluid flow equations. Jawad et al. [17] stated that, due to the nonlinearity in fluid flow problems, there is more than one solution; however, seeking all possible solutions challenges researchers. Further, several of the approaches are ineffective because solutions are close together. Rohni [18] reported in her Ph.D. thesis that "the multiplicity of solutions in fluid dynamics and heat transfer is important to be able to be computed, since solutions arising from bifurcations often interact with one another producing otherwise inexplicable phenomena". Moreover, she claimed that, in order to notice all possible solutions of any fluid, the flow problem is still a challenge. Mishra and DebRoy [19] described that multiple solutions have many important applications when these are related to heat transfer, because the final qualities and structure of many products of material processing in the industries can be improved by the concept of multiple solutions. Lund et al., [20] considered nanofluid, with the base fluid as sodium alginate and solid particles of copper and silver, over the shrinking sheet and obtained dual solutions and reported that the existence of multiple solutions depends upon how the researchers set the ranges of different applied physical parameters. Furthermore, they specified that the dual solution depended on the range of the suction parameter. Khashi'ie et al. [21] highlighted the importance of the multiple solutions and the stability analysis during the examination of the mixed convection flow of the micropolar fluid. They stated "if the problem has non-unique solutions but the researchers manage to find one solution only, there is a probability that the solution is the lower branch solution (unstable/not real). This will lead to the misinterpretation of the flow and heat transfer characteristics". These papers include some significant studies on multiple solutions [22–24]. Here, it is worth highlighting that, when multiple solutions appeared in any fluid flow problem, the study of stability analysis should be considered. The first step in performing the stability analysis is, according to Rana et al. [25], to change the governing equation to unsteady form by adding a new time dependent variable for similarity. Lund et al., [26] found the dual solution of the Williamson fluid and claimed that there existed an infinite number of the values of the smallest eigenvalue by performing the stability of solutions. Waini et al., [27] reported that the stability of any solution dependent upon the sign of the value of the smallest eigenvalue. If the value of eigenvalue is positive, it indicated the stable solution, while the negative value of the smallest eigenvalue indicated the unstable solution [28].

The pointwise prime objectives of this study are, as follows:

1. Reconsidering the problem of Aurangzaib et al. [29].
2. To find all possible multiple solutions.
3. To perform the stability analysis that has not been considered by the Aurangzaib et al. [29].
4. Indicating the stable and unstable solutions by doing stability analysis, which cannot be experientially seen, due to that mathematical analysis is necessary.

The authors tried their best to find all possible solutions due to the importance of multiple solutions and lack of available literature in which triple solutions of any fluid flow problem were noticed. This paper is divided into six sections; Section 1 is for the brief introduction of micropolar fluid and multiple solutions. Mathematical formulation, derivation of stability analysis, and methodology are kept in Section 2, Section 3, and Section 4 respectively. Section 5 is for the result and discussion and Section 6 is for the conclusion.

2. Mathematical Formulation

An incompressible laminar boundary layer two-dimensional flow of the micropolar fluid over the exponentially shrinking sheet has been considered. The corresponding velocities of x and y-axes are u and v. The shrinking velocity is assumed to be $u_w(x) = -U_w e^{\frac{2x}{l}}$. The temperature of the sheet is taken to be $T_w(x) = T_\infty + T_0 e^{\frac{x}{2l}}$, as shown in Figure 1. The $N = N(x,y)$ is supposed as the angular velocity. The respective boundary layer movement equation, along with micro rotations and the heat transfer equations can be expressed as vectors in accordance with the abovementioned assumptions.

$$\nabla \cdot V = 0 \tag{1}$$

$$\rho \frac{dV}{dt} = -\nabla P + (\mu + \kappa)\nabla^2 V + \kappa(\nabla \times N) \tag{2}$$

$$\rho j \frac{dN}{dt} = \gamma \nabla^2 N - \kappa(2N - \nabla \times V) \tag{3}$$

$$\rho c_p \frac{dT}{dt} = k\nabla^2 T \tag{4}$$

in which velocity vector is $V \equiv [u(x,y), v(x,y), 0]$, the micro-rotation vector is N, ρ stands for fluid density, μ for viscosity coefficient, κ for vertex viscosity, j is the density of micro-rotation, and γ stands for micropolar constant. We get following boundary layer equations according to the scale analysis.

$$\frac{\partial u}{\partial x} + \frac{\partial v}{\partial y} = 0 \tag{5}$$

$$u\frac{\partial u}{\partial x} + v\frac{\partial u}{\partial y} = \left(\vartheta + \frac{\kappa}{\rho}\right)\frac{\partial^2 u}{\partial y^2} + \frac{\kappa}{\rho}\frac{\partial N}{\partial y} \tag{6}$$

$$u\frac{\partial N}{\partial x} + v\frac{\partial N}{\partial y} = \frac{\gamma}{\rho j}\frac{\partial^2 N}{\partial y^2} - \frac{\kappa}{\rho j}\left(2N + \frac{\partial u}{\partial y}\right) \tag{7}$$

$$u\frac{\partial T}{\partial x} + v\frac{\partial T}{\partial y} = \alpha\frac{\partial^2 T}{\partial y^2} \tag{8}$$

Subject to these boundary conditions

$$v = v_w(x); \quad u = u_w(x); \quad N = -m\frac{\partial u}{\partial y}; \quad T = T_w(x); \quad at \quad y = 0$$

$$u \to 0; \quad N \to 0; \quad T \to T_\infty \quad as \quad y \to \infty \tag{9}$$

Now, we look for similarity transformation variables in order to transform Equations (6)–(8) with boundary conditions (9)

$$u = U_w e^{\frac{x}{l}} f'(\eta); v = -\sqrt{\frac{\vartheta U_w}{2l}} e^{\frac{x}{2l}}(f(\eta) + \eta f'(\eta)); N = U_w e^{\frac{3x}{2l}}\sqrt{\frac{U_w}{2\vartheta l}} h(\eta);$$

$$\theta(\eta) = \frac{(T - T_\infty)}{(T_w - T_\infty)}; \eta = y\sqrt{\frac{U_w}{2\vartheta l}} e^{\frac{x}{2l}} \tag{10}$$

By applying Equation (10) in Equations (6)–(9), we have the following system of similarity transformed ordinary differential equations

$$(1+K)f''' + ff'' - 2f'^2 + Kh' = 0 \tag{11}$$

$$\left(1 + \frac{K}{2}\right)h'' + fh' - 3f'h - K(2h + f'') = 0 \tag{12}$$

$$\frac{1}{Pr}\theta'' + f\theta' - f'\theta = 0 \tag{13}$$

Subject to boundary conditions

$$\begin{aligned} f(0) = S; \quad f'(0) = -1; \quad h(0) = -nf''(0); \quad \theta(0) = 1 \\ f'(\eta) \to 0; \quad h(\eta) \to 0; \quad \theta(\eta) \to 0; \quad as\ \eta \to \infty \end{aligned} \tag{14}$$

where prime denotes the differentiation with respect to η, the micropolar material parameter is $K = \frac{\kappa}{\mu}$, Prandtl number is $Pr = \frac{\vartheta}{\alpha}$, and suction is $S = -\frac{v_w}{\sqrt{\vartheta U_w/2l}}$.

The physical quantities of interest include skin friction, the stress of local couples, and the local number of Nusselt, which are described as

$$C_f = \frac{\left[(\mu + \kappa)\frac{\partial u}{\partial y} + \kappa N\right]_{y=0}}{\rho u_w^2}; M_x = \frac{-\gamma\left[\frac{\partial N}{\partial y}\right]_{y=0}}{\rho x u_w^2}; N_u = \frac{-x\left[\frac{\partial T}{\partial y}\right]_{y=0}}{(T_w - T_\infty)} \tag{15}$$

By applying similarity transformation variables (10) in Equation (15), we have

$$\begin{aligned} C_f(Re_x)^{\frac{1}{2}}\sqrt{2l/x} = (1 + (1-m)K)f''(0), M_x Re_x = \left(1 + \frac{K}{2}\right)h'(0), \\ N_u(Re_x)^{-\frac{1}{2}}\sqrt{2l/x} = -\theta'(0) \end{aligned} \tag{16}$$

where $Re_x = xu_w/\vartheta$ is the local Reynolds number.

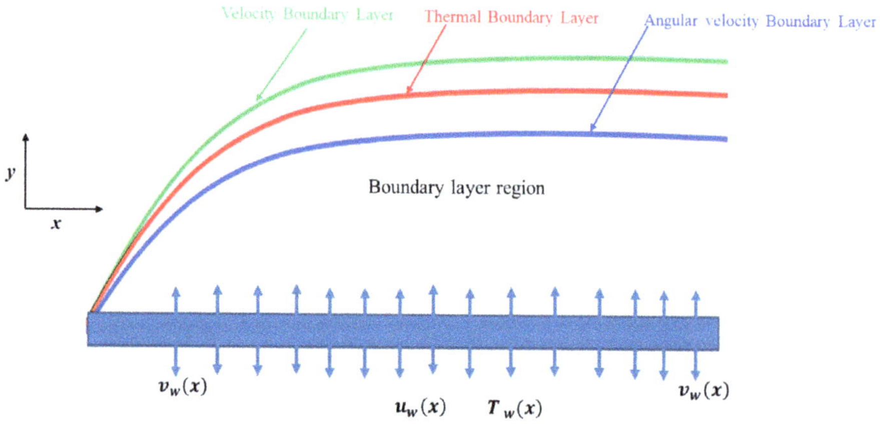

Figure 1. Flow of problem and coordinate system.

3. Stability Analysis

According to Nasir et al., [28] and Rana et al., [25], we need to introduce the unsteady form of Equations (6)–(8) in order to perform stability test,

$$\frac{\partial u}{\partial t} + u\frac{\partial u}{\partial x} + v\frac{\partial u}{\partial y} = \left(\vartheta + \frac{\kappa}{\rho}\right)\frac{\partial^2 u}{\partial y^2} + \frac{\kappa}{\rho}\frac{\partial N}{\partial y} \tag{17}$$

$$\frac{\partial N}{\partial t} + u\frac{\partial N}{\partial x} + v\frac{\partial N}{\partial y} = \frac{\gamma}{\rho j}\frac{\partial^2 N}{\partial y^2} - \frac{\kappa}{\rho j}\left(2N + \frac{\partial u}{\partial y}\right) \tag{18}$$

$$\frac{\partial T}{\partial t} + u\frac{\partial T}{\partial x} + v\frac{\partial T}{\partial y} = \frac{k^*}{\rho c_p}\frac{\partial^2 T}{\partial y^2} \tag{19}$$

where time is denoted by t. The similarity transformations are written according to Roşca and Pop [30], as,

$$\psi = \sqrt{2\vartheta L a}e^{\frac{x}{2L}}f(\eta, \tau); \ N = ae^{\frac{3x}{2L}}\sqrt{\frac{a}{2\vartheta L}}h(\eta, \tau); \theta(\eta, \tau) = \frac{(T - T_\infty)}{(T_w - T_\infty)};$$

$$\eta = y\sqrt{\frac{a}{2\vartheta L}}\ e^{\frac{x}{2L}} \ ; \ \tau = \frac{a}{2L}e^{\frac{x}{L}}.t \tag{20}$$

By applying these similarity transformations, we reduced our Equations (17)–(19) into the following form

$$(1 + K)\frac{\partial^3 f}{\partial \eta^3} + f\frac{\partial^2 f}{\partial \eta^2} - 2\left(\frac{\partial f}{\partial \eta}\right)^2 + K\frac{\partial h}{\partial \eta} - \frac{\partial^2 f}{\partial \tau \partial \eta} = 2\tau\left(\frac{\partial f}{\partial \eta}\frac{\partial^2 f}{\partial \eta \partial x} - \frac{\partial^2 f}{\partial \eta^2}\frac{\partial f}{\partial x}\right) \tag{21}$$

$$\left(1 + \frac{K}{2}\right)\frac{\partial^2 h}{\partial \eta^2} + f\frac{\partial h}{\partial \eta} - 3h\frac{\partial f}{\partial \eta} - 2Kh - K\frac{\partial^2 f}{\partial \eta^2} - \frac{\partial h}{\partial \tau} = 2\tau\left(\frac{\partial f}{\partial \eta}\frac{\partial h}{\partial x} - \frac{\partial h}{\partial \eta}\frac{\partial f}{\partial x}\right) \tag{22}$$

$$\frac{1}{Pr}\frac{\partial^2 \theta}{\partial \eta^2} + f\frac{\partial \theta}{\partial \eta} - \theta\frac{\partial f}{\partial \eta} - \frac{\partial \theta}{\partial \tau} = 2\tau\left(\frac{\partial f}{\partial \eta}\frac{\partial \theta}{\partial x} - \frac{\partial \theta}{\partial \eta}\frac{\partial f}{\partial x}\right) \tag{23}$$

with the following boundary conditions

$$f(0, \tau) = S; \ \frac{\partial f(0, \tau)}{\partial \eta} = -1; \ h(0, \tau) = -m\frac{\partial^2 f(0, \tau)}{\partial \eta^2}; \ \theta(0, \tau) = 1$$

$$\frac{\partial f(\eta, \tau)}{\partial \eta} \to 0; \ h(\eta, \tau) \to 0; \ \theta(\eta, \tau) \to 0 \qquad as \ \eta \to \infty \tag{24}$$

and then we applied perturbed with the disturbance (see Roşca and Pop [30]) in Equations (21)–(24) with the following functions

$$f(\eta, \tau) = f_0(\eta) + e^{-\varepsilon\tau}F(\eta, \tau)$$
$$h(\eta, \tau) = h_0(\eta) + e^{-\varepsilon\tau}G(\eta, \tau) \tag{25}$$
$$\theta(\eta, \tau) = \theta_0(\eta) + e^{-\varepsilon\tau}H(\eta, \tau)$$

where $F(\eta, \tau)$, $G(\eta, \tau)$, and $H(\eta, \tau)$ are small relative to $f_0(\eta)$, $h_0(\eta)$, and $\theta_0(\eta)$, respectively. Further, ε is the unknown eigenvalue. By substituting Equation (25) in Equations (21)–(23) by keeping $\tau = 0$, we have following the linearized eigenvalue problems

$$(1 + K)F_0''' + f_0F_0'' + F_0f_0'' - 4f_0'F_0' + KG_0' + \varepsilon F_0' = 0 \tag{26}$$

$$\left(1 + \frac{K}{2}\right)G_0'' + f_0G_0' + F_0h_0' - 3h_0F_0' - 3h_0F_0' - 2KG_0 - KF_0'' + \varepsilon G_0 = 0 \tag{27}$$

$$\frac{1}{Pr}H_0'' + f_0 H_0' + F_0 \theta_0' - f_0' H_0 - F_0' \theta_0 + \varepsilon H_0 = 0 \tag{28}$$

With the boundary conditions

$$F_0(0) = 0, \quad F_0'(0) = 0, \quad G_0(0) = -nF_0''(0), \quad H_0(0) = 0$$
$$F_0'(\eta) \to 0, \quad G_0(\eta) \to 0, \quad H_0(\eta) \to 0, \quad as \ \eta \to \infty \tag{29}$$

We have to solve above linearized Equations (10)–(13) with new relax boundary conditions in order to find the values of smallest eigenvalue. In this particular problem, we have relaxed $H_0(\eta) \to 0$ as $\eta \to \infty$ into $H_0'(0) = 1$, see [31–33].

4. Numerical Methods

The governing ODEs are highly non-linear and, therefore, we adopt the numerical approach in order to solve Equations (11)–(14) and Equations (26)–(19). In this study, two methods have been employed, namely shooting method and Three-stage Lobatto III-A formula, which were used in many research articles of same authors previously [refer to 17,20,23,26]. The descriptions regarding these methods are explained below.

4.1. Shooting Method

The shooting technique along with the Runge Kutta method of the fourth order is employed in order to obtain the numerical solutions of Equations (11)–(13) subject to the boundary conditions. Shooting method helps to reduce the third order ODEs (11)–(13) into the first-order ODEs, such that

$$p = f', q = p', \ r = h'; \ q' = \frac{1}{1+K}\left\{2p^2 - fq - Kr\right\} \tag{30}$$

$$r = h; \ r' = \frac{2}{2+K}\left\{3ph - fr + K(2h + q)\right\} \tag{31}$$

$$s = \theta'; \ s' = Pr\{p\theta - fs\} \tag{32}$$

with conditions

$$f(0) = S; \ p(0) = -1; \ q(0) = \alpha_1; \ h(0) = -n\alpha_1; r(0) = \alpha_2; \ \theta(0) = 1; s(0) = \alpha_3$$

where $\alpha_1, \alpha_2,$ and α_3 are called as unknown initial conditions. These three missing values $\alpha_1, \alpha_2,$ and α_3 have to be obtained by using different shoots; this process of shoots will be continue until the profiles of the $f'(\eta) \to 0; \ h(\eta) \to 0;$ and $\theta(\eta) \to 0$ are satisfied the boundary condition $\eta \to \infty$. Maple (18) software has been used to convert the system of the third order ODEs into the system of the first order ODEs, for this process shootlib function is built-in Maple. Using RK method solves the system of the first order ODEs. Further, a detailed discussion about the shooting method with Maple software can be seen in the paper of Meade et al. [34].

4.2. Three-Stage Lobatto III-A Formula

Three-stage Lobatto III-A formula is built in BVP4C function with aid of C^1 piece-wise cubic polynomial in the finite difference code. According to Lund et al., [35] and Raza et al., [36], "this collocation polynomial and formula offers a C^1 continuous solution in which mesh error control and selection are created on the residual of the continuous solution. The tolerance of relative error is fixed 10^{-5} for the current problem. The suitable mesh determination is created and returned in the field sol.x. The bvp4c returns solution, called as sol.y., as a construction. In any case, values of the solution are gotten from the array named sol.y relating to the field sol.x". In addition, Figure 2 explains the algorithm of the method for stability analysis of the solutions.

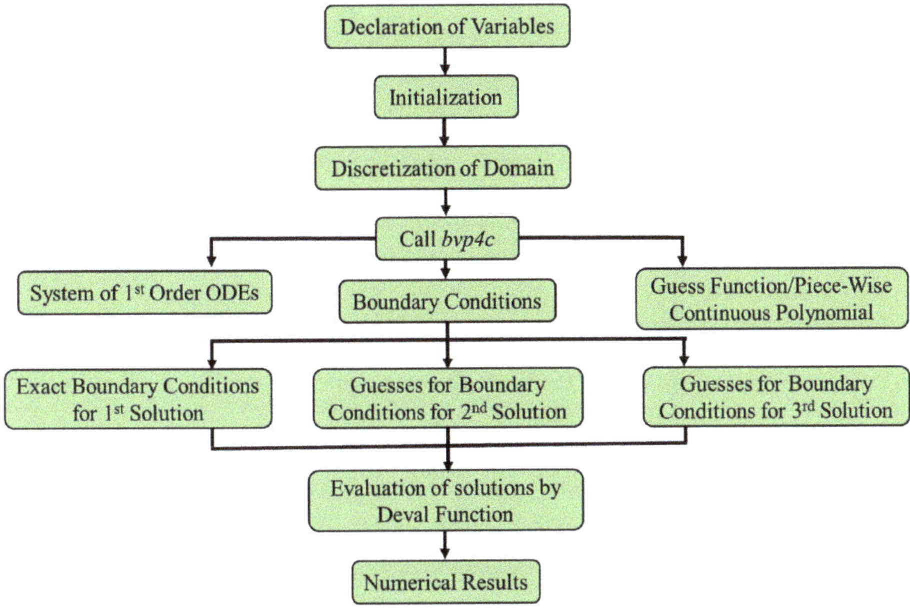

Figure 2. Description of three-stage Lobatto III-A formula.

5. Result and Discussion

Aurangzaib et al., [29] solved these Equations (11)–(14) and found dual solutions, which was the main contribution of authors. Aurangzaib et al., [29] have given some strong statements during the investigation of critical points. Before going to detail, we check the accuracy of our method (shooting method) by comparing our results with previously published literature in Table 1 and found excellent agreement with them. Moreover, we also qualitatively compared our results with Aurangzaib et al., [29] in Figure 3 and found in the good agreement. This gives trust in our numerical calculation and urges us to additionally contemplate this problem. Aurangzaib et al., [29] found dual solutions and stated in the result and discussion section that "the present study shows that for $K = 0.1$, i.e., for micropolar fluid, the similarity solutions exist when $S \geq 2.3231$ and no similarity solution exists for $S < 2.3231$". Firstly, we would like to clarify that there exist triple solutions not dual solutions; secondly, there is a range of multiple solutions and single solution. From Figure 4, a conclusion can be made that there exists multiple similarity solutions when $S \geq 2.3224$ and only a single similarity solution exists for $S < 2.3224$ when $K = 0.1$. However, the range of multiple similarity solution is $2.3769 \leq S$ and there also exists only a single solution when $S < 2.3769$ when $K = 0.2$ (see Figure 4). Furthermore, skin friction increases as suction is increased in the third solution. It is worth mentioning here that there is no range of no solution. This is one of the big reasons that insist us to reconsider and re-examine the whole problem by reproducing all of the results because there are triple solutions in order to provide true knowledge to the readers and researchers.

Table 1. The comparison of values of heat transfer rate for different values of *Pr*.

Pr	*M*	Ishak [33]	Pramanik [37]	Raju et al. [38]	Present Results
1	0	0.9548	0.9547	0.954734	0.954955
2	0	1.4715	1.4714	1.471426	1.471421
3	0	1.8691	1.8691	1.869134	1.869044
10	0	3.6603	3.6603	3.660312	3.660354

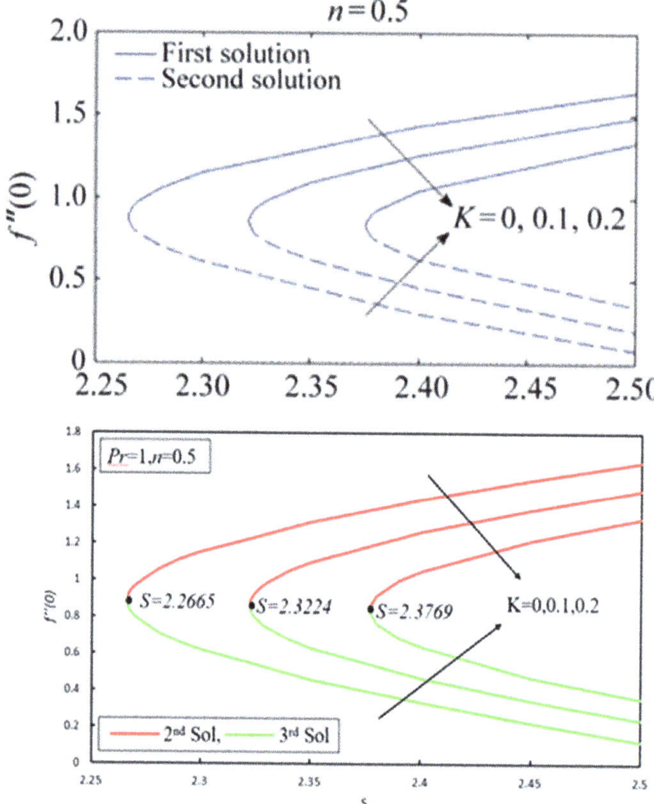

Figure 3. Comparison between upper graph: existing results obtained by Aurangzaib et al., [29] and lower graph: present results.

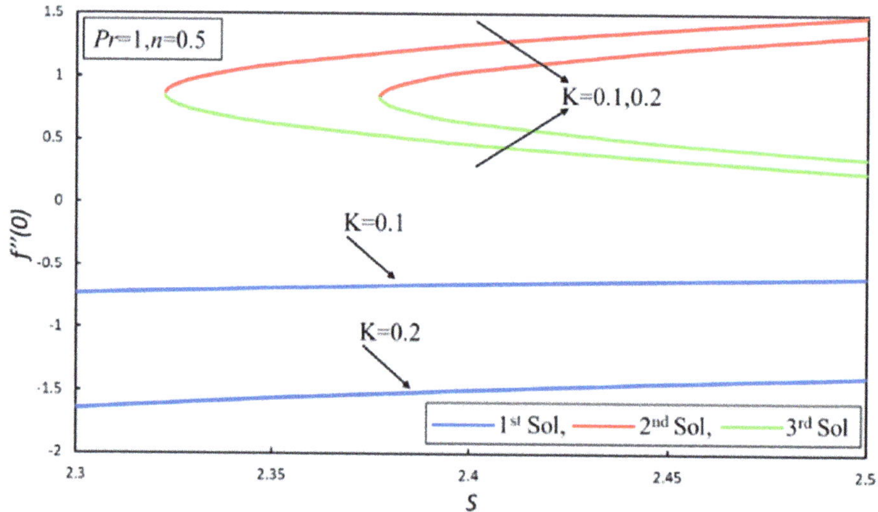

Figure 4. Variation of $f''(0)$ for different fixed values of K with suction parameter S.

Figure 5 illustrates the effect of micropolar parameter on the $h'(0)$. The effect of local couple stress enhanced as the suction increases in the first and third solutions because increasing suction creates additional resistance in the flowing fluid inside the boundary layer. However, increments in the material parameter produce more coupling of stress. Figure 6 shows the nature of the heat transfer rate for various values of the suction. It has been examined that the heat transfer rate increases in the first and second solutions for the higher values of the suction parameter, while the third solution shows opposite compliance.

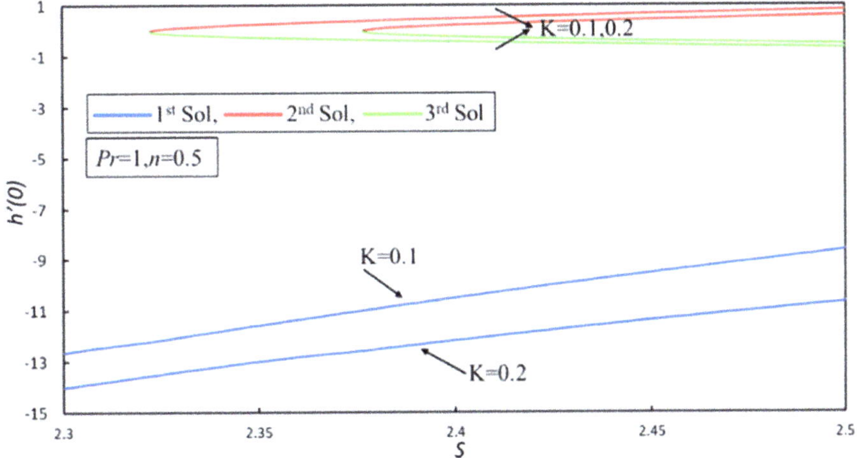

Figure 5. Variation of $h'(0)$ for different fixed values of K with suction parameter S.

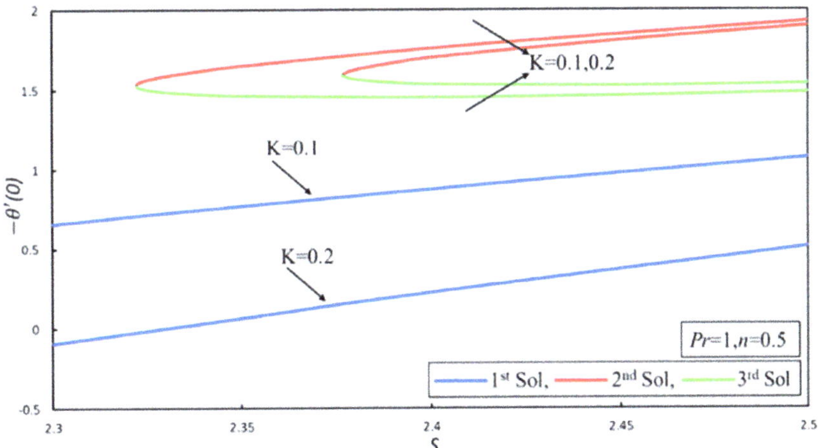

Figure 6. Variation of $-\theta'(0)$ for different fixed values of K with suction parameter S.

Finally, we plot Figures 7–10 to show the existence of triple solutions of velocity, microrotation, and temperature profiles for different values of material parameter K. In Figure 7, the dual nature of behavior has been noticed in the first solution. Velocity profile increases in the second solution when K is increased; the physical material parameter reduces the effect of drag force due to that thickness of the momentum boundary layer enhanced. On the other hand, the opposite trend has been observed in the third solution. Dual behavior can be noticed microrotation profile in Figure 8 for all solutions. The thickness of thermal boundary layer increases in the first and second solutions as material parameter K is increased, as in Figure 9, since the non-Newtonian parameter produces more viscosity, decreases the velocity of profiles, and forces fluid flow to stay on the hotter surface, as a result temperature of fluid increases and the boundary layer becomes thicker. However, the opposite behavior is noticed for the third solution. The Prandtl effect on the temperature distribution is depicted in Figure 10. It is observed that the temperature of fluid diminishes for the higher values of the Prandtl number for all solutions. Physically, it can be explained, as the Prandtl number ($Pr = \frac{\mu c_p}{k}$) has an inverse relationship with the thermal conductivity and, consequently, diminishes the thickness of the thermal boundary layer.

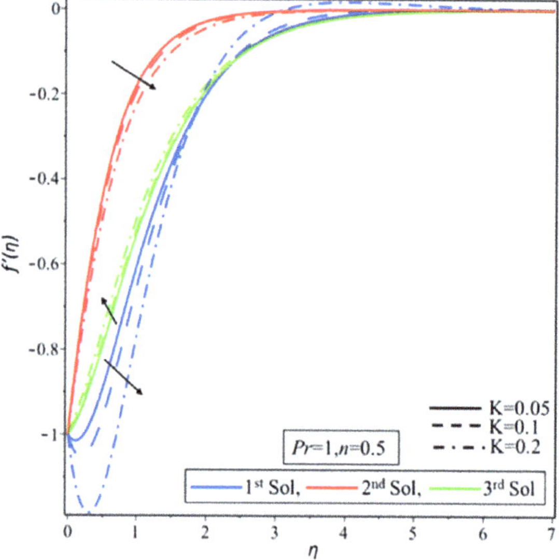

Figure 7. Effect of K on $f'(\eta)$ when $S = 2.5$.

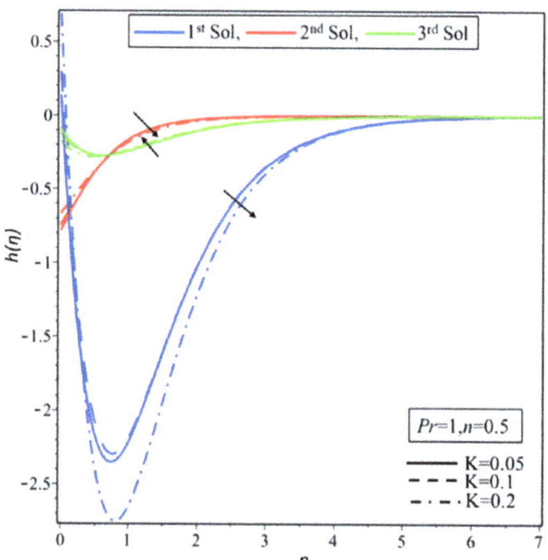

Figure 8. Effect of K on $h(\eta)$ when $S = 2.5$.

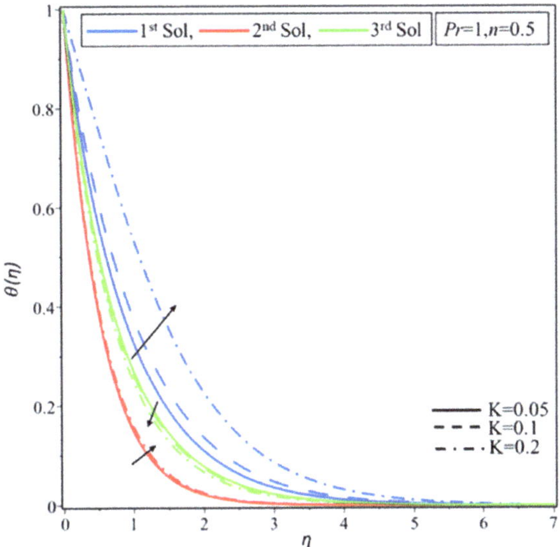

Figure 9. Effect of K on $\theta(\eta)$ when S = 2.5.

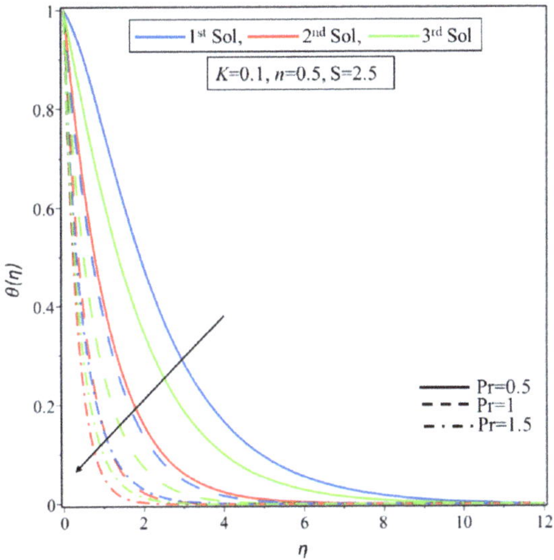

Figure 10. Effect of Pr on $\theta(\eta)$.

Table 2 shows the smallest eigenvalues ε for the selected values of S and K. The positive smallest eigenvalue makes the initial disturbance decay and, in this way, the flow becomes stable. Conversely, the negative smallest eigenvalue outcomes in an initial growth of disturbance, in this manner, the flow is unstable. It is seen from Table 2 that ε is negative for the second and the third solutions, while positive for the first solution. Thus, the second and the third solution are not stable, and the first solution is stable. From this discussion, it can be concluded that the first solution of Aurangzaib et al., [29] is not stable and not physically realizable; therefore, in this stage, it could be said that the first solution is actually the second or the third solution.

Table 2. Smallest eigen values ε at several values of S and K when $Pr = 1$.

K	S	ε		
		1st solution	2nd solution	3rd solution
0.1	2.3224	1.28061	0	0
-	2.4	1.0662	−0.06382	−0.13406
0.2	2.3769	1.36201	0	0
-	2.4	1.1364	−0.10482	−0.17482

6. Conclusions

The micropolar fluid over the shrinking surface has been considered. The system of governing equations has been transformed into the system of ODEs by using appropriate exponential similarity transformation. The system of ODEs is reduced to IVPs by employing the shooting method before solving IVPs by the Runge Kutta method. The pointwise conclusions of this study are given below:

1. Triple solutions appear.
2. There are ranges of multiple solutions and no solutions that depend upon the suction parameter.
3. According to stability analysis, the first solution is stable, which can be experimentally seen.
4. The results of Aurangzaib et al., [29] are unstable.
5. The thickness of thermal boundary layer increases in the first and the second solutions as material parameter K is increased.
6. Increments in the material parameter produce more couple stress.

Author Contributions: Conceptualization, L.A.L. and Z.O.; Funding acquisition, D.B.; Investigation, I.K.; Methodology, L.A.L., and I.K.; Project administration, I.K.; Software, L.A.L., I.K., Z.O., D.B., and K.S.N.; Supervision, I.K.; Review and Revision: D.B. and K.S.N.; Formal Analysis: D.B. and K.S.N. All authors have read and agreed to the published version of the manuscript.

Funding: This research is also supported by the Universiti Utara Malaysia.

Acknowledgments: This project was supported by the Deanship of Scientific Research at Prince Sattam Bin Abdulaziz University under the research project No. 2020/01/16436.The first author is thankful to his all teachers, especially Prof. Mazhar Ali Lund (U. B Govt. Degree college Dadu) who taught him free of cost in 2007–2009.

Conflicts of Interest: The authors declare no conflict of interest.

References

1. Eringen, A. Simple microfluids. *Int. J. Eng. Sci.* **1964**, *2*, 205–217. [CrossRef]
2. Eringen, A.C. Theory of micropolar fluids. *J. Math. Mech.* **1966**, *16*, 1–18. [CrossRef]
3. Lukaszewicz, G. *Micropolar Fluids: Theory and Applications*; Springer Science & Business Media: Berlin/Heidelberg, Germany, 1999.
4. Eringen, A.C. *Microcontinuum Field Theories: II. Fluent Media (Vol. 2)*; Springer Science & Business Media: Berlin/Heidelberg, Germany, 2001.
5. Lok, Y.Y.; Ishak, A.; Pop, I. Oblique stagnation slip flow of a micropolar fluid towards a stretching/shrinking surface: A stability analysis. *Chin. J. Phys.* **2018**, *56*, 3062–3072. [CrossRef]
6. Sheremet, M.; Pop, I.; Ishak, A. Time-dependent natural convection of micropolar fluid in a wavy triangular cavity. *Int. J. Heat Mass Transf.* **2017**, *105*, 610–622. [CrossRef]
7. Bhattacharyya, K.; Mukhopadhyay, S.; Layek, G.; Pop, I. Effects of thermal radiation on micropolar fluid flow and heat transfer over a porous shrinking sheet. *Int. J. Heat Mass Transf.* **2012**, *55*, 2945–2952. [CrossRef]
8. Ishak, A.; Nazar, R.; Pop, I. Dual solutions in mixed convection boundary layer flow of micropolar fluids. *Commun. Nonlinear Sci. Numer. Simul.* **2009**, *14*, 1324–1333. [CrossRef]
9. Ramzan, M.; Farooq, M.; Hayat, T.; Chung, J.D. Radiative and Joule heating effects in the MHD flow of a micropolar fluid with partial slip and convective boundary condition. *J. Mol. Liq.* **2016**, *221*, 394–400. [CrossRef]

10. Turkyilmazoglu, M. Mixed convection flow of magnetohydrodynamic micropolar fluid due to a porous heated/cooled deformable plate: Exact solutions. *Int. J. Heat Mass Transf.* **2017**, *106*, 127–134. [CrossRef]

11. Shah, Z.; Islam, S.; Ayaz, H.; Khan, S. Radiative Heat and Mass Transfer Analysis of Micropolar Nanofluid Flow of Casson Fluid Between Two Rotating Parallel Plates With Effects of Hall Current. *J. Heat Transf.* **2018**, *141*, 022401. [CrossRef]

12. Lund, L.A.; Ching, D.L.C.; Omar, Z.; Khan, I.; Nisar, K.S. Triple Local Similarity Solutions of Darcy-Forchheimer Magnetohydrodynamic (MHD) Flow of Micropolar Nanofluid Over an Exponential Shrinking Surface: Stability Analysis. *Coatings* **2019**, *9*, 527. [CrossRef]

13. Lund, L.A.; Omar, Z.; Khan, U.; Khan, I.; Baleanu, D.; Nisar, K.S. Stability Analysis and Dual Solutions of Micropolar Nanofluid over the Inclined Stretching/Shrinking Surface with Convective Boundary Condition. *Symmetry* **2020**, *12*, 74. [CrossRef]

14. Dero, S.; Rohni, A.M.; Saaban, A. MHD Micropolar Nanofluid Flow over an Exponentially Stretching/Shrinking Surface: Triple Solutions. *J. Adv. Res. Fluid Mech. Therm. Sci.* **2019**, *56*, 165–174.

15. Lund, L.A.; Omar, Z.; Khan, I. Mathematical analysis of magnetohydrodynamic (MHD) flow of micropolar nanofluid under buoyancy effects past a vertical shrinking surface: Dual solutions. *Heliyon* **2019**, *5*, e02432. [CrossRef] [PubMed]

16. Lund, L.A.; Omar, Z.; Dero, S.; Khan, I. Linear stability analysis of MHD flow of micropolar fluid with thermal radiation and convective boundary condition: Exact solution. *Heat Transf.-Asian Res.* **2019**, *49*, 461–476. [CrossRef]

17. Raza, J.; Rohni, A.M.; Omar, Z. Rheology of micropolar fluid in a channel with changing walls: Investigation of multiple solutions. *J. Mol. Liq.* **2016**, *223*, 890–902. [CrossRef]

18. Rohni, A.M. Multiple Similarity Solutions of Steady and Unsteady Convection Boundary Layer Flows in Viscous Fluids and Nanofluids. Ph.D. Thesis, Universiti Sains Malaysia, Penang, Malaysia, 2013.

19. Mishra, S.; Debroy, T. A computational procedure for finding multiple solutions of convective heat transfer equations. *J. Phys. D Appl. Phys.* **2005**, *38*, 2977–2985. [CrossRef]

20. Lund, L.A.; Omar, Z.; Khan, I.; Dero, S. Multiple solutions of Cu-C6H9NaO7 and Ag-C6H9NaO7 nanofluids flow over nonlinear shrinking surface. *J. Cent. South Univ.* **2019**, *26*, 1283–1293. [CrossRef]

21. Khashi'Ie, N.S.; Arifin, N.M.; Nazar, R.; Hafidzuddin, E.H.; Wahi, N.; Pop, I. Mixed Convective Flow and Heat Transfer of a Dual Stratified Micropolar Fluid Induced by a Permeable Stretching/Shrinking Sheet. *Entropy* **2019**, *21*, 1162. [CrossRef]

22. Raza, J.; Rohni, A.M.; Omar, Z. A Note on Some Solutions of Copper-Water (Cu-Water) Nanofluids in a Channel with Slowly Expanding or Contracting Walls with Heat Transfer. *Math. Comput. Appl.* **2016**, *21*, 24. [CrossRef]

23. Lund, L.A.; Omar, Z.; Khan, I.; Raza, J.; Bakouri, M.; Tlili, I. Stability Analysis of Darcy-Forchheimer Flow of Casson Type Nanofluid Over an Exponential Sheet: Investigation of Critical Points. *Symmetry* **2019**, *11*, 412. [CrossRef]

24. Dero, S.; Uddin, M.J.; Rohni, A.M. Stefan Blowing and Slip Effects on Unsteady Nanofluid Transport Past a Shrinking Sheet: Multiple Solutions. *Heat Transfer-Asian Res.* **2019**, *48*, 2047–2066. [CrossRef]

25. Rana, P.; Shukla, N.; Gupta, Y.; Pop, I. Analytical prediction of multiple solutions for MHD Jeffery–Hamel flow and heat transfer utilizing KKL nanofluid model. *Phys. Lett. A* **2019**, *383*, 176–185. [CrossRef]

26. Lund, L.A.; Omar, Z.; Khan, I. Analysis of dual solution for MHD flow of Williamson fluid with slippage. *Heliyon* **2019**, *5*, e01345. [CrossRef] [PubMed]

27. Waini, I.; Ishak, A.; Pop, I. Unsteady flow and heat transfer past a stretching/shrinking sheet in a hybrid nanofluid. *Int. J. Heat Mass Transf.* **2019**, *136*, 288–297. [CrossRef]

28. Nasir, N.A.A.M.; Ishak, A.; Pop, I. Stagnation point flow and heat transfer past a permeable stretching/shrinking Riga plate with velocity slip and radiation effects. *J. Zhejiang Univ. A* **2019**, *20*, 290–299. [CrossRef]

29. Aurangzaib; Uddin, S.; Bhattacharyya, K.; Khan, I. Micropolar fluid flow and heat transfer over an exponentially permeable shrinking sheet. *Propuls. Power Res.* **2016**, *5*, 310–317. [CrossRef]

30. Roşca, A.V.; Pop, I. Flow and heat transfer over a vertical permeable stretching/shrinking sheet with a second order slip. *Int. J. Heat Mass Transf.* **2013**, *60*, 355–364. [CrossRef]

31. Rahman, M.; Roşca, A.; Pop, I. Boundary layer flow of a nanofluid past a permeable exponentially shrinking/stretching surface with second order slip using Buongiorno's model. *Int. J. Heat Mass Transf.* **2014**, *77*, 1133–1143. [CrossRef]

32. Harris, S.D.; Ingham, D.B.; Pop, I. Mixed Convection Boundary-Layer Flow Near the Stagnation Point on a Vertical Surface in a Porous Medium: Brinkman Model with Slip. *Transp. Porous Media* **2008**, *77*, 267–285. [CrossRef]

33. Ishak, A. MHD boundary layer flow due to an exponentially stretching sheet with radiation effect. *Sains Malays.* **2011**, *40*, 391–395.

34. Meade, D.B.; Haran, B.S.; White, R.E. The shooting technique for the solution of two-point boundary value problems. *Maple Tech. Newsl.* **1996**, *3*, 1–8.

35. Lund, L.A.; Omar, Z.; Khan, I. Quadruple solutions of mixed convection flow of magnetohydrodynamic nanofluid over exponentially vertical shrinking and stretching surfaces: Stability analysis. *Comput. Methods Programs Biomed.* **2019**, *182*, 105044. [CrossRef] [PubMed]

36. Raza, J.; Mebarek-Oudina, F.; Chamkha, A. Magnetohydrodynamic flow of molybdenum disulfide nanofluid in a channel with shape effects. *Multidiscip. Model. Mater. Struct.* **2019**, *15*, 737–757. [CrossRef]

37. Pramanik, S. Casson fluid flow and heat transfer past an exponentially porous stretching surface in presence of thermal radiation. *Ain Shams Eng. J.* **2014**, *5*, 205–212. [CrossRef]

38. Raju, C.S.K.; Sandeep, N.; Sugunamma, V.; Babu, M.J.; Reddy, J.R. Heat and mass transfer in magnetohydrodynamic Casson fluid over an exponentially permeable stretching surface. *Eng. Sci. Technol. Int. J.* **2016**, *19*, 45–52. [CrossRef]

MDPI
St. Alban-Anlage 66
4052 Basel
Switzerland
Tel. +41 61 683 77 34
Fax +41 61 302 89 18
www.mdpi.com

Crystals Editorial Office
E-mail: crystals@mdpi.com
www.mdpi.com/journal/crystals